THE LIBRARY
ST. MARY'S COLLEGE OF MARYLAND
ST. MARY'S CITY, MARYLAND 20686

# Reductions in Organic Synthesis

ACS SYMPOSIUM SERIES 641

# Reductions in Organic Synthesis

## Recent Advances and Practical Applications

**Ahmed F. Abdel-Magid,** EDITOR
*R. W. Johnson Pharmaceutical Research Institute*

Developed from a symposium sponsored
by the Division of Organic Chemistry
at the 210th National Meeting
of the American Chemical Society.

American Chemical Society, Washington, DC

Library of Congress Cataloging-in-Publication Data

Reductions in organic synthesis: recent advances and practical applications / Ahmed F. Abdel-Magid, editor.

p. cm.—(ACS symposium series, ISSN 0097–6156; 641)

"Developed from a symposium sponsored by the Division of Organic Chemistry at the 210th National Meeting of the American Chemical Society, Chicago, Illinois, August 20–24, 1995."

Includes bibliographical references and indexes.

ISBN 0–8412–3381–0

1. Reduction (Chemistry)—Congresses. 2. Organic Compounds—Synthesis—Congresses.

I. Abdel-Magid, Ahmed F., 1947– . II. American Chemical Society. Division of Organic Chemistry. III. American Chemical Society. Meeting (210th: 1995: Chicago, Ill.) IV. Series.

QD281.R4R43 1996
547'.23—dc20                                              96–26217
                                                              CIP

This book is printed on acid-free, recycled paper.

Copyright © 1996

American Chemical Society

All Rights Reserved. The appearance of the code at the bottom of the first page of each chapter in this volume indicates the copyright owner's consent that reprographic copies of the chapter may be made for personal or internal use or for the personal or internal use of specific clients. This consent is given on the condition, however, that the copier pay the stated per-copy fee through the Copyright Clearance Center, Inc., 222 Rosewood Drive, Danvers, MA 01923, for copying beyond that permitted by Sections 107 or 108 of the U.S. Copyright Law. This consent does not extend to copying or transmission by any means—graphic or electronic—for any other purpose, such as for general distribution, for advertising or promotional purposes, for creating a new collective work, for resale, or for information storage and retrieval systems. The copying fee for each chapter is indicated in the code at the bottom of the first page of the chapter.

The citation of trade names and/or names of manufacturers in this publication is not to be construed as an endorsement or as approval by ACS of the commercial products or services referenced herein; nor should the mere reference herein to any drawing, specification, chemical process, or other data be regarded as a license or as a conveyance of any right or permission to the holder, reader, or any other person or corporation, to manufacture, reproduce, use, or sell any patented invention or copyrighted work that may in any way be related thereto. Registered names, trademarks, etc., used in this publication, even without specific indication thereof, are not to be considered unprotected by law.

PRINTED IN THE UNITED STATES OF AMERICA

# Advisory Board

## ACS Symposium Series

Robert J. Alaimo
Procter & Gamble Pharmaceuticals

Mark Arnold
University of Iowa

David Baker
University of Tennessee

Arindam Bose
Pfizer Central Research

Robert F. Brady, Jr.
Naval Research Laboratory

Mary E. Castellion
ChemEdit Company

Margaret A. Cavanaugh
National Science Foundation

Arthur B. Ellis
University of Wisconsin at Madison

Gunda I. Georg
University of Kansas

Madeleine M. Joullie
University of Pennsylvania

Lawrence P. Klemann
Nabisco Foods Group

Douglas R. Lloyd
The University of Texas at Austin

Cynthia A. Maryanoff
R. W. Johnson Pharmaceutical Research Institute

Roger A. Minear
University of Illinois at Urbana–Champaign

Omkaram Nalamasu
AT&T Bell Laboratories

Vincent Pecoraro
University of Michigan

George W. Roberts
North Carolina State University

John R. Shapley
University of Illinois at Urbana–Champaign

Douglas A. Smith
Concurrent Technologies Corporation

L. Somasundaram
DuPont

Michael D. Taylor
Parke-Davis Pharmaceutical Research

William C. Walker
DuPont

Peter Willett
University of Sheffield (England)

# Foreword

THE ACS SYMPOSIUM SERIES was first published in 1974 to provide a mechanism for publishing symposia quickly in book form. The purpose of this series is to publish comprehensive books developed from symposia, which are usually "snapshots in time" of the current research being done on a topic, plus some review material on the topic. For this reason, it is necessary that the papers be published as quickly as possible.

Before a symposium-based book is put under contract, the proposed table of contents is reviewed for appropriateness to the topic and for comprehensiveness of the collection. Some papers are excluded at this point, and others are added to round out the scope of the volume. In addition, a draft of each paper is peer-reviewed prior to final acceptance or rejection. This anonymous review process is supervised by the organizer(s) of the symposium, who become the editor(s) of the book. The authors then revise their papers according to the recommendations of both the reviewers and the editors, prepare camera-ready copy, and submit the final papers to the editors, who check that all necessary revisions have been made.

As a rule, only original research papers and original review papers are included in the volumes. Verbatim reproductions of previously published papers are not accepted.

ACS BOOKS DEPARTMENT

# Contents

Preface .................................................................................................... ix

1. Sixty Years of Hydride Reductions ............................................... 1
   Herbert C. Brown and P. Veeraraghavan Ramachandran

2. New Developments in Chiral Ruthenium (II) Catalysts
   for Asymmetric Hydrogenation and Synthetic Applications ........... 31
   Jean Pierre Genet

3. Tartrate-Derived Ligands for the Enantioselective LiAlH$_4$
   Reduction of Ketones: A Comparison of TADDOLates
   and BINOLates ............................................................................ 52
   Albert K. Beck, Robert Dahinden, and Florian N. M. Kühnle

4. Electrophilic Assistance in the Reduction of Six-Membered
   Cyclic Ketones by Alumino- and Borohydrides ............................. 70
   Jacqueline Seyden-Penne

5. Recent Advances in Asymmetric Reductions
   with B-Chlorodiisopinocampheylborane ....................................... 84
   P. Veeraraghavan Ramachandran and Herbert C. Brown

6. The Practical Enantioselective Reduction of Prochiral Ketones .... 98
   Anthony O. King, David J. Mathre, David M. Tschaen,
   and Ichiro Shinkai

7. Diphenyloxazaborolidine for Enantioselective Reduction
   of Ketones ................................................................................... 112
   George J. Quallich, James F. Blake, and Teresa M. Woodall

8. Diastereo- and Enantioselective Hydride Reductions of Ketone
   Phosphinyl Imines to Phosphinyl Amines: Synthesis
   of Protected Amines and Amino Acids ........................................ 127
   Robert O. Hutchins, Qi-Cong Zhu, Jeffrey Adams,
   Samala J. Rao, Enis Oskay, Ahmed F. Abdel-Magid,
   and MaryGail K. Hutchins

9. Remote Acyclic Diastereocontrol in Hydride Reductions
   of 1,n-Hydroxy Ketones .................................................. 138
   Bruce E. Maryanoff, Han-Cheng Zhang, Michael J. Costanzo,
   Bruce D. Harris, and Cynthia A. Maryanoff

10. Synthesis, Characterization, and Synthetic Utility of Lithium
    Aminoborohydrides: A New Class of Powerful, Selective,
    Air-Stable Reducing Agents ........................................... 153
    Gayane Godjoian, Gary B. Fisher, Christian T. Goralski,
    and Bakthan Singaram

11. Sodium Borohydride and Carboxylic Acids: A Novel
    Reagent Combination .................................................... 167
    Gordon W. Gribble

12. Use of Sodium Triacetoxyborohydride in Reductive
    Amination of Ketones and Aldehydes ............................ 201
    Ahmed F. Abdel-Magid and Cynthia A. Maryanoff

INDEXES

Author Index ..................................................................... 219

Affiliation Index ................................................................ 219

Subject Index .................................................................... 219

# Preface

REDUCTION IS A VERY IMPORTANT FIELD of research in organic synthesis, judging by the number of publications in this field. A recent literature search on reductive amination resulted in about 300 references that used this reaction in the period from 1990 to 1994. Considering that reductive amination is only a small aspect of reduction in organic synthesis, one can imagine the magnitude of research conducted in the vast area of reduction. The research in this field ranges from theoretical and mechanistic studies to practical applications in the synthesis of a wide variety of molecules. As the complexity of the synthetic targets that face the synthetic organic chemist increases, the need for mild and selective reduction procedures becomes obvious. Recent advances in reduction methods allow not only the means to carry out a chemical conversion but also the capacity to do so with high degrees of chemo-, regio-, and stereoselectivity.

The symposium upon which this book is based, "Reduction in Organic Chemistry," was held at the 210th national meeting of the American Chemical Society (ACS) in Chicago, August 20–25, 1995. The symposium was held to satisfy a wide interest and to provide a forum to present some of the recent advances in reduction by top researchers in the field. The success of that symposium, as judged by its large attendance, shows clearly the need for this book as a means of gathering several of the recent developments in reduction into one volume. The ACS Symposium Series, which is designed for rapid publication in areas that are changing and developing rapidly, is ideal for this format. It allows the expansion of the scope of the symposium to cover additional subjects of interest.

This book contains 12 chapters on different aspects of reduction written by accomplished researchers with extensive records in their fields. The authors were selected to present readers with diverse experiences from the United States and Europe and to cover both academic and industrial research. The chapters cover topics on boron- and aluminum-based hydride reagents, along with and catalytic hydrogenation reduction procedures. The book's opening chapter will be of particular interest to readers. It is an overview of hydride reduction from its infancy to today's latest advances with 200 references cited, written by the person who started and developed hydride reduction over the past 60 years, Herbert C. Brown, the 1979 Nobel Prize laureate.

The goal of producing this book is to be helpful to professional synthetic organic chemists in industry and academic institutions as well as graduate students. The book provides a quick reference for current topics of reduction and may be used as an educational tool. The literature citations at the end of the chapters should be an added means of fully exploring and expanding the literature search of a specific topic. Also included are some of the most recent advances in process development and practical applications of reduction procedures that may appeal to professionals in process research.

**Acknowledgments**

The idea of organizing a symposium on reduction was suggested to me by my colleague, Cynthia Maryanoff, the current chair-elect of the ACS Division of Organic Chemistry. I thank her for that and for continued support and advice throughout the organization of the symposium and the production of this book. I thank each of the authors whose outstanding contributions to this book made it such a worthy project and a valuable addition to the literature. I appreciate their professional courtesy, patience, and kindness which made this an enjoyable experience.

I thank Dieter Seebach and his coworkers Albert K. Beck, Robert Dahinden, and Florian N. M. Kunle of Eidgenssischen Technischen Hochschule; Jean Pierre Genet of Ecole Nationale Supérieure de Chimie de Paris; Jacqueline Seyden-Penne of Université Paris Sud; George J. Quallich of Pfizer, Inc.; Anthony O. King and Ichiro Shinkai of Merck Research Laboratories; and Bakthan Singaram of the University of California—Santa Cruz for their contributions to the book. I also thank Stephen L. Buchwald of the Massachusetts Institute of Technology and Mark J. Burk of Duke University for their participation in the symposium. I am especially thankful to those who contributed as speakers in the symposium as well as authors in the book: Herbert C. Brown of Purdue University, Bruce E. Maryanoff of the R. W. Johnson Pharmaceutical Research Institute, Robert O. Hutchins of Drexel University, and Gordon W. Gribble of Dartmouth College.

I appreciate the financial support to the symposium by the following companies: FMC Corporation (Lithium Division), Merck Research Laboratories, the R. W. Johnson Pharmaceutical Research Institute, SmithKline Beecham, and Wyeth-Ayerst Research. I also thank the ACS Division of Organic Chemistry for sponsoring the symposium. Last, but not least, I thank the ACS Books Department staff, particularly Michelle Althius, for her outstanding efforts in the production of this book.

AHMED F. ABDEL-MAGID
R. W. Johnson Pharmaceutical Research Institute
Spring House, PA 19477

# Chapter 1

# Sixty Years of Hydride Reductions

Herbert C. Brown and P. Veeraraghavan Ramachandran

H. C. Brown and R. B. Wetherill Laboratories of Chemistry, Purdue University, West Lafayette, IN 47907−1393

A survey of hydride reductions in organic chemistry from its beginnings has been made. Persuaded by Alfred Stock's book entitled "The Hydrides of Boron and Silicon" that he received as a graduation gift in 1936 from his classmate (now his wife) Sarah Baylen, the senior author undertook research with Professor H. I. Schlesinger and Dr. A. B. Burg, exploring the chemistry of diborane. His Ph.D. research, begun in 1936 involved a study of the reaction of diborane with aldehydes and ketones, and other compounds with a carbonyl group. This development initiated the hydride era of organic reductions. Necessities of WWII research led to the discovery of sodium borohydride and the discovery of the alkali metal hydride route to diborane. The systematic study to modify sodium borohydride and lithium aluminum hydride led to a broad spectrum of reagents for selective reductions. It is now possible to selectively reduce one functional group in the presence of another. The study of the reduction characteristics of sodium borohydride led to the discovery of hydroboration and the versatile chemistry of organoboranes. An examination of the hydroboration of α-pinene led to the discovery of an efficient asymmetric hydroboration agent, diisopinocampheylborane, $Ipc_2BH$. This led to the devlopment of a general asymmetric synthesis and to the discovery of efficient reagents for asymmetric reduction. Research progressed, one discovery leading to another, opening up a whole new continent of chemistry.

We are nearing the close of a century that witnessed unprecedented scientific and technological progress that was probably unimaginable even half a century ago. While the overall advancement of science and technology is phenomenal, in reality it has taken place over several generations, made possible by the untiring dedication of the scientists involved to their research. In fact, every decade achieved significant advances, permitting the next generation to move forward in their own pursuit of knowledge. Later developments might cause some of the work of the earlier workers to appear trivial. However, one can only admire the tenacity of the pioneers whose steadfastness has led us to where we are today.

In chemistry, the invention and perfection of new sophisticated instruments and methodologies have facilitated the analysis of reaction intermediates and products. Newer industries catering to the needs of chemists have decreased the necessity of having to prepare many of the starting materials and reagents.

The senior author has had the rare good fortune to carry out research on one topic for sixty years, developing it from its very beginnings. As in any scientific research his sixty-year career has been a combination of serendipities and the systematic research that followed the initial observations. It is the capability of the scientist to observe and infer when one stumbles upon the unexpected that makes research so fascinating.

Hydride reductions have come a long way since the observation that diborane rapidly reduces aldehydes and ketones, the development of the alkali metal hydride route to diborane, and the discoveries of sodium borohydride (SBH) and lithium aluminum hydride (LAH). The area is so vast that it is impossible to condense all of the literature on reductions involving these and other modified hydride reagents in a short review such as this. Several monographs and books and tens of reviews have appeared in this area (1-13). We have given references to several of the reviews that have appeared pertaining to each area in the corresponding sections of this chapter. Any of the earlier reviews that has not been mentioned or the original work not included is due solely to the limitations of space. Discussion of the applications of most of the reagents described herein are made in two recent multi-volume series (14, 15). We would call the reader's attention to these reviews for a comprehensive knowledge of hydride reagents for reduction.

We shall attempt to take the readers through a chronological tour of the development of the hydride reduction area so that they can appreciate how research can progress, where one observation can open new major areas of study.

## Beginnings

**Pre-Borohydride Era.** At the beginning of this century, the reduction of an aldehyde, ketone or carboxylic acid ester was carried out by the generation of hydrogen from zinc dust, sodium amalgum, or iron and acetic acid (16). Later, sodium in ethanol (17) or zinc and sodium hydroxide in ethanol (18) were used for this purpose. In the second quarter of this century, independent research by Verley (19), Meerwein (20) and Ponndorf (21) led to the M-P-V reduction (22) whereby the reduction of an aldehyde or ketone was achieved with the aluminum alkoxides of *sec*-alcohols.

*All of these procedures were made obsolete by a reaction carried out in search of a solvent to purify sodium borohydride!*

## Diborane for Carbonyl Reductions

The interest of chemists in structural theory and their curiosity in unravelling the mysterious electron-deficient structure of a simple compound, such as diborane (23) led Schlesinger to the synthesis of borane-carbonyl (24) and to the examination of reactions of diborane with aldehydes and ketones. This study initiated the hydride era of organic reductions. It was soon discovered that aldehydes and ketones react rapidly with diborane even at low temperatures in the absence of solvents to produce dialkoxy derivatives, which can be rapidly hydrolyzed to the corresponding alcohols (eq 1-2) (25).

$$4 R_2CO + B_2H_6 \longrightarrow 2 (R_2CHO)_2BH \xrightarrow{6 H_2O} 4 R_2CHOH + 2 H_2 + 2 B(OH)_3 \quad (1)$$

$$4 RCHO + B_2H_6 \longrightarrow 2 (RCH_2O)_2BH \xrightarrow{6 H_2O} 4 RCH_2OH + 2 H_2 + 2 B(OH)_3 \quad (2)$$

However, the lack of availability of diborane hindered progress in the application of this "easy" procedure for reductions. The situation was changed by the necessities of war research.

## World War II Research and Preparation of Sodium Borohydride - A Historical Perspective.

A request from the National Defense Research Committee (NDRC) to investigate the synthesis of volatile compounds of uranium having low molecular weight, but without the corrosive properties of $UF_6$, led Schlesinger's group to extend the method of preparation of other metal borohydrides, such as aluminum and beryllium borohydride (26-28) (which happened to be the most volatile compounds of these metals), to uranium borohydride (eq 3-6) (29).

$$3\ LiEt + 2\ B_2H_6 \longrightarrow 3\ LiBH_4 + BEt_3 \quad (3)$$

$$3\ BeMe_2 + 4\ B_2H_6 \longrightarrow 3\ Be(BH_4)_2 + 2\ BMe_3 \quad (4)$$
$$\text{sp } 91°C$$

$$AlMe_3 + 2\ B_2H_6 \longrightarrow Al(BH_4)_3 + BMe_3 \quad (5)$$
$$\text{bp } 45\ °C$$

$$UF_4 + 2\ Al(BH_4)_3 \longrightarrow U(BH_4)_4 + 2\ AlF_2(BH_4) \quad (6)$$

But the need for the then rare species diborane hampered progress. This problem was circumvented by a series of reactions to achieve a practical procedure for the preparation of uranium borohydride. We discovered that lithium hydride readily reacts with boron trifluoride-etherate in ethyl ether (EE) to produce diborane, which was subsequently transformed into $U(BH_4)_4$ (eq 7-10) (30-32).

$$6\ LiH + 8\ BF_3 \xrightarrow{EE} B_2H_6\uparrow + 6\ LiBF_4\downarrow \quad (7)$$

$$2\ LiH + B_2H_6 \xrightarrow{EE} 2\ LiBH_4 \quad (8)$$

$$AlCl_3 + 3\ LiBH_4 \longrightarrow Al(BH_4)_3\uparrow + 3\ LiCl \quad (9)$$

$$UF_4 + 2\ Al(BH_4)_3 \longrightarrow U(BH_4)_4 + 2\ AlF_2(BH_4) \quad (10)$$

The lack of an available supply of LiH during the war became an impediment. The corresponding reaction with sodium hydride failed in EE, though it was discovered several years later that other solvents, such as THF and diglyme, not available in 1940, facilitate the reaction. The necessity to synthesize sodium borohydride led to the following sequence (eq 11-15) (33).

$$NaH + B(OMe)_3 \xrightarrow{reflux} NaBH(OMe)_3 \quad (11)$$

$$6\ NaBH(OMe)_3 + 8\ BF_3 \longrightarrow B_2H_6\uparrow + 6\ NaBF_4 + 6\ B(OMe)_3 \quad (12)$$

$$2\ NaBH(OMe)_3 + B_2H_6 \longrightarrow 2\ NaBH_4 + 2\ B(OMe)_3 \quad (13)$$

$$AlCl_3 + 3\ NaBH_4 \longrightarrow Al(BH_4)_3\uparrow + 3\ NaCl \quad (14)$$

$$UF_4 + 2\ Al(BH_4)_3 \longrightarrow U(BH_4)_4 + 2\ AlF_2(BH_4) \quad (15)$$

The problem of handling $UF_6$ had been mastered, so that the NDRC was no longer interested in $U(BH_4)_4$. Fortunately, the Signal Corps was interested in exploiting the feasibility of sodium borohydride for the field generation of hydrogen. The demand led to a more practical preparation of sodium borohydride (34), by the treatment of sodium hydride with methyl borate at 250 °C (eq 16).

$$4 \text{ NaH} + \text{B(OCH}_3)_3 \xrightarrow{250\ °C} \text{NaBH}_4 + 3 \text{ NaOCH}_3 \quad (16)$$

During the search for a proper solvent to separate the two solid products, we observed that sodium borohydride dissolved in acetone with a vigorous reaction. However, it could not be recovered by the removal of the solvent (eq 17-18). *The discovery of sodium borohydride as a hydrogenating agent was thus made!*

$$\text{NaBH}_4 + 4 \text{ (CH}_3)_2\text{CO} \longrightarrow \text{NaB[OCH(CH}_3)_2]_4 \quad (17)$$
$$\text{NaB[OCH(CH}_3)_2]_4 + 2 \text{ H}_2\text{O} \longrightarrow 4 \text{ (CH}_3)_2\text{CHOH} + \text{NaB(OH)}_4 \quad (18)$$

The difference in solubility of sodium borohydride and sodium methoxide in ammonia is exploited to purify the former. It proved more convenient to use isopropylamine. This is the basis of the current industrial process for preparing millions of pounds of sodium borohydride per year.

**Lithium Aluminum Hydride.**

The procedure developed for the preparation of borohydrides was extended for the synthesis of the corresponding aluminum derivatives. Thus lithium aluminum hydride was synthesized from lithium hydride and aluminum chloride in EE, using a small quantity of LAH to facilitate the reduction (eq 19-21) (*35*).

$$4 \text{ LiH} + \text{AlCl}_3 \longrightarrow \text{LiAlH}_4 + 3 \text{ LiCl} \quad (19)$$
$$3 \text{ LiAlH}_4 + \text{AlCl}_3 \longrightarrow 4 \text{ AlH}_3 + 3 \text{ LiCl} \quad (20)$$
$$4 \text{ AlH}_3 + 4 \text{ LiH} \longrightarrow 4 \text{ LiAlH}_4 \quad (21)$$

**Selective Reductions**

**Two Extremes of a Broad Spectrum.**

The discovery of SBH and LAH revolutionized the reduction of functional groups in organic chemistry. These reagents are not speciality chemicals any more. Lithium aluminum hydride is capable of reducing practically all organic functional groups, such as aldehydes, ketones, acid chlorides, carboxylic esters, acids and anhydrides, lactones, nitriles, amides, oximes, azo and hydrazo compounds, sulfoxides, etc. (*36*). It reduces alkyl and aryl halides to the corresponding hydrocarbon (*37*). Although LAH reduces a *tert*-amide to the corresponding amine, 1-acylaziridines are reduced to the corresponding aldehydes (*38*). On the other hand, sodium borohydride is a remarkably gentle reducing agent. It readily reduces only aldehydes, ketones and acid chlorides (*39*). There was a need to find a family of reducing agents with capabilities between these two extremes and our systematic research led to the development of a series of reagents that lie between and even beyond these extremes (Table 1).

We modified these reagents, either by decreasing the reducing power of LAH or by increasing that of SBH to make a complete spectrum of reagents for selective reductions. We studied (1) the influence of solvents, (2) the effect of cations, (3) the effect of introducing substituents of varying steric and electronic requirements in the complex, (4) the development of acidic reducing agents, such as alane and borane, and (5) the effect of introducing substituents into these acidic reagents.

A standard set of 56 functional groups were chosen for testing the capability of the modified reagents in tetrahydrofuran (THF) at 0 °C (*39*).

*This study marked the beginning of the Selective Reductions Era.*
Others have also made important contributions to this area, as discussed below.

Table 1. Two extremes in selective reductions

|         | NaBH$_4$ | LiAlH$_4$ |
|---------|----------|-----------|
| RCHO    | +        | +         |
| R$_2$CO | +        | +         |
| RCOCl   | +        | +         |
| RCO$_2$R' | –      | +         |
| RCO$_2$H | –       | +         |
| RCONR'$_2$ | –     | +         |
| RCN     | –        | +         |
| RNO$_2$ | –        | +         |
| RCH=CH$_2$ | –     | –         |

**Effect of Solvents.** The very high reactivity of LAH restricted the choice of solvents to hydrocarbons, ethers and *tert*-amines. It is generally used in EE, THF and diglyme, in all of which it is a powerful reducing agent. Sodium borohydride is a better reducing agent in water and alcohols. This reagent is not soluble in EE and is sparingly soluble in THF. Although it is soluble in diglyme, the reducing power is curtailed to the extent that even acetone is not rapidly reduced in this solvent.

**Effect of Cations.** Changing the cations from lithium to sodium in the metal aluminum hydride did not alter the reactivity very much (*40*). While sodium borohydride is soluble in diglyme (DG) and triglyme, lithium borohydride is soluble in simple ether solvents, such as EE and THF. Also there is a marked difference in the reactivity of these two borohydrides (*41*). Thus, sodium borohydride reduces esters only very sluggishly, whereas lithium borohydride reduces them rapidly (eq 22).

$$\text{RCH}_2\text{OH} \xleftarrow{\text{LiBH}_4} \text{RCOOR'} \xrightarrow{\text{NaBH}_4} \text{No or slow reaction} \quad (22)$$

A convenient procedure to convert SBH to LBH using lithium halide in simple ether solvents was developed (*42*). The addition of one equiv of lithium chloride or lithium bromide to a one molar solution of sodium borohydride generates lithium borohydride with precipitation of sodium halide. The reagent can be used without the removal of the salt ( eq 23) (*43*).

$$\text{RCOOEt} \xrightarrow{\text{NaBH}_4\text{-LiBr}} \text{RCH}_2\text{OH} \quad (23)$$

Kollonitsch and coworkers achieved rapid reduction of esters by sodium borohydride in the presence of Li, Mg, Ca, Ba, and Sr salts (*44,45*). Preparation of zinc borohydride from sodium borohydride (eq 24) and the reactions have been reported (*46,47*). Nöth (*48*) and Yoon (*49*) carried out a systematic study of the preparative method and concluded that the reagent is not the simple Zn(BH$_4$)$_2$, but a complex mixture of borohydrides.

$$2 \text{ NaBH}_4 + \text{ZnCl}_2 \longrightarrow \text{"Zn(BH}_4)_2\text{"} + 2 \text{ NaCl} \quad (24)$$

Addition of 0.33 molar equiv of aluminum chloride to sodium borohydride in diglyme results in a clear solution that has much enhanced reducing power, approaching that of lithium aluminum hydride (eq 25) (*50*). The reduction of an unsaturated ester, ethyl oleate, using this reagent mixture (51,52) led to the discovery of hydroboration!

$$3 \text{ NaBH}_4 + \text{AlCl}_3 \rightleftharpoons \text{Al(BH}_4)_3 + 3 \text{ NaCl} \tag{25}$$

Nöth reported that the $^{11}$B NMR of such solutions in diglyme indicate the presence of several species, such as NaBH$_4$, NaB$_2$H$_7$, NaAlCl$_3$BH$_4$ and NaAlCl$_3$H (53).

**Effect Of Substituents**
**Alkoxyaluminohydrides.** A systematic study of the reaction of lithium aluminum hydride in ethereal solvents with *pri-*, *sec-* and *tert-*alcohols using hydride analysis (54) and $^{27}$Al NMR spectroscopy (Ramachandran, P. V.; Gong, B., unpublished data) reveals that an equilibrium exists between various alkoxy derivatives. Both the *pri-* and *sec-*alcohols provide the tetraalkoxy derivative with four equiv of the alcohol (eq 26-28). However, the *tert-*alcohol does not react past the trialkoxyaluminohydride stage (eq 29). While methanol and ethanol provide the corresponding trialkoxyaluminohydride derivative cleanly, 2-propanol provides only the tetraalkoxy derivative irrespective of the molar equiv of the alcohol used.

$$\text{LiAlH}_4 + 4 \text{ MeOH} \longrightarrow \text{LiAl(MeO)}_4 + 4 \text{ H}_2 \uparrow \tag{26}$$

$$\text{LiAlH}_4 + 4 \text{ EtOH} \longrightarrow \text{LiAl(EtO)}_4 + 4 \text{ H}_2 \uparrow \tag{27}$$

$$\text{LiAlH}_4 + 4 \text{ } i\text{-PrOH} \longrightarrow \text{LiAl}(i\text{-PrO})_4 + 4 \text{ H}_2 \uparrow \tag{28}$$

$$\text{LiAlH}_4 + 3 \text{ } t\text{-BuOH} \longrightarrow \text{LiAl}(t\text{-BuO})_3\text{H} + 3 \text{ H}_2 \uparrow \tag{29}$$

Lithium tri-*tert*-butoxyaluminohydride proved to be exceptionally stable in the solid form or in ether solvents (54, 55). Substitution of the hydride with alkoxy groups decreases the reducing power of the substituted LAH considerably. The reagent reduces aldehydes, ketones and acid chlorides (56). Lactones and epoxides react slowly, whereas carboxylic acids and esters do not react with the exception of aryl esters. This reagent is capable of reducing nitriles (eq 30), *tert*-amides (eq 31) and aromatic acid chlorides (eq 32) to aldehydes in excellent yield (57).

$$\text{RCOCl} + \text{LiAl}(t\text{-BuO})_3\text{H} \longrightarrow \text{RCHO} + \text{LiCl} + \text{Al}(t\text{-BuO})_3 \tag{30}$$
$$\text{RCN} + \text{LiAl}(t\text{-BuO})_3\text{H} \longrightarrow \longrightarrow \text{RCHO} \tag{31}$$
$$\text{RCONR'}_2 + \text{LiAl}(t\text{-BuO})_3\text{H} \longrightarrow \longrightarrow \text{RCHO} \tag{32}$$

The corresponding sodium tri-*tert*-butoxyaluminohydride is capable of reducing aliphatic acid chlorides as well (58).
Lithium trimethoxyaluminohydride (LTMA) (59) and lithium triethoxyaluminohydride (LTEA) (60) are powerful reducing agents, closely resembling LAH, but more selective. LTEA reduces aromatic and aliphatic nitriles, and *tert*-amides to the corresponding aldehdydes. The difference in the reducing characteristics of LTBA and LTMA is shown in eq 33.

$$\text{Ph}\diagup\diagdown\diagup\text{OH} \xleftarrow{\text{LTMA}} \text{Ph}\diagup\diagdown\overset{\overset{\text{O}}{\|}}{\text{H}} \xrightarrow{\text{LTBA}} \text{Ph}\diagup\diagdown\diagup\text{OH} \tag{33}$$

LTBA can be very selective, distinguishing between the carbonyl groups of aldehydes and ketones (eq 34) (61).

$$\text{[structure with CHO] } \xrightarrow{\text{LTBA}} \text{[structure with OH]} \qquad (34)$$

The increased steric requirements of the alkoxy groups of the reagent aids in a more stereoselective reduction of certain bicyclic ketones.

Sodium bis(2-methoxyethoxy)aluminum hydride (SMEAH) (Red-Al, Vitride, Alkadride) is a stable dialkoxyaluminum hydride that resembles LAH in its reducing capabilities, but possesses unique properties, such as higher solubility in ether solvents and aromatic hydrocarbons, and thermal stability (*62*) (eq 35).

$$NaAlH_4 + 2\ MeOCH_2CH_2OH \longrightarrow NaAlH_2(OCH_2CH_2OMe)_2 \qquad (35)$$

It shows selective behavior in the reactions of epoxides. Unsymmetrical epoxides are opened with preferential attack at the least substituted carbon (eq 36) (*63*).

$$H_{15}C_7\text{-epoxide-}CH_2OH\ R \xrightarrow{H^-} H_{15}C_7\text{-}CH(OH)\text{-}CR(OH)\text{-} + H_{15}C_7\text{-}C(OH)(R)\text{-}CH_2OH \qquad (36)$$

|  |  |  |  |  |
|---|---|---|---|---|
| R = H, LAH | 4 | : | 1 |  |
| R = H, Red-Al | 100 | : | 1 |  |
| R = Me, Red-Al | 1 | : | 100 |  |

Recently Harashima prepared a series of substituted Red-Al with even higher selectivity than Red-Al itself (eq 37) (*64*).

$$NaAlH_2(OCH_2CH_2OMe)_2 + ROH \longrightarrow NaAlHOR(OCH_2CH_2OMe)_2 + H_2 \qquad (37)$$

**Monoalkoxyaluminumtrihydride.** Our systematic study of the reaction of LAH and SAH with a series of alcohols, phenols, diols, triols, pri- and sec-amines using simultaneous hydride and $^{27}Al$ NMR analysis has identified several new trialkoxy and dialkoxy species derived from both hydride reagents. Most importantly, we observed that the reaction of SAH with tricyclohexylcarbinol provides a stable trihydrido species (eq 38). The $^{27}Al$ NMR spectra of this product in THF reveals a quartet at $\delta$ 107 ppm. With a second equiv of the carbinol, it forms a solid dihydroaluminum compound and adds no more of the carbinol (Ramachandran, P. V.; Gong, B., unpublished data).

$$NaAlH_4 + Chx_3COH \longrightarrow NaAlH_3OCChx_3 + H_2 \qquad (38)$$

Reductions by alkoxyaluminum hydrides have been thoroughly reviewed by Malek (*12, 13*).

**Alkoxyborohydrides.** Unlike the aluminohydrides, the alkoxyborohydrides cannot be synthesized by the treatment of sodium borohydride with the corresponding alcohols. They are prepared by the treatment of the borate esters with the corresponding metal hydrides in the absence of solvents (eq 39). However, they undergo rapid disproportionation in solvents (*65*).

$$4 \text{ (MeO)}_3\text{B} + 4 \text{ NaH} \xrightarrow{\text{neat}} 4 \text{ Na(MeO)}_3\text{BH} \xrightarrow{\text{THF}} \text{NaBH}_4 + 3 \text{ NaB(OMe)}_4 \quad (39)$$

Although trimethoxy- and triethoxyborohydrides disproportionate, the corresponding triisopropoxyborohydrides are stable (eq 40-42) (66, 67). They are mild reducing agents, similar to SBH or LTBA.

$$\text{NaH} + (i\text{-PrO})_3\text{B} \xrightarrow[170 \text{ h}]{\text{THF, reflux}} \text{Na}(i\text{-PrO})_3\text{BH} \quad (40)$$

$$\text{NaH} + (i\text{-PrO})_3\text{B} \xrightarrow[130 \, ^\circ\text{C, 1h}]{\text{Triglyme}} \text{Na}(i\text{-PrO})_3\text{BH} \quad (41)$$

$$\text{KH} + (i\text{-PrO})_3\text{B} \xrightarrow[1 \text{ h}]{\text{THF, RT}} \text{K}(i\text{-PrO})_3\text{BH} \quad \text{(KIPBH)} \quad (42)$$

**Aminoborohydrides.** Although sodium aminoborohydrides have been known for quite some time (68, 69), recently Singaram and coworkers described an efficient synthesis of lithium aminoborohydrides (eq. 43) (70). Unlike the alkoxyborohydrides, the aminoborohydrides are very powerful reducing agents that are capable of performing virtually all of the transformations for which LAH is currently used. Yet, the reagents are stable to air, similar to SBH.

$$\text{BH}_3 \cdot \text{SMe}_2 + \text{R}_2\text{NH} \xrightarrow[\text{SMe}_2]{} \text{R}_2\text{NH} \cdot \text{BH}_3 \xrightarrow{n\text{-BuLi}} \text{LiR}_2\text{NBH}_3 + n\text{-BuH} \quad (43)$$

A series of lithium aminoborohydrides of varying steric and electronic requirements have been synthesized. The chemistry of these reagents are reviewed by Singaram and coworkers in this book.

**Acyloxyborohydrides.** The treatment of SBH with carboxylic acids provides the corresponding acyloxyborohydrides (71). Gribble and coworkers showed the applicability of acyloxyborohydrides, especially sodium triacetoxyborohydride for reductions (eq. 44) (72). These reagents have extended the scope of SBH. They are selective in reducing aldehydes in the presence of ketones. Morover, α- and β-hydroxy ketones are reduced cleanly to *anti*-diols.

$$\text{NaBH}_4 + 3 \text{ RCOOH} \longrightarrow \text{NaBH(OCOR)}_3 + 3 \text{ H}_2 \quad (44)$$

Tetra-*n*-butylammonium- (73) and tetramethylammonium triacetoxyborohydride (eq 45-46) are more selective than the sodium counterpart for reductions. Evans described the synthesis of tetramethylammonium triacetoxyborohydride for the stereoselective reduction of β–hydroxy ketones to the corresponding *anti*-diols (74).

$$\text{NaBH}_4 + \text{Me}_4\text{NOH} \longrightarrow \text{Me}_4\text{NBH}_4 \quad (45)$$

$$\text{Me}_4\text{NBH}_4 + 3 \text{ AcOH} \longrightarrow \text{Me}_4\text{N(AcO)}_3\text{BH} \quad (46)$$

These derivatives are discussed in detail elsewhere in this book by Gribble and also by Abdel-Magid.

**Sulfurated Borohydride.** Lalancette and coworkers reported the synthesis of a sulfurated borohydride by the treatment of SBH with sulfur at room temperature in appropriate organic solvents (eq 47) (75). This reagent is capable of reducing oximes to the corresponding amines with yields depending on the steric requirement of the oxime (76).

$$NaBH_4 + 3\,S \longrightarrow NaBH_2S_3 + H_2 \quad (47)$$

**Alkylaluminohydrides.** The syntheses and reactions of lithium $n$-butyl- (77) and lithium $tert$-butyl(diisobutyl)aluminum hydrides (78) have been reported (eq 48).

$$i\text{-}Bu_2AlH + RLi \longrightarrow Li[i\text{-}Bu_2RAlH] \quad (R = n\text{-}Bu, t\text{-}Bu) \quad (48)$$

**Alkylborohydrides.**

**Trialkylborohydrides.** The addition of metal hydrides to trialkylboranes provide the corresponding borohydrides (eq 49). Although these compounds were prepared during WW-II research (33), the exceptional reducing power of these borohydrides were discovered during a study of the carbonylation of organoboranes catalyzed by LTBA (79).

$$LiH + Et_3B \xrightarrow[25\,^\circ C]{THF} LiEt_3BH \text{ (Super-Hydride)} \quad (49)$$

Our initial aim in the selective reductions project was to increase the reducing power of SBH to bring it closer to LAH in the spectrum of reagents. However, we encountered a borohydride, lithium triethylborohydride, that is far more powerful than LAH. Due to the superior hydridic qualities, the trialkylborohydrides have been termed "Super Hydrides". Increase in the steric bulk of the alkyl groups of these trialkylborohydrides make these reagents more selective than LAH without compromising their reductive capability.

Alternate methods for their synthesis were also discovered (80-82), especially for hindered trialkylborohydrides, such as tri-$sec$-butylborohydrides (eq 50).

$$LiAlH(OMe)_3 + sec\text{-}Bu_3B \xrightarrow[25\,^\circ C]{THF} \underset{\text{(L-Selectride)}}{Li\text{-}sec\text{-}Bu_3BH} + Al(OMe)_3 \quad (50)$$

Corey had shown that treatment of trialkylboranes with $t$-butyllithium provides the corresponding trialkylborohydrides (82). This was applied in the synthesis of Selectrides (eq 51) (80).

$$t\text{-}BuLi + Sia_3B \xrightarrow[-78\,^\circ C]{THF} LiSia_3BH \text{ (LS-Selectride)} \quad (51)$$

**Super Hydride.** Lithium triethylborohydride (Super Hydride) is used for reductive dehalogenation. It exhibits enormous nucleophilic power in $S_N2$ displacement reactions with alkyl halides, $10^4$ fold more powerful than $LiBH_4$ (eq 52) (83).

$$\begin{array}{c} R\text{-}Br \xrightarrow[RT,\,2\,min]{LiEt_3BH,\,THF} \\ R'\text{-}Br \xrightarrow[RT,\,3\,h]{LiEt_3BH,\,THF} \end{array} \longrightarrow \text{alkane} \quad (52)$$

It has been observed that the dehalogenation can be achieved by *in situ* generated Super Hydride, using lithium hydride and catalytic amounts of $Et_3B$ (eq 53) (*84*).

$$\text{ArCH}_2\text{Cl} \xrightarrow[\text{cat. Et}_3\text{B}]{\text{LiH}} \text{ArCH}_3 \tag{53}$$

It is stereospecific in the reductive opening of epoxides (eq. 54) (*85*).

$$\text{cyclohexene oxide} \xrightarrow{\text{LiEt}_3\text{BH}} \text{cyclohexanol} \quad \begin{array}{l}99\% \text{ yield}\\ \geq 99\% \text{ } tert\text{-}\end{array} \tag{54}$$

The exceptional nature of Super Hydride and its applications in organic reductions has been reviewed in detail (*86*).

Yoon and coworkers reported the preparation of potassium triethylborohydride (eq 55). This reagent is milder than the corresponding lithium analog (*87*).

$$\text{KH} + \text{Et}_3\text{B} \xrightarrow[25\,°\text{C, 24 h}]{\text{THF}} \text{KEt}_3\text{BH} \tag{55}$$

They also reported the synthesis of a bulky Super Hydride from triphenylborane (eq 56) (*88*).

$$\text{KH} + \text{Ph}_3\text{B} \xrightarrow[25\,°\text{C, 6h}]{\text{THF}} \text{KPh}_3\text{BH} \tag{56}$$

This reagent is capable of discriminating in the selective reduction of a mixture of 2- and 4-heptanones (eq 57) (*88*).

$$\begin{array}{c}\text{2-heptanone}\\ +\\ \text{4-heptanone}\end{array} \xrightarrow[-78\,°\text{C}]{\text{KPh}_3\text{BH}} \begin{array}{l}\text{2-heptanol} \quad 94\%\\ +\\ \text{4-heptanol} \quad 6\%\end{array} \tag{57}$$

**Selectrides.** Hindered trialkylborohydrides, such as K- (*66, 89*) and L-Selectrides (*90*) reduce cyclic and bicyclic ketones to the corresponding alcohols with remarkable stereoselectivity. LS-Selectride (*91*) achieves the best selectivity of them all (eq 58).

$$\text{ketone} \xrightarrow{H^-} \text{cis-OH} + \text{trans-OH}$$

|  | cis : trans |
|---|---|
| $NaBH_4$ | = 13 : 87 |
| L-Selectride | = 96.5 : 3.5 |
| LS-Selectride | = 99.5 : 0.5 |

(58)

**Mono- and Dialkylborohydrides.** We established convenient procedures for the preparation of mono- and dialkylborohydrides from the corresponding borinates and boronates, respectively, by treatment with LAH (eq 59-60) (*92, 93*).

$$RB(OR')_2 + LiAlH_4 \xrightarrow[0\,°C,\ 15\ min.]{n\text{-pentane-EE}} LiRBH_3 + AlH(OR')_2 \downarrow \quad (59)$$

R = Me, *n*-Bu, *t*-Bu, Ph, etc.

$$R_2BOR' + LiAlH_4 \xrightarrow[0\,°C,\ 15\ min.]{n\text{-pentane-EE}} LiR_2BH_2 + AlH_2OR' \quad (60)$$

R = Me, *n*-Bu, *t*-Bu, Ph, etc.

The monohydrodialkoxyalane produced from the reaction of the boronate (eq 59) precipitates out, whereas the dihydromonoalkoxyalane produced from the borinate (eq 60) is soluble in pentane-EE. Hence we used lithium monoethoxyaluminum hydride for the syntheses of dialkylborohydrides (eq 61) (*93*).

$$R_2BOR'_2 + LiAl(OEt)H_3 \xrightarrow[0\,°C,\ 15\ min.]{n\text{-pentane-EE}} LiR_2BH_2 + AlH(OEt)OR' \downarrow \quad (61)$$

R = Me, *n*-Bu, *t*-Bu, Ph, etc.

The boronates necessary for these reactions can be obtained via hydroboration reactions or from the alkylmetals or alkyl Grignard reagents as shown in eq 62 and 63 (*95*). These procedures provide an efficient general route to synthesize different types of alkylborohydrides, including lithium methylborohydride, that are not available via the routes discussed thus far.

$$RMgX + B(OMe)_3 \longrightarrow RB(OMe)_2 + MgX(OMe) \downarrow \quad (62)$$

$$RLi + B(OMe)_3 \longrightarrow RB(OMe)_2 + LiOMe \quad (63)$$

These procedures are applicable for the syntheses of optically active borohydrides (R = chiral) as well. The capability of these borohydrides in asymmetric reduction has not been explored. However, efficient procedures to convert these alkylborohydrides into the corresponding chiral alkylboranes have been established (*vide infra*).

**Cyanoborohydride.** Wittig synthesized the first cyanoborohydride by treating $LiBH_4$ with HCN (*96*). The corresponding reaction of sodium borohydride with hydrogen cyanide provides a white crystalline solid, sodium cyanoborohydride, which is a much milder and more selective reagent than the parent reagent (eq 64) (*97*). An improved process for the preparation of sodium cyanoborohydride involves the reaction of sodium cyanide with borane-THF (eq 64) (*98*). The stability of this reagent in acid solutions down to pH 3, and its solubility in THF, water, methanol, HMPA, DMF, sulfolane, etc. make it a unique reagent.

$$NaBH_4 + HCN \xrightarrow[H_2]{THF} NaBH_3CN \leftarrow BH_3 \cdot THF + NaCN \quad (64)$$

NaBH$_3$CN reduces alkyl halides and tosylates to the corresponding alkanes in the presence of a variety of reactive functional groups, such as aldehyde, ketone, epoxide, cyano, ester, carboxylic acid, amide, etc. (eq 65) (*99*).

$$Ph\text{-epoxide-CH}_2Br \xrightarrow[70\,°C,\,12\,h]{NaBH_3CN} Ph\text{-epoxide-CH}_3 \quad (65)$$

This reagent is widely used for the reduction of ketoximes to hydroxylamines (eq 66) (*100, 101*) and for the reductive amination of aldehydes and ketones (eq 67) (*100*).

$$\text{(sugar)}=N-OH \xrightarrow{NaBH_3CN} \text{(sugar)}-NHOH \quad (66)$$

$$\underset{R_2}{\overset{R_1}{>}}=O + H-N\underset{R_4}{\overset{R_3}{<}} \rightleftharpoons \underset{R_2}{\overset{R_1}{>}}=\overset{+}{N}\underset{R_4}{\overset{R_3}{<}} \xrightarrow{NaBH_3CN} \underset{R_2}{\overset{R_1}{>}}-N\underset{R_4}{\overset{R_3}{<}} \quad (67)$$

An unhindered ketone is selectively aminated in the presence of a relatively hindered ketone (eq 68) (*102*).

$$\text{diketosteroid} \xrightarrow[MeOH]{NH_4OAc,\,NaBH_3CN} \text{amino-ketosteroid} \quad (68)$$

Another important application of sodium cyanoborohydride is the reduction of tosylhydrazones to the corresponding alkanes (eq 69) (*103*).

$$\underset{R_2}{\overset{R_1}{>}}=O \longrightarrow \underset{R_2}{\overset{R_1}{>}}=NNHTs \xrightarrow{NaBH_3CN} \underset{R_2}{\overset{R_1}{>}} \quad (69)$$

The applications of sodium cyanoborohydride in organic syntheses have been reviewed earlier (*98, 104-105*).

## Acidic Reducing Agents

So far discussions were made of complex borohydrides and aluminohydrides whose application in reduction involve transfer of the hydride moiety to an electron deficient center of the substrate (*106,107*). In other words, the reagents are nucleophilic. Accordingly, substituents that enhance the electron deficiency at the reaction site increased the rate of reduction. For example, SBH in diglyme reduces chloral and acetyl chloride much more rapidly than aldehydes and ketones, for example, pivalaldehyde.

Rate of reaction with NaBH$_4$

On the other hand, diborane and alane are electron deficient molecules and hence behave as Lewis acids. Consequently, reduction involving these molecules are expected to involve an electrophilic attack on the center of highest electron density. Thus, pivalaldehyde is reduced much faster than chloral.

Rate of reaction with BH$_3$

**Alane.** Aluminum hydride, AlH$_3$, can be prepared by the treatment of LAH with AlCl$_3$ in EE. However, AlH$_3$ in EE is unstable and polymerizes relatively rapidly (*108*). This reagent is conveniently prepared by the addition of the calculated amount of 100% sulfuric acid to a standardized solution of LAH or SAH in THF (eq 70) (*109*).

$$2 \text{ LiAlH}_4 + \text{H}_2\text{SO}_4 \xrightarrow{\text{THF}} \text{Li}_2\text{SO}_4 \downarrow + 2 \text{ AlH}_3 + 2 \text{ H}_2 \uparrow \quad (70)$$

Apparently, coordination of the AlH$_3$ with the THF prevents the rapid association of the AlH$_3$ that is observed in EE.

AlH$_3$ is used for the selective reduction of carboxylic acid esters to the corresponding alcohols in the presence of halogen and nitro substituents (*110*). Another application of this reagent is for the reduction of *tert*-amides to the corresponding amines in excellent yields. This becomes especially important since this reduction is compatible for amides with unsaturation present. Utilization of borane-THF for this purpose results in concurrent hydroboration.

**Alane-Amine Complex.** Wiberg and coworkers reported a 1:1 and 1:2 complex of alane with trimethylamine in 1952 (*108*). Recently, Marlett and Park described the reducing power of AlH$_3$-amine complexes (*111*). The utility of alane-triethylamine complex was systematically studied by us (*112*). This complex permits the convenient use of alane in organic synthesis with high efficiency. SAH is preferred for the preparation of alane since the salt formed, NaCl, has very little solubility in THF and can be easily removed by filtration (eq 71-72).

$$NaAlH_4 + HCl \longrightarrow AlH_3 + NaCl\downarrow + H_2\uparrow \qquad (71)$$
$$AlH_3 + Et_3N \longrightarrow H_3Al\cdot NEt_3 \qquad (72)$$

**Dialkylalanes.** The preparation and reactions of dialkylalanes have been reviewed before (113). One of the most widely used dialkylalane is diisobutylaluminum hydride, DIBAL-H (114). A representative application of DIBAL-H is the reduction of α,β-unsaturated esters to the corresponding allylic alcohols (eq 73) (115).

$$\text{(73)}$$

**Borane.** Originally we carried out all of the reactions involving diborane in the gas phase in vacuum lines (25). Then we discovered that diborane can be conveniently generated by the treatment of SBH in diglyme with boron-trifluoride-etherate (eq 74) (116) and reactions were carried out by bubbling the gaseous diborane into solutions of the compound in EE, THF, or DG. Other methods of preparation are discussed in two early reviews (117, 118). We soon discovered that borane can be conveniently used as a monomer by complexing with a suitable ligand. Borane is now commonly used as a complex in THF, $H_3B\cdot THF$, or a dimethyl sulfide complex, $H_3B\cdot SMe_2$, used in THF or other solvents.

$$3\ NaBH_4 + 4\ BF_3 \xrightarrow{DG} 2\ B_2H_6\uparrow + 3\ NaBF_4 \qquad (74)$$

**Borane-Tetrahydrofuran.** The reactivity of borane depends on the complexing agent also, since the mechanism of reaction involves prior dissociation and formation of free borane (119). Borane-THF is prepared by passing gaseous diborane through THF (120, 121). Although a 4 $M$ solution can be prepared, it loses some of the borane upon storage and THF is slowly cleaved to give $n$-BuO–B< moieties. Hence it is currently marketed as a 1$M$ solution. This reagent is capable of reducing aldehydes, ketones, lactones, carboxylic acids, *tert*-amides, and nitriles (121). Acid chlorides, epoxides, and esters are reduced slowly. Borane-THF can tolerate a variety of functional groups. One of the important application of borane-THF has been in the rapid and quantitative reduction of carboxylic acids to the corresponding alcohols under remarkably mild conditions in the presence of various functional groups (eq 75) (122).

$$X\text{-}C_6H_4\text{-}COOH \xrightarrow{BH_3\cdot THF} X\text{-}C_6H_4\text{-}CH_2OH \qquad X = Cl, Br, I, NO_2, CN, COOEt, \text{etc.} \qquad (75)$$

The remarkable difference in the behavior of the borohydride and borane reagents toward the nature of the substrate has been exploited in the synthesis of both ($R$)- and ($S$)-mevalonolactone from a common precursor ( eq 76) (123).

$$\text{HO—CH(COOH)(COOMe)} \xrightarrow[\text{2. BH}_3\cdot\text{THF}]{\begin{array}{c}\text{LiBH}_4\\ \text{1. Ac}_2\text{O}\end{array}} \begin{array}{c}\text{HO—CH(COOH)(CH}_2\text{OH)}\\ \text{HO—CH(CH}_2\text{OH)(COOMe)}\end{array} \longrightarrow \begin{array}{c}(R)\text{-lactone}\\ (S)\text{-lactone}\end{array} \quad (76)$$

Recently, Arase and coworkers reported a lithium borohydride catalyzed selective reduction of the carbonyl group of conjugated and unconjugated alkenones with borane-THF (eq 77) (*124*). This methodology provides an efficient synthesis of allylic alcohols and other enols.

$$\text{CH}_2=\text{CHCH}_2\text{CH}_2\text{C(O)CH}_3 \xrightarrow[\substack{1\%\text{ LiBH}_4\\-50\,^\circ\text{C}}]{\text{BH}_3\cdot\text{THF}} \text{CH}_2=\text{CHCH}_2\text{CH}_2\text{CH(OH)CH}_3 \quad (77)$$

**Borane-Dimethyl Sulfide Complex.** Although borane can be used conveniently as borane-THF, the low concentration and the necessity to add trace amounts of SBH to stabilize the reagent (diminishing the cleavage of THF) made the introduction of alternate complexes of borane desirable. Adams and co-workers introduced the dimethyl sulfide complex (*125*) for hydroboration (*126*) and subsequent research proved this to be as efficient as borane-THF. The reagent can be stored as a neat material (10 *M*) and reactions can be carried out at much higher concentrations in a variety of aprotic solvents. Alternately, one can utilize commercial solutions of 2*M* BMS in THF. Such solutions exhibit long-term stability. The reagent is capable of reducing acids, esters, amides, nitriles etc. The hydroborations and reductions are made possible by free borane produced by dissociation. Our systematic study has shown that distillation of the free dimethyl sulfide from THF solutions aids in achieving fast reaction rates. This phenomenon was taken advantage in the facile reduction of carboxylic esters, and amides (eq 78) (*127*).

$$\text{RCOOEt} \xrightarrow{\text{BH}_3\cdot\text{SMe}_2} \text{RCH}_2\text{OH} \quad (78)$$

*pri*-Amides are reduced by one equiv of hydride from BMS whereas *sec*- and *tert*-amides need one equiv of the borane to complex with the amine product formed. This requirement for excess diborane is avoided by using one equiv of boron trifluoride-etherate. Thus, we have achieved efficient reduction of both aliphatic and aromatic *pri*-, *sec*-, and *tert*-amides (eq 79) (*127*).

$$\text{RCONR}'_2 \xrightarrow[\text{BF}_3\cdot\text{EE}]{2/3\ \text{BH}_3\cdot\text{SMe}_2} \text{RCH}_2\text{NR}'_2 \quad (79)$$

BMS reduces nitriles via the formation of borazines (eq 80). This mechanism alters the stoichiometry of the reaction so that three equiv of hydride are required to give a quantitative yield of the amine product (*127*).

$$3 \text{ RCN} + 3 \text{ BH}_3 \cdot \text{SMe}_2 \longrightarrow [\text{cyclic intermediate}] + 3 \text{ Me}_2\text{S} \longrightarrow 3 \text{ RCH}_2\text{NH}_2 \quad (80)$$

The applications of BMS in hydroborations, reductions and in organic syntheses have been reviewed on several occasions (128-133).

Several other sulfide complexes of borane have been reported. Adams observed that tetrahydrothiophene is a weaker base toward $BH_3$ than dimethyl sulfide. However, contrary to expectations, the complex is less reactive than BMS (126). We have shown that borane-1,4-thioxane complex (134) avoids the stench of the volatile dimethylsulfide component and more readily provides borane than BMS. We have since developed several new sulfide complexes for hydroboration (Brown, H. C.; Zaidlewicz, M., unpublished results). Several solid borane-sulfide complexes have also been reported in the literature (135).

**Borane-amines.** Although amine-boranes have been known for several decades (136, 137), they have not yet found significant use in organic reductions and syntheses as one might expect. This may be due to the strong coordination of the nitrogen lone pair and the boron, since dissociation appears to be a key step for reaction. We have now synthesized several new amine-boranes that are strong enough to form a stable complex, but weak enough to liberate free borane for hydroborations and reductions (Brown, H. C.; Zaidlewicz, M.; Dalvi, P. V., unpublished results).

One of the borane-amines currently available that deserve special attention is pyridine-borane (136). This reagent is capable of reducing aldehyde oximes to the corresponding hydroxylamines (eq 81) (138) and is also used for reductive aminations (139).

$$\underset{R_2}{\overset{R_1}{>}}=O \longrightarrow \underset{R_2}{\overset{R_1}{>}}=N-OH \xrightarrow{BH_3 \cdot \text{Pyridine}} \underset{R_2}{\overset{R_1}{>}}-NHOH \quad (81)$$

**Alkylboranes.** The synthesis of alkylboranes can be achieved from olefins, acetylenes, and dienes directly by hydroboration or from lithium alkylborohydrides by treatment with a proper acidic reagent.

**From Olefins via Hydroboration.** The hydroboration of simple alkenes with borane generally proceeds directly to the formation of the trialkylborane, $R_3B$. However, in a number of instances, it has proved possible to control the hydroboration reaction to achieve the synthesis of monoalkylboranes, $RBH_2$, dialkylboranes, $R_2BH$, and cyclic boranes (eq 82-85) (140-142).

$$2 \, \text{alkene} + BH_3 \longrightarrow (\text{Sia})_2BH \quad \text{Sia}_2BH \quad (82)$$

$$\text{>=<} + BH_3 \longrightarrow \text{>-<}-BH_2 \quad ThxBH_2 \qquad (83)$$

$$2 \text{ (α-pinene)} + BH_3 \longrightarrow (\text{pinanyl})_2BH \quad Ipc_2BH \qquad (84)$$

$$\text{(cyclooctadiene)} + BH_3 \longrightarrow H-B\text{(bicyclic)} \quad \text{9-BBN} \qquad (85)$$

These boranes have found unique applications in organic syntheses which have been discussed in several reviews and books (*128-133*). Of particular interest is diisopinocampheylborane, derived by the hydroboration of α-pinene. This reagent achieved the first non-enzymatic asymmetric reaction achieving very high ee by the hydroboration of a prochiral *cis*-olefin. *This reaction marked the beginning of non-enzymatic asymmetric synthesis in high ee.*

The α-pinene-boron moiety exhibits remarkable efficacy as a chiral auxiliary (Figure 1) (*143-146*).

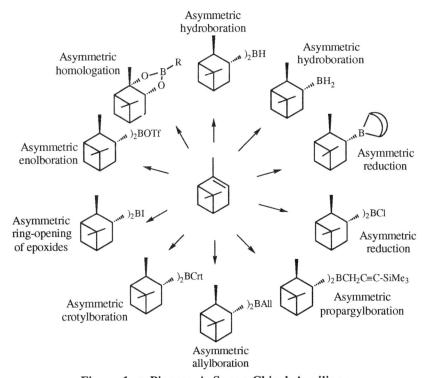

Figure 1. α-Pinene: A Super Chiral Auxiliary

**From Borohydrides.** While the boranes discussed above are prepared via direct hydroboration, there are several alkylboranes, methylborane for example, that can not be obtained via this route. We discovered an efficient general synthesis of mono- and dialkylboranes from the corresponding borohydrides (eq 86-87), which in turn can be obtained from the corresponding borinates and boronates, respectively (eq 59-60). This procedure allows the syntheses of boranes, such as methylborane and *t*-butylborane, etc. that are inaccessible via hydroboration.

$$LiR_2BH_2 + HCl \longrightarrow R_2BH + LiCl + H_2 \tag{86}$$
$$LiRBH_3 + HCl \longrightarrow RBH_2 + LiCl + H_2 \tag{87}$$

We examined convenient procedures for the generation of the boranes from the borohydrides (*147*). A major advantage of this procedure is the synthesis of optically pure boranes from optically pure borinates and boronates for a general synthesis of optically pure organic molecules (eq 88) (*148*).

$$LiR^*BH_3 + HCl \longrightarrow R^*BH_2 + LiCl + H_2 \tag{88}$$

Masamune applied this procedure for the synthesis of an optically active borolane (eq 89) (*149*).

(89)

The simple synthesis of the R*B< moiety through hydroboration with R*BH$_2$ and the ability to substitute the boron atom by other atoms and groups with complete retention makes possible a general asymmetric synthesis, as indicated in Figure 2. (Reactions that have been demonstrated experimentally are shown by arrows with solid lines.)

We have reviewed our general asymmetric synthetic procedures earlier (Figure 2) (*143-146*) and they will not be discussed here.

**Alkoxyboranes.** Due to their decreased Lewis acidity, dialkoxyboranes, such as 4,4,6-trimethyl-1,3,2-dioxaborinane (*150*) are very poor reducing agents. However, the acidity can be increased by using appropriate diols, such as catechol. Thus, the treatment of borane-THF and catechol readily provides catecholborane, a very mild reducing and hydroborating agent (eq 90) (*151*). The reducing characteristics of catecholborane have been explored in detail (*152*). This reagent has been a favorite for transition metal catalyzed hydroborations (*153*) and oxazaborolidine catalyzed asymmetric reductions (*154*).

(90)

**Haloboranes.** Reagents that are stronger Lewis acids than borane were synthesized by substituting hydrogen with halogen, such as bromine and chlorine. The synthesis was conveniently achieved by redistribution (eq 91-92) (*155*).

$$2 \ BX_3 \cdot SMe_2 + BH_3 \cdot SMe_2 \longrightarrow 3 \ BHX_2 \cdot SMe_2 \ (X = Cl, Br) \tag{91}$$
$$BX_3 \cdot SMe_2 + 2 \ BH_3 \cdot SMe_2 \longrightarrow 3 \ BH_2X \cdot SMe_2 \ (X = Cl, Br) \tag{92}$$

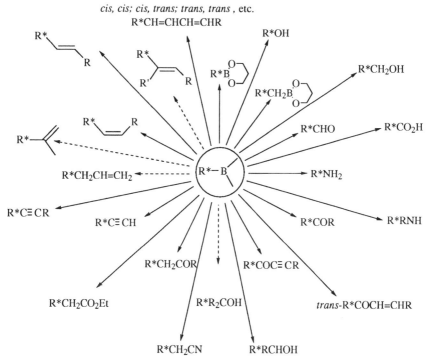

**Figure 2. A General Asymmetric Synthesis via Asymmetric Hydroboration**

Dichloroborane and monochloroborane are conveniently synthesized as the etherates by treating lithium borohydride with a stoichiometric quantity of boron trichloride in EE as the solvent (eq 93-94) (156).

$$\text{LiBH}_4 + \text{BCl}_3 \xrightarrow{\text{EE}} 2\text{ BH}_2\text{Cl}\cdot\text{EE} + \text{LiCl} \quad (93)$$

$$\text{LiBH}_4 + 3\text{ BCl}_3 \xrightarrow{\text{EE}} 4\text{ BHCl}_2\cdot\text{EE} + \text{LiCl} \quad (94)$$

These chloroborane reagents have found unique applications for selective reductions. For example, dichloroborane-methylsulfide reduces many types of azides to the corresponding amines in high yield (eq 95) (157).

$$\text{RN}_3 + \text{HBCl}_2\cdot\text{SMe}_2 \xrightarrow[\substack{\text{RT, 1 h} \\ \text{reflux, 1h}}]{\text{CH}_2\text{Cl}_2} \text{RNHBCl}_2 \xrightarrow[\text{2. KOH}]{\text{1. H}_3\text{O}^+} \text{RNH}_2 \quad (95)$$
$$\text{SMe}_2 + \text{N}_2$$

This reaction allows the selective reduction of an azide in the presence of an olefin by dichloroborane, or the hydroboration of an olefin in the presence of an azide by borane-THF (eq 96) (Salunkhe, A. M., unpublished results).

$$R\text{-}CH_2CH_2)_3B \xleftarrow{BH_3 \cdot THF} R\overset{}{=} + RN_3 \xrightarrow{HBCl_2 \cdot SMe_2} R\overset{}{=} + RNH_2 \quad (96)$$

Monochloroborane-dimethyl sulfide is an efficient reagent for the chemo- and regioselective opening of epoxides (eq 97, *158*).

$$R\overset{O}{\triangle} \xrightarrow[THF, -60\,°C]{H_2BCl \cdot SMe_2} R\underset{}{\overset{Cl}{\curlywedge}}OH + R\underset{}{\overset{OH}{\curlywedge}}Cl \quad (97)$$

R = Ph,    94 : 6
R = n-pent,  6 : 94

α-Oxysubstituted epoxides are cleaved to provide the corresponding regio- and stereodefined *anti*-chlorohydrin (eq 98) (*159*). The reaction is believed to proceed with anchimeric assistance from the side-chain oxygen.

$$R\overset{O}{\triangle}\text{-}OR' \xrightarrow[-60\,°C]{H_2BCl \cdot SMe_2} R\underset{OH}{\overset{Cl}{\curlywedge}}OR' \quad (98)$$

**Alkylchloroboranes.** Chloroborane-dimethylsulfide hydroborates two equiv of unhindered olefins to provide the corresponding dialkylchloroborane. One such compound, *B*-chlorodiisopinocampheylborane (Aldrich: DIP-Chloride) has been shown to be an excellent asymmetric reducing agent (eq 99) (*160*). This reagent is discussed by us in another chapter in this book.

$$2 \xrightarrow[Me_2S]{H_2BCl \cdot SMe_2} (\text{Ipc})_2BCl \xrightarrow[EE, -25\,°C]{Ph\overset{O}{\curlywedge}} Ph\overset{OH}{\curlywedge} \quad (99)$$
                                                                98% ee

When the olefin is hindered, such as 2,3-dimethyl-2-butene (thexylene), the hydroboration stops at the monoalkylchloroborane stage providing thexylchloroborane-dimethylsulfide (eq 100) (*161*). This reagent reveals certain unique properties. For example, it reduces carboxylic acids partially, efficiently and quantitatively, to the corresponding aldehydes (eq 101) (*162*).

$$\overset{}{>}=\overset{}{<} \xrightarrow{BH_2Cl \cdot SMe_2} \overset{}{>}\!\!-\!\!\overset{}{<}\text{-}BHCl \cdot SMe_2 \quad (ThxBHCl \cdot SMe_2) \quad (100)$$

$$RCOOH \xrightarrow{ThxBHCl \cdot SMe_2} RCHO \quad (101)$$

**Chemoselectivity and Stereoselectivity in Reductions**

Chemoselectivity is important in organic reductions. Many of the modified reagents discussed in this review are capable of achieving chemoselective reduction.

Stereoselectivity in acyclic and cyclic reductions is achieved by controlling the steric requirements of the reagents. These subjects including the mechanistic aspects of acyclic and cyclic stereoselection (Cram, Cornforth, anti-Cram, Karabatsos, and Felkin models) have been reviewed in detail by Greeves (163).

Table 2 summarizes the steric control achieved by several reagents discussed in this review for the reduction of 2-methylcyclohexanone.

Table 2. Diastereoselective Reduction of 2-Methylcyclohexanone

| Reagent | % cis | % trans |
|---|---|---|
| LAH | 25 | 65 |
| SBH | 31 | 69 |
| LBH | 30 | 70 |
| LTMA | 31-72[a] | 69-28[a] |
| LTBA | 27 | 73 |
| L-Selectride | 99.3 | 0.7 |
| K-Selectride | >99 | <1 |
| LiSia$_3$BH | 99.7 | 0.3 |
| KIPBH | 92 | 8 |
| Diborane | 74 | 26 |
| Sia$_2$BH | 79 | 21 |
| Chx$_2$BH | 94 | 6 |
| 9-BBN | 40 | 60 |
| ThxBHCl•SMe$_2$ | 95 | 5 |

[a]Depends on the solvent.

One of the interesting applications of borohydride reductions in organic syntheses has been the stereoselctive reduction of β-diketones or β-hydroxy ketones to diols, an especially valuable transformation due to the importance of such polyols in antibiotics and other natural products. Hence methodologies for the selective synthesis of either the *syn*- or *anti*-diol from the corresponding hydroxy ketone have been developed. Naraska (164) and Prasad (165) developed methods for the synthesis of essentially pure *syn* diols and Evans (166) developed procedures for the preparation of essentially pure *anti*-diols (eq 102).

(102)

## Asymmetric Reduction

We have discussed the beginnings of diborane, sodium borohydride, lithium aluminum hydride, and many modifications thereof. Ever since chemists became familiar with these reagents, they have modified them with different chiral auxiliaries to achieve asymmetric reduction. Although the initial attempts by Bothner-By (*167*) and Landor and coworkers (*168*) to modify LAH did not achieve much asymmetry in the reduction of ketones, subsequent research has led to several quite successful asymmetric reducing agents.

**LAH Modified Reagents.** Several alcohols, amines, amino alcohols, diols and triols have been used to modifiy LAH to prepare asymmetric reducing agents. This subject has been reviewed several times (*169-173*). The following are some of the most successful reagents developed by Mosher (*174*), Mukaiyama (*175*), Terashima (*176*), Fujisawa (*177*), Vigneron (*178*) and Noyori (*179*).

LiAlH$_4$ + Darvon Alcohol (*174*)    LiAlH$_4$ + Diamine (*175*)

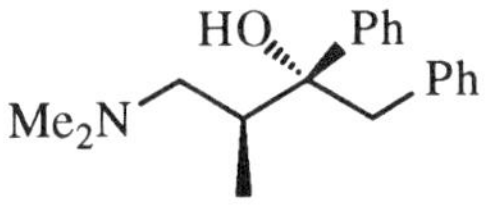

Darvon alcohol

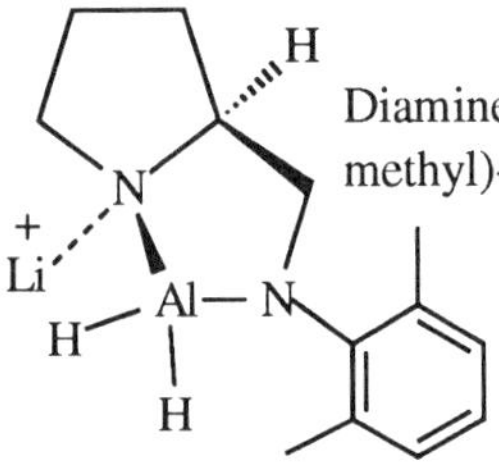

Diamine = (S)- -(2,6-Xylidino-methyl)-pyrrolidine

LiAlH$_4$ + MEP + NEA (*176*)    LiAlH$_4$ + Aminobutanol (*177*)

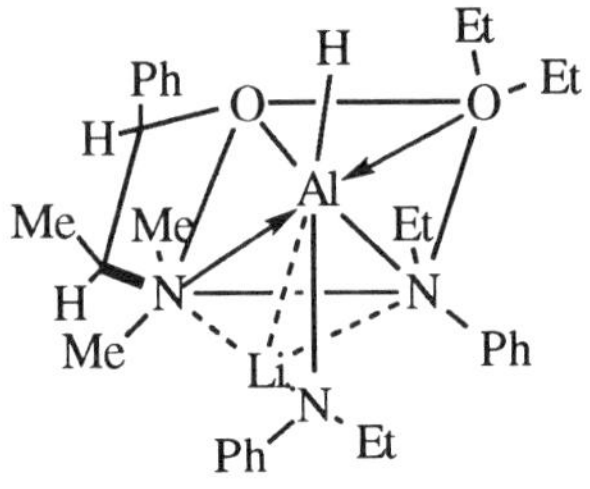

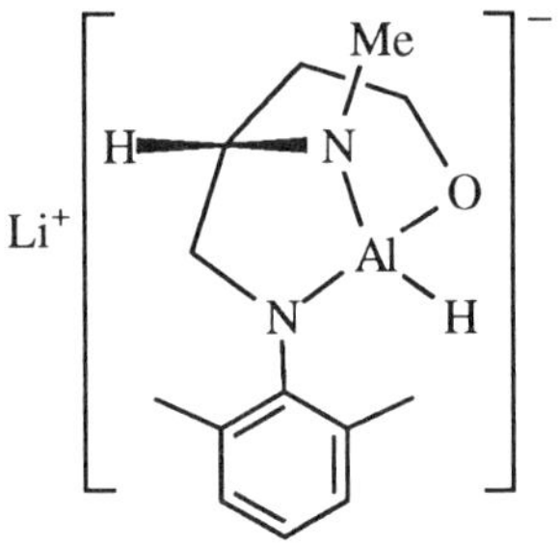

LiAlH$_4$ + MEP + ArOH (*178*)    LiAlH$_4$ + Binaphthol + EtOH (*179*)

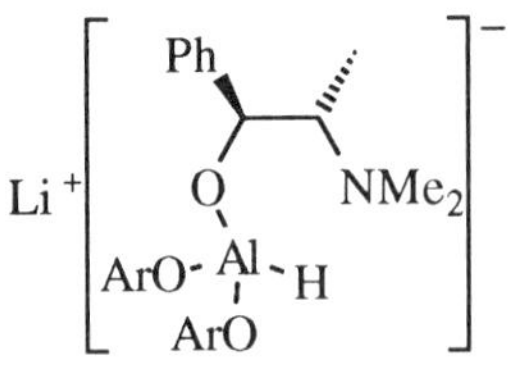

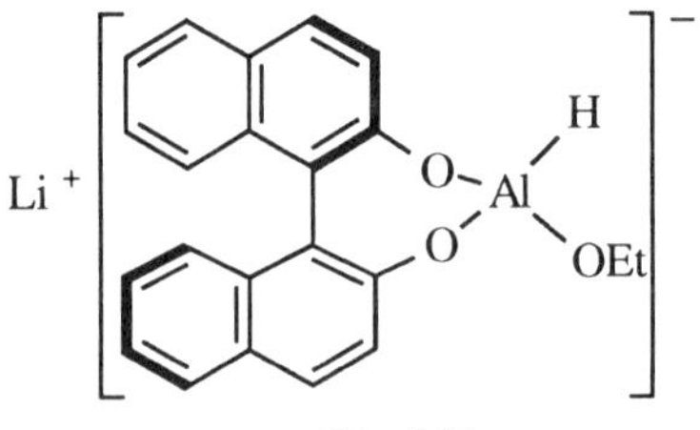

ArOH = 3,5-dimethyl phenol

Binal-H

**SAH modified Reagents.** We synthesized the corresponding reagents by treating sodium aluminum hydride with Darvon alcohol, N-methylephedrine, menthol, and binaphthol and carried out the asymmetric reduction of prochiral ketones. In all of the cases studied, we obtained results similar to or slightly better than those obtained with the lithium analog (Ramachandran, P. V.; Gong, B., unpublished data).

**Borohydride Reagents.** Attempts to modify SBH with several optically active acids, including amino acids, have not led to any highly successful reducing agent thus far. This subject has been reviewed previously (*173, 180*). Morrison (*181*) and Hirao (*182*) and their coworkers modified SBH with two equiv of carboxylic acid and 1,2:5,6-di-*O*-isopropylidene-D-glucofuranose. Yamada and coworkers prepared a reagent from SBH and *N*-benzyloxycarbonylproline that achieves high ee for the reduction of imines (*183*) Modifications of SBH with mandelic (*184*), lactic (*185*), tartaric (*186, 187*), camphanic or malic acids (*187*) have also been reported.

$NaBH_4$ + *i*-Pr-COOH + DIPGF

(DIPGF)

$NaBH_4$ + 3 [proline-COOBnz structure]
COOBnz

However, Soai and coworkers transformed lithium borohydride using *N, N*-dibenzoylcysteine (DBC) and *tert*-butanol into a reagent capable of reducing aromatic β-keto esters to the corresponding hydroxy esters in very high ee (*188*).

$LiBH_4$ + DBC + *t*-BuOH    DBC = $(PhCONH-CH(COOH)-CH_2-S-)_2$

We have shown that the same results can be achieved by preparing the reagent from sodium borohydride and catalytic amounts of lithium borohydride or lithium chloride (Ramachandran, P. V.; Teodorovic´, A. V., unpublished data).

We synthesized several chiral dialkylmonoalkoxy- and alkyldialkoxyborohydrides from the corresponding borinates or boronates by treatment with KH and tested them for the reduction of ketones (*189*). We identified potassium 9-*O*-(1,2;5,6-di-*O*-isopropylidene-α-*D*-glucofuranosyl)-9-boratabicyclo-[3.3.1]nonane (K-Glucoride) as an unusually efficient reagent for the reduction of hindered aromatic ketones, and α-keto esters (*190*). Hutchins and coworkers applied this reagent for the reduction of phosphinamides (*191*) and Cho and coworkers used it for the reduction of imines (*192*).

K⁺ [K-Glucoride structure]    K-Glucoride

**Chiral Super Hydrides.** Lithium (*B*-isopinocampheyl-9-borabicyclo[3.3.1]nonyl) hydride (Alpine-Hydride) prepared from α-pinene is a poor reagent for the reduction of ketones (*193*). Midland synthesized the corresponding borohydride from nopol benzyl ether (NB-Enantride) and it proved to be an excellent reagent for the reduction of straight chain aliphatic ketones (*194*). Several analogs of this reagent have been prepared that are moderately successful (*195*) (Weissman, S. A.; Ramachandran, P. V. *Tetrahedron Lett.*, in press).

$$Li^+ \left[ \begin{array}{c} \text{structure with B-H and R group on bicyclic} \end{array} \right]^-$$

R = H (Alpine-Hydride)
R = OMe
R = NEt$_2$
R = N(i-Bu)$_2$
R = CH$_3$ (Eapine-Hydride)
R = CH$_2$OMe
R = CH$_2$OBnz (NB-Enantride)
R = CH$_2$NEt$_2$
R = CH$_2$N(i-Bu)$_2$

**Oxazaborolidines.** Although several chemists attempted the modifications of borane with different chiral auxiliaries (*180*), it was the systematic study of Hirao, Itsuno and coworkers using amino alcohols, derived from amino acids, that led to superior borane-modified reagents for asymmetric reduction (*196*). Itsuno and coworkers revealed the catalytic nature of the aminoalcohol-borane system (*197*). Corey and coworkers identified the catalyst as oxazaborolidines (*198*). A flood of oxazaborolidines that achieve moderate to good enantiomeric excess have been reported in the literature since then and continues to be reported. This subject has been reviewed before (*199*) and is also the subject of a chapter by Quallich in this book.

Buono and coworkers have shown that oxazaphospholidines also act as asymmetric catalytsts for reductions with borane. However the ee achieved is much poorer than those of the boron counterpart (*200*).

## Conclusions

Persuaded by Alfred Stock's book entitled "The Hydrides of Boron and Silicon" which he received as a graduation gift in 1936 from his classmate Sarah Baylen (now his wife), the senior author undertook research with Professor H. I. Schlesinger and Dr. A. B. Burg, exploring the chemistry of diborane.

The Ph.D. study of the reaction of diborane with aldehydes and ketones opened up the hydride era of organic reductions. The study of this reaction led to the discovery of alkali metal borohydrides. Study of the reducing characteristics of the aluminohydrides and borohydrides led to the addition of a wide variety of reducing agents for selective reductions to the organic chemist's arsenal. The capabilities of several of the reagents that fill in the spectrum between and beyond sodium borohydride and lithium aluminumhydride are summarized in Table 3.

Study of the reducing characteristics of sodium borohydride led to the discovery of hydroboration. Hydroboration provided both simple synthetic routes to organoboranes and a wide variety of organoborane reagents. Investigations established that these compounds have a most versatile chemistry. This study provided in 1961 the first non-enzymatic asymmetric synthesis in high ee and opened another significant area of research to chemists. Clearly, a Major New Continent of Chemistry was discovered sixty years ago. The landmarks of the sixty-year hydride reduction projects are the following.

1. Beginnings
2. Volatile Compounds of Uranium
3. Alkali Metal Route to Diborane
4. Alkali Metal Borohydrides
5. Selective Reductions
6. Hydroborations
7. Versatile Organoboranes
8. Asymmetric Hydroboration
9. Asymmetric Synthesis Made Easy
10. α-Pinene: Superior Chiral Auxiliary
11. Asymmetric Reductions
12. Asymmetric Allyl- and Crotylboration
13. Asymmetric Enolborations
14. Asymmetric Opening of Meso Epoxides

Table 3. Broad spectrum of selective reducing agents

| | KIPBH | NaBH$_4$ | LTBA | LiBH$_4$ | Al(BH$_4$)$_3$ | BH$_3$·THF | Sia$_2$BH | 9-BBN | AlH$_3$ | LTMA | LiAlH$_4$ | LiEt$_3$BH |
|---|---|---|---|---|---|---|---|---|---|---|---|---|
| RCHO | + | + | + | + | + | + | + | + | + | + | + | + |
| R$_2$CO | + | + | + | + | + | + | + | + | + | + | + | + |
| RCOCl | + | + | + | + | + | − | − | + | + | + | + | + |
| RCO$_2$R' | − | − | ± | + | + | ± | − | ± | + | + | + | + |
| RCO$_2$H | − | − | − | − | + | + | + | ± | + | + | + | + |
| RCONR'$_2$ | − | − | − | − | − | + | + | + | + | + | + | − |
| RCN | − | − | − | − | + | − | ± | + | + | + | + | + |
| RNO$_2$ | − | − | − | − | − | − | − | + | + | + | + | + |
| RCH=CH$_2$ | − | − | − | − | − | + | + | + | − | − | − | − |

It will require a new generation of chemists to continue this exploration and apply the riches of the New Continent for the good of Mankind.

## Acknowledgment

The financial assistance from the United States Army Research Office for this program for the past 40+ years which made most of the study reported in this chapter possible is gratefully acknowledged.

## Literature Cited

1. Brown, H. C. *Boranes in Organic Synthesis*; Cornell University Press: Ithaca, NY, 1972.
2. Brown, H. C.; Krishnamurthy, S. *Tetrahedron* **1979**, *35*, 567.
3. Pizey, J. S. *Synthetic Reagents*; Ellis Horwood: Chichester, England, 1974, Vol 1; pp 101-204.
4. Seyden-Penne, J. *Reductions by the Alumino- and Borohydrides in Organic Synthesis*; VCH-Lavoier: Paris, 1991.
5. Hajos, A. *Complex Hydrides and Related Reactions in Organic Synthesis*; Elsevier: Amsterdam, 1979.
6. Walker, E. R. H. *Chem. Soc. Rev.* **1976**, *5*, 23.
7. Hudlicky, M. *Reductions in Organic Chemistry*; Horwood: Chichester, 1984.
8. Brown, W. G. *Org. React.* **1951**, *6*, 469.
9. Muetterties, E. L. *Boron Hydride Chemistry*; Academic: New York, NY, 1975.
10. House, H. O. *Modern Synthetic Reactions*; Benjamin/Cummings: Reading, MA, 1972.

11. Gaylord, N. G. *Reduction with Complex Metal Hydrides*; Interscience: New York, NY, 1956.
12. Malek, J. *Org. React.* **1985**, *34*, 1.
13. Malek, J. *Org. React.* **1988**, *36*, 249.
14. Trost, B. M.; Fleming, I. Eds. *Comprehensive Organic Chemistry*; Pergamon: Elmsford, NY, 1991.
15. Paquette, L. A. Ed. *Encyclopedia of Reagents for Organic Synthesis*; John Wiley: Chichester, England, 1995.
16. Bouis, J.; Carlet, H. *Liebigs Ann. Chem.* **1862**, *124*, 352.
17. Thoms, H.; Mannich, C. *Ber. Dtsch. Chem. Ges.* **1903**, *36*, 2544.
18. Wiselogle, F. Y.; Sonneborn III, H. **1943**, *Org. Synth. Coll. Vol. I,* 317.
19. Verley, A. *Bull. Soc. Chim. Fr.* **1925**, *37*, 537.
20. Meerwein, H.; Schmidt, R. *Liebigs. Ann. Chem.* **1925**, *444*, 221.
21. Ponndorf, W. *Z. Angew. Chem.* **1926**, *39*, 138.
22. Wilds, A. L. *Org. React.* **1944**, *2*, 178.
23. Stock, A. *Hydrides of Boron and Silicon*; Cornell University Press: Ithaca, NY 1933.
24. Burg, A. B.; Schlesinger, H. I. *J. Am. Chem. Soc.* **1937**, *59*, 780.
25. Brown, H. C.; Schlesinger, H. I.; Burg, A. B. *J. Am. Chem. Soc.* **1939**, *61*, 673.
26. Schlesinger, H. I.; Sanderson, R. T.; Burg, A. B. *J. Am. Chem. Soc.* **1940**, *62*, 3421.
27. Schlesinger, H. I.; Burg, A. B. *J. Am. Chem. Soc.* **1940**, *62*, 3425.
28. Schlesinger, H. I.; Brown, H. C.; Abraham, B. A.; Davidson, A. E.; Finholt, A. E.; Lad, R. A.; Knight, J.; Schwartz, A. M. *J. Am. Chem. Soc.* **1953**, *75*, 191.
29. Schlesinger, H. I.; Brown, H. C. *J. Am. Chem. Soc.* **1953**, *75*, 219.
30. Schlesinger, H. I.; Brown, H. C.; Gilbreath, J. R. Katz, J. J. *J. Am. Chem. Soc.* **1953**, *75*, 195.
31. Schlesinger, H. I.; Brown, H. C.; Hoekstra, H. R.; Rapp. L. R. *J. Am. Chem. Soc.* **1953**, *75*, 199.
32. Schlesinger, H. I.; Brown, H. C.; Hyde, E. K. *J. Am. Chem. Soc.* **1953**, *75*, 209.
33. Schlesinger, H. I.; Brown, H. C.; Sheft, I.; Ritter, D. M. *J. Am. Chem. Soc.* **1953**, *75*, 192.
34. Schlesinger, H. I.; Brown, H. C.; Finholt, A. E. *J. Am. Chem. Soc.* **1953**, *75*, 205.
35. Finholt, A. E.; Bond, Jr., A. C.; Schlesinger, H. I. *J. Am. Chem. Soc.* **1947**, *69*, 1199.
36. Chaikin, S. W.; Brown, W. G. *J. Am. Chem. Soc.* **1949**, *71*, 122.
37. Brown, H. C.; Krishnamurthy, S. *J. Org. Chem.* **1969**, *34*, 3918.
38. Brown, H. C.; Tsukamoto, A. *J. Am. Chem. Soc.* **1961**, *83*, 4549.
39. Brown, H. C.; Weissman, P. M.; Yoon, N. M. *J. Am. Chem. Soc.* **1966**, *88*, 1458.
40. Cha, J. S.; Brown, H. C. *J. Org. Chem.* **1993**, *58*, 4727.
41. Nystrom, R. F.; Chaikin, S. W.; Brown, W. G. *J. Am. Chem. Soc.* **1949**, *71*, 3245.
42. Brown, H. C.; Choi, Y. M.; Narasimhan, S. *Inorg. Chem.* **1981**, *20*, 4454.
43. Yoon, N. M.; Cha, J. S. *J. Kor. Chem. Soc.* **1977**, *21*, 108.
44. Kollonitsch, J.; Fuchs, P.; Gabor, V. *Nature* **1954**, *173*, 125.
45. Kollonitsch, J.; Fuchs, P.; Gabor, V. *Nature* **1956**, *175*, 346.
46. Gensler, W. J.; Johnson, F.; Sloan, A. D. B. *J. Am. Chem. Soc.* **1960**, *82*, 6074.
47. Ranu, B. C. *Synlett* **1993**, 885.
48. Noth, H.; Wiberg, E.; Winter, L. P. *Z. Anorg. Allg. Chem.* **1969**, *370*, 209.
49. Yoon, N. M.; Lee, H. J.; Kim, H. K.; Kang, J. *J. Kor. Chem. Soc.* **1976**, *20*, 59.

50. Brown, H. C.; Subba Rao, B. C. *J. Am. Chem. Soc.* **1956**, *78*, 2582.
51. Brown, H. C.; Subba Rao, B. C. *J. Org. Chem.* **1957**, *22*, 1136.
52. Brown, H. C.; Subba Rao, B. C. *J. Am. Chem. Soc.* **1959**, *81*, 6423.
53. Noth, H. *Proc. Hydride Symposium II* Metalgesellschaft AG, Frankfurt, Germany, p.51.
54. Brown, H. C.; McFarlin, R. F. *J. Am. Chem. Soc.* **1958**, *80*, 5372.
55. Brown, H. C.; Shoaf, C. J. *J. Am. Chem. Soc.* **1964**, *86*, 1079.
56. Brown, H. C.; Weissman, P. M. *Isr. J. Chem.* **1963**, *1*, 430.
57. Brown, H. C.; Subba Rao, B. C. *J. Am. Chem. Soc.* **1958**, *80*, 5377.
58. Cha, J. S.; Brown, H. C. *J. Org. Chem.* **1993**, *58*, 4732.
59. Brown, H. C.; Garg, C. P. *J. Am. Chem. Soc.* **1964**, *86*, 1085.
60. Brown, H. C.; Tsukamoto, A. *J. Am. Chem. Soc.* **1964**, *86*, 1089.
61. Semmelhack, M. F.; Stauffer, R. D. *J. Org. Chem.* **1975**, *40*, 3619.
62. Bazant, V.; Capka, M.; Cerny, M.; Chvalosky, V.; Kochloefl, K.; Kraus, M.; Malek, J. *Tetrahedron Lett.* **1968**, 3303.
63. Viti, S. M. *Tetrahedron Lett.* **1982**, *23*, 4541.
64. Harashima, S.; Oda, O.; Amemiya, S.; Kojima, K. *Tetrahedron* **1991**, *47*, 2773.
65. Brown, H. C.; Mead, E. J.; Shoaf, C. J. *J. Am. Chem. Soc.* **1956**, *78*, 3616.
66. Brown, C. A. *J. Am. Chem. Soc.* **1973**, *95*, 4100.
67. Brown, C. A.; Krishnamurthy, S.; Kim, S. C. *J. Chem. Soc. Chem. Commun.* **1973**, 373.
68. Aftandilian, V. D.; Miller, H. C.; Muetterties, E. L. *J. Am. Chem. Soc.* **1961**, *83*, 2471.
69. Hutchins, R. O.; Learn, K.; El-Telbany, F.; Stercho, Y. P. *J. Org. Chem.* **1984**, *49*, 2438.
70. Fisher, G. B.; Fuller, J. C.; Harrison, J.; Alvarez, S. G.; Burkhardt, E. R.; Goralski, C. T.; Singaram, B. *J. Org. Chem.* **1994**, *59*, 2438.
71. Brown, H. C. Subba Rao, B. C. *J. Am. Chem. Soc.* **1960**, *82*, 681.
72. Gribble, G. W.; Ferguson, W. S. *J. Chem. Soc. Chem. Commun.* **1975**, 535.
73. Nutaitis, C. F.; Gribble, G. W. *Tetrahedron Lett.* **1983**, *24*, 4287.
74. Evans, D. A.; Chapman, K. T.; *Tetrahedron Lett.* **1986**, *27*, 5939.
75. Lalancette, J. M.; Frêche, A.; Monteux, R. *Can. J. Chem.* **1968**, *46*, 2754.
76. Lalancette, J. M.; Brindle, J. R. *Can. J. Chem.* **1970**, *48*, 6378.
77. Kim, S.; Ahn, K. H.; Chung, Y. W. *J. Org.Chem.* **1982**, *47*, 4581.
78. Kim, S.; Ahn, K. H.; *J. Org.Chem.* **1984**, *49*, 1717.
79. Khuri, A. Ph. D. Thesis, Purdue University, West Lafayette, IN, 1960.
80. Brown, C. A.; Krishnamurthy, S. *J. Organometal. Chem.* **1978**, *156*, 111.
81. Binger, P.; Benedikt, G.; Rotermund, G. W. Köster, R. *Liebigs Ann. Chem.* **1968**, *717*, 21.
82. Corey, E. J.; Albonico, S. M.; Koelliker, U.; Schaaf, T. K.; Varma, R. K. *J. Am. Chem. Soc.* **1971**, *93*, 1491.
83. Brown, H. C.; Krishnamurthy, S. *J. Org. Chem.* **1983**, *48*, 3085.
84. Brown, H. C.; Ramachandran, P. V.; Teodorovic, A.; Swaminathan, S. *Tetrahedron Lett.* **1991**, *32*, 6691.
85. Krishnamurthy, S.; Schubert, R. M.; Brown, H. C. *J. Am. Chem. Soc.* **1973**, *95*, 8486.
86. Krishnamurthy, S. *Aldrichim. Acta* **1974**, *7*, 55.
87. Yoon, N. M. Yang, H. S.; Hwang, Y. S. *Bull. Kor. Chem. Soc.* **1987**, *8*, 285.
88. Yoon, N. M.; Kim, K. E.; Kang, J. *J. Org. Chem.* **1986**, *51*, 226.
89. Ganem, B. *J. Org. Chem.* **1975**, *40*, 156.
90. Brown, H. C.; Krishnamurthy, S. *J. Am. Chem. Soc.* **1972**, *94*, 7159.
91. Krishnamurthy, S.; Brown, H. C. *J. Am. Chem. Soc.* **1976**, *98*, 3383.
92. Singaram, B.; Cole, T. E.; Brown, H. C. *Organometallics* **1984**, *3*, 774.
93. Singaram, B.; Cole, T. E.; Brown, H. C. *Organometallics* **1984**, *3*, 1520.

94. Snyder, H. R.; Kuck, J. A.; Johnson, J. R. *J. Am. Chem. Soc.* **1938**, *60*, 105.
95. Srebnik, M.; Cole, T. E.; Ramachandran, P. V.; Brown, H. C. *J. Org. Chem.* **1989**, *54*, 6085.
96. Wittig, G. *Liebigs Ann. Chem.* **1951**, *573*, 209.
97. Wade, R. C.; Sullivan, E. A.; Berchied, Jr. J. R.; Purcell, K. F. *Inorg. Chem.* **1970**, *9*, 2146.
98. Sullivan, E. A. *The Alembic* **1991**, *44*, 1.
99. Hutchins, R. O.; Kandasamy, D.; Maryanoff, C. A.; Masilamani, D.; Mayanoff, B. E. *J. Org. Chem.* **1977**, *42*, 82.
100. Borch, R. F.; Bernstein, M. D.; Durst, H. D. *J. Am. Chem. Soc.* **1971**, *93*, 2897.
101. Tronchet, J. M. J.; Bizzozero, N.; Geoffroy, M. *Carbohydr. Res.* **1989**, *191*, 138.
102. Boutique, M. -H.; Jacquesy, R. *Bull. Soc. Chim. Fr.* **1973**, 750.
103. Hutchins, R. O.; Milewski, C. A.; Maryanoff, B. E. *J. Am. Chem. Soc.* **1971**, *93*, 1793.
104. Lane, C. F. *Synthesis* **1975**, *135*.
105. Hutchins, R. O.; Natale, N. R. *Org. Prep. Proc. Int.* **1979**, *11*, 201.
106. Brown, H. C.; Wheeler, O. H.; Ichikawa, K. *Tetrahedron* **1957**, *1*, 214.
107. Trevoy, W.; Brown, W. G. *J. Am. Chem. Soc.* **1949**, *71*, 1675.
108. Wiberg, E.; Graf, H.; Schmidt, M.; Uson, R. *Z. Naturforsch.* **1952**, *76*, 578.
109. Brown, H. C.; Yoon, N. M. *J. Am. Chem. Soc.* **1966**, *88*, 1464.
110. Yoon, N. M.; Brown, H. C. *J. Am. Chem. Soc.* **1968**, *90*, 2927.
111. Marlett.; E. M.; Park, W. S. *J. Org. Chem.* **1990**, *55*, 2968.
112. Cha, J. S.; Brown, H. C. *J. Org. Chem.* **1993**, *58*, 3974.
113. Mole, T.; Jeffery, E. A. *Organoaluminium Compounds*; Elsevier: Amsterdam, 1972.
114. Zweifel, G.; Miller, J. A. *Org. React.* **1984**, *32*, 375.
115. Daniewski, A. R.; Wojceichowska, W. *J. Org. Chem.* **1982**, *47*, 2993.
116. Zweifel, G.; Brown, H. C. *Org. React.* **1963**, *13*, 1.
117. Lane, C. F. *Chem. Rev.* **1976**, *76*, 773.
118. Long, L. H. *Prog. Inorg. Chem.* **1972**, *15*, 1.
119. Brown, H. C.; Chandrasekharan, J.; Wang, K. K. *Pure Appl. Chem.* **1983**,*55*, 1387.
120. Elliott, J. R.; Roth, W. L.; Roedel, G. F.; Boldebuck, E. M. *J. Am. Chem. Soc.* **1952**, *74*, 5211.
121. Brown, H. C.; Heim, P. Yoon, N. M. *J. Am. Chem. Soc.* **1970**, *92*, 1637.
122. Yoon, N. M.; Pak, C. S.; Brown, H. C.; Krishnamurthy, S.; Stocky, T. P. *J. Org. Chem.* **1973**, *38*, 2786.
123. Huang, F. C.; Lee, L. F. H.; Mittal, R. S. D.; Ravikumar, P. R.; Chan, J. A.; Sih, C. J. *J. Am. Chem. Soc.* **1975**, *97*, 4144.
124. Arase, A.; Hoshi, M.; Yamaki, T.; Nakanishi, H. *J. Chem. Soc. Chem. Commun.* **1994**, 855.
125. Coyle, T. D.; Kaesz, H. D.; Stone, F. G. A. *J. Am. Chem. Soc.* **1959**, *81*, 2989.
126. Braun, L. M.; Braun, R. A.; Crissman, H. R.; Opperman, M.; Adams, R. M. *J. Org. Chem.* **1971**, *36*, 2388.
127. Brown, H. C.; Choi, Y. M.; Narasimhan, S. *J. Org. Chem.* **1982**, *47*, 3153.
128. Brown, H. C. *Hydroboration*; Benjamin: Reading, MA, 1962.
129. Brown, H. C. *Organic Syntheses via Boranes*; Wiley: New York, NY, 1975.
130. Mikhailov, B. M.; Bubnov, Y. N. *Organoborane Compounds in Organic Synthesis*; Harwood: London, 1984.
131. Onak, T.; *Organoborane Chemistry*; Academic: New York, NY, 1975.
132. Cragg, G. M. L. *Organoboranes in Organic Synthesis*; Marcel Decker, New York, NY 1973.
133. Hutchins, R. O.; Cistone, F. *Org. Prep. Proc. Int.* **1981**, *13*, 225.

134. Brown, H. C.; Mandal, A. K. *J. Org. Chem.* **1992**, *57*, 4970.
135. Follet, M. *Chem. Ind.* **1986**, 123.
136. Hutchins, R. O.; Learn, K.; Nazer, B.; Pytlewski, D.; Pelter, A. *Org. Proc. Prep. Int.* **1984**, *16*, 335.
137. Lane, C. F. *Aldrichim. Acta.* **1973**, *6*, 51.
138. Kawase, M.; Kikugawa, Y. *J. Chem. Soc. Perkin Trans. I* **1979**, 643.
139. Pelter, A.; Rosser, R. M.; Mills, S. *J. Chem. Soc. Perkin Trans. I* **1984**, 717.
140. Brown, H. C.; Zweifel, G. *J. Am. Chem. Soc.* **1961**, *83*, 1241.
141. Knights, E. F.; Brown, H. C. *J. Am. Chem. Soc.* **1968**, *90*, 5280.
142. Zweifel, G.; Brown, H. C. *J. Am. Chem. Soc.* **1963**, *85*, 2066.
143. Brown, H. C.; Ramachandran, P. V. *Pure Appl. Chem.* **1991**, *63*, 307.
144. Brown, H. C.; Ramachandran, P. V. *Pure Appl. Chem.* **1994**, *66*, 201.
145. Brown, H. C.; Ramachandran, P. V. In *Advances in Asymmetric Synthesis* Vol. 1. Hassner, A. Ed. JAI Press: Greenwich, CT, 1995.
146. Brown, H. C.; Ramachandran, P. V. *J. Organometal. Chem.* **1995**, *500*, 1.
147. Cole, T. E.; Bakshi, R. K.; Srebnik, M.; Singaram, B.; Brown, H. C. *Organometallics* **1986**, *5*, 2303.
148. Brown, H. C.; Singaram, B.; Cole, T. E. *J. Am. Chem. Soc.* **1985**, *107*, 460.
149. Masamune, S.; Kim, B. M.; Peterson, J. S.; Sato, T.; Veenstra, S. J.; Imai, T. *J. Am. Chem. Soc.* **1985**, *107*, 4549.
150. Woods, W. G.; Strong, P. L. *J. Am. Chem. Soc.* **1966**, *88*, 4667.
151. Brown, H. C.; Gupta, S. K. *J. Am. Chem. Soc.* **1971**, *93*, 1816.
152. Lane, C. F.; Kabalka, G. *Tetrahedron* **1976**, *32*, 981.
153. Männig, D.; Nöth, H. *Angew. Chem. Int. Ed. Engl.* **1985**, *24*, 878.
154. Corey, E. J.; Bakshi, R. K. *Tetrahedron Lett.* **1990**, *31*, 611.
155. Brown, H. C.; Ravindran, N. *J. Org. Chem.* **1977**, *42*, 2533.
156. Brown, H. C.; Ravindran, N. *J. Am. Chem. Soc.* **1976**, *98*, 1785
157. Salunkhe, A. M.; Brown, H. C. *Tetrahedron Lett.* **1995**, *36*, 7987.
158. Bovicelli, P.; Minicone, E.; Ortaggi, G. *Tetrahedron Lett.* **1991**, *32*, 3719.
159. Bovicelli, P.; Lupattelli, P.; Bersani, M. T. Minicone, E. *Tetrahedron Lett.* **1992**, *33*, 6181.
160. Brown, H. C.; Ramachandran, P. V. *Acc. Chem. Res.* **1992**, *25*, 16.
161. Brown, H. C.; Nazer, B.; Cha, J. S.; Sikorski, J. A. *J. Org. Chem.* **1986**, *51*, 5264.
162. Brown, H. C.; Cha, J. S.; Yoon, N. M.; Nazer, B. *J. Org. Chem.* **1987**, *52*, 5400.
163. Greeves, N. In ref. 14. Vol. 8.; pp 1-24.
164. Narasaka, K.; Pai, F. -C. *Tetrahedron* **1984**, *40*, 2233.
165. Chen, K.-M.; Hardtmann, G. E.; Prasad, K.; Repic, O.; Shapiro, M. J. *Tetrahedron Lett.* **1987**, *28*, 155.
166. Evans, D. A.; Chapman, K. T.; Carreira, E. M. *J. Am. Chem. Soc.* **1988**, *110*, 3560.
167. Bothner-by, A. A. *J. Am. Chem. Soc.* **1951**, *73*, 846.
168. Landor, S. R.; Miller, B. J.; Tatchell, A. R. *Proc. Chem. Soc.* **1964**, 227.
169. Grandbois, E. R.; Howard, S. I.; Morrison, J. D. In *Asymmetric Synthesis*; Morrison, J. D. Ed. Academic: Orlando, FL, 1983, Vol. 2; pp. 71-90.
170. Haubenstock, H. *Top. Curr. Chem.* **1983**, *14*, 231.
171. Nishizawa, M.; Noyori, R. In ref. 14, Vol. 8, pp 159-182.
172. Morrison, J. D.; Mosher, H. S. *Asymmetric Organic Reactions*; Prentice-Hall: New York, NY, 1971.
173. Brown, H. C.; Park, W. S.; Cho, B. T.; Ramachandran, P. V. *J. Org. Chem.* **1987**, *52*, 5406.
174. Yamaguchi, S.; Mosher, H. S. *J. Org. Chem.* **1973**, *38*, 1870.
175. Mukaiyama, T.; Asami, M.; Hanna, J.; Kobayashi, S. *Chem. Lett.* **1977**, 783.
176. Terashima, S.; Tanno, N.; Koga, K. *J. Chem. Soc. Chem. Commun.* **1980**, 1026.

177. Sato, T.; Gotoh, Y.; Wakabayashi, Y.; Fujisawa, T. *Tetrahedon Lett.* **1983**, 4123.
178. Vigneron, J. P.; Jacquet, I. *Tetrahedron* **1976**, *32*, 939.
179. Noyori, R.; Tomino, I.; Tanimoto, Y. Nishizawa, M. *J. Am. Chem. Soc.* **1984**, *106*, 6709.
180. Midland, M. M. In *Asymmetric Synthesis*; Morrison, J. D. Ed. Academic: Orlando, FL, 1983, Vol. 2; pp. 45-69.
181. Morrison, J. D.; Grandbois, E. R.; Howard, S. I. *J. Org. Chem.* **1980**, *45*, 4229.
182. Hirao, A.; Nakahama, S.; Mochizuki, H.; Itsuno, S.; Yamazaki, N. *J. Org. Chem.* **1980**, *45*, 4231.
183. Yamada, K.; Takeda, M.; Iwakuma, T. *J. Chem. Soc. Perkin Trans. I* **1983**, 265.
184. Nasipuri, D.; Sarkar, A.; Konar, S. K.; Ghosh, A. *Ind. J. Chem.* **1982**, *21B*, 212.
185. Bianchi, G.; Achilli, F.; Gamba, A.; Vercesi, D. *J. Chem. Soc. Perkin Trans. I* **1988**, 417.
186. Yatagai, M.; Ohnuki, T. *J. Chem. Soc. Perkin Trans. I* **1990**, 1826.
187. Polyak, F. D.; Solodin, I. V.; Dorofeeva, T. V. *Syn. Commun.* **1991**, *21*, 1137.
188. Soai, K.; Oyamada, H.; Yamanoi, T. *J. Chem. Soc. Chem. Commun.* **1984**, 413.
189. Brown, H. C.; Park, W. S.; Cho, B. T. *J. Org. Chem.* **1986**, *51*, 3278.
190. Brown, H. C.; Park, W. S.; Cho, B. T. *J. Org. Chem.* **1988**, *53*, 1231.
191. Hutchins, R. O.; Abdel-Majid, A.; Stercho, Y. P.; Wambsgans, A. *J. Org. Chem.* **1987**, *52*, 702.
192. Cho, B. T.; Chun, Y. S. *J. Chem. Soc. Perkin Trans. I.* **1990**, 3200.
193. Krishnamurthy, S.; Vogel, F.; Brown, H. C. *J. Org. Chem.* **1977**, *42*, 2534.
194. Midland, M. M.; Kazubski, A.; Woodling, R. E. *J. Org. Chem.* **1991**, *56*, 1068.
195. Ramachandran, P. V.; Brown, H. C.; Swaminathan, S. *Tetrahedron Asym.* **1990**, *1*, 433.
196. Itsuno, S.; Nakano, M.; Miyazaki, K.; Masuda, H.; Ito, K.; Hiao, A.; Nakahama, S. *J. Chem. Soc. Perkin Trans. I.* **1985**, 2039.
197. Itsuno, S.; Sakurai, Y.; Ito, K.; Hirao, A.; Nakahara, S. *Bull. Chem. Soc. Jpn.* **1987**, *60*, 395.
198. Corey, E. J.; Bakshi, R. K.; Shibata, S. *J. Am. Chem. Soc.* **1987**, *109*, 5551.
199. Singh, V. K. *Synthesis* **1992**, 605.
200. Brunel, J. M.; Pardigon, O.; Faure, B.; Buono, G. *J. Chem. Soc. Chem. Commun.* **1992**, 287.

## Chapter 2

# New Developments in Chiral Ruthenium (II) Catalysts for Asymmetric Hydrogenation and Synthetic Applications

Jean Pierre Genet

Laboratoire de Synthèse Organique Associé au Centre National de la Recherche Scientifique, Ecole Nationale Supérieure de Chimie de Paris, 11 rue Pierre et Marie Curie, 75231 Paris, France

This chapter covers catalytic asymmetric synthesis effected by chiral ruthenium (II) complexes. New general and useful methods for the synthesis of diphosphine ruthenium (II) complexes : (P*P)RuX$_2$ (X = carboxylato, 2-methylallyl, halide), P*P =, BIPHEMP, MeO-BIPHEP, BINAP DIPAMP, DIOP, CHIRAPHOS, DUPHOS etc., as well as hydrido and dinuclear chiral ruthenium complexes are reviewed. These catalysts were evaluated in asymmetric hydrogenation reactions of prochiral substrates. Allylic alcohols, α- and β- acylamino acrylic acids, enamides, α- and β-keto esters, diketones were easily reduced to give the corresponding saturated products in good yields. High efficiency is displayed by Ru catalysts having atropisomeric ligands and C$_2$ symmetric bis phospholanes (e.g. Me-DUPHOS, Et-DUPHOS and iPr-BPE). Chirally labile compounds capable of undergoing *in situ* racemization were hydrogenated with high diastereoselectivity (*syn* or *anti*) and high enantioselectivity. This reaction provides a powerful tool, dynamic kinetic resolution, for the synthesis of chiral compounds. Applications of these processes, particularly significant in the preparation of synthetic intermediates and pharmaceuticals, are summarized.

In the last few years the demand for optically pure compounds has grown rapidly. Asymmetric synthesis using transition metal catalysis is an ideal method for preparing optically active materials. The development of this method started in the 1970's with chiral rhodium (I) catalysts which have been used with great success in homogeneous hydrogenation of prochiral olefins. To date synthetic organic chemists have designed a number of selective catalysts. Many reviews have been published on asymmetric synthesis in general *(1-12)*. A wide range of compounds contain a hydrogen atom at the stereogenic centre. As this hydrogen atom can be introduced into an appropriate prochiral substrate by a hydrogenation reaction, asymmetric hydrogenation provides a major route to highly enantiomerically-pure compounds. Spectacular enantioselectivities (up to 99 %) were obtained with rhodium (I) catalysts. There are many reviews covering the field *(13-16)*.
In this chapter we limit the topic to the homogeneous asymmetric reactions catalyzed by chiral ruthenium catalysts. What follows is a survey of the present state of the

field, with special emphasis on development of new chiral ruthenium catalysts containing a large variety of chiral diphosphines.

**Chiral Ligands.** Increases in stereoselectivity can be related to their structure. They differ in size and position of the chiral information (*17-18*). The most efficient phosphines are diphosphines ; the chirality can either be located on the phosphorous atom (DIPAMP) or on the carbon skeleton of the diphosphine (CHIRAPHOS, SKEWPHOS, BPPM, NORPHOS, CBD, PROPHOS, DUPHOS, etc.). One other important class of ligands is atropisomeric ligands such as BINAP, BIPHEMP, MeO-BIPHEP, BICHEP. Some representative chiral ligands used for the preparation of chiral ruthenium catalysts are shown scheme 1.

**scheme 1.** Chiral diphosphines for asymmetric hydrogenation with Ru(II) catalysts

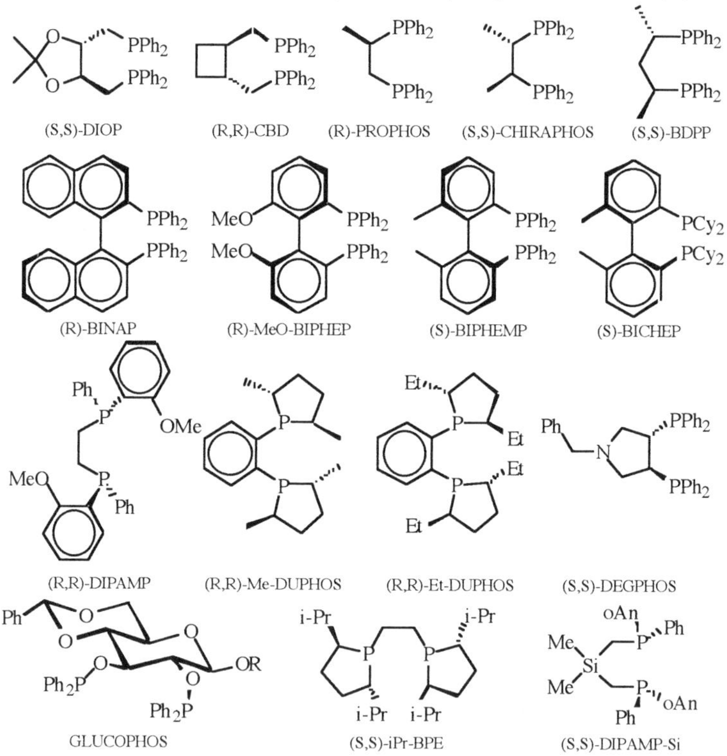

Abbreviations of representative ligands : DIOP : (S,S)-2,3-O-isopropylidene-2,3-dihydroxy-1,4 bis (diphenylphosphinobutane) ; CBD : (R,R)-*trans*-1,2-bis(diphenylphosphinomethyl) cyclobutane ; PROPHOS : (R)-1,2-bis (diphenylphosphino) propane ; CHIRAPHOS : (S,S)-2,3-bis diphenylphosphinobutane ; BDPP : (S,S)-2,4-bis (diphenylphosphino) pentane (SKEWPHOS) ; BINAP : (R)-2,2'-bis (diphenylphosphino)1,1'-binaphtyl ; MeO-BIPHEP : (R)-2,2'-bis (diphenylphosphino)-6,6' dimethoxy ; BIPHEMP : (S)-2,2'-bis (diphenylphosphino)-6,6'-dimethyl-1,1'-diphenyl ; BICHEP : (S)-2,2'-bis (dicyclohexylphosphino)-6,6'-dimethyl-1,1'-biphenyl ; DIPAMP : (R,R)-1,2-bis [(O-methoxyphenyl)-phenylphosphino] ethane ; DUPHOS : (R,R)-1,2-bis (phospholano) benzene ; DEGPHOS : (S,S)-1-benzyl-3,4-bis (diphenylphosphino) pyrrolidine ; iPr-BPE : (S,S)1,2-bis (*trans*-2,5 diisopropylphospholano) ethane.

## 2. GENET  Chiral Ruthenium (II) Catalysts for Asymmetric Hydrogenation

**Preparation of chiral ruthenium (II) catalysts.** There are general methods for the preparation of chiral Rh(I) catalysts. In contrast to rhodium, there are few reports concerning chiral ruthenium catalysts. The first chiral binuclear Ru (II) catalyst was prepared by James in 1975 *(19)* from [RuCl$_2$(PPh$_3$)$_3$] using the phosphine exchange method. This catalyst has been found effective for catalytic hydrogenation of unsaturated carboxylic acids with moderate enantioselectivities (60 %).

Since this time, spectacular enantioselectivities were obtained in numerous hydrogenation reactions when Saburi and Ikariya *(20)* introduced the bimetallic complex[Ru$_2$Cl$_4$(BINAP$_2$)]NEt$_3$ **2** and Noyori *(21-23)* the mononuclear BINAPRu(OAc)$_2$ **3** catalyst. However, the procedure for the synthesis of the ruthenium complex **2** from polymeric starting material **1** is not general, requires the presence of triethylamine, and is conducted in toluene at reflux (scheme 2).

**scheme 2.** Synthesis of BINAP Ru(II) catalysts

(CODRuCl$_2$)n  $\xrightarrow[\text{Toluene (reflux, 12 h)}]{\text{BINAP / NEt}_3}$  [RuCl$_2$(BINAP)]$_2$NEt$_3$
**1**  **2**

[RuCl$_2$(BINAP)]$_2$NEt$_3$  $\xrightarrow[\text{tBuOH (reflux, 12h)}]{\text{AcONa}}$  BINAPRu(OAc)$_2$  $\xrightarrow{\text{HX}}$  "BINAPRuX$_2$"
**2**  **3**  **4**

The development of ruthenium chemistry for homogenous asymmetric catalysis requires mild and reliable synthesis of chiral ruthenium catalysts. In 1991, a major improvement was made by Genet and co-workers *(24)* who found that P*P Ru(2-methylallyl)$_2$ complexes are produced from CODRu(2-methylallyl)$_2$ (commercially available from Acros Organics). This procedure is general and several chiral phosphines have been used. ( P*P = BINAP, BPPM, NORPHOS, DEGPHOS, PROPHOS, BDPP, CBD, BINAP, BIPHEMP, MeO-BIPHEP, GLUCOPHOS, DUPHOS, etc...) scheme 3.

**scheme 3.** General synthesis of chiral diphosphine Ru(2-methylallyl)$_2$ complexes

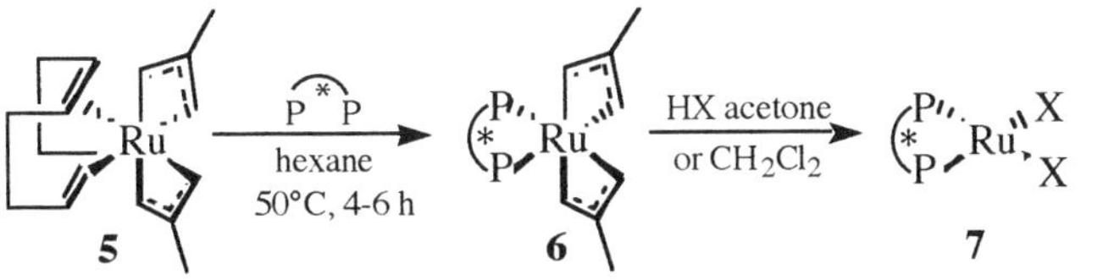

**5** $\xrightarrow[\text{hexane } 50°C, 4-6 h]{\text{P*P}}$ **6** $\xrightarrow{\text{HX acetone or CH}_2\text{Cl}_2}$ **7**

50-85 % yields

Our general synthetic method allows the production of Ru (II) catalysts containing tertiary biphosphines, having the phosphorus atom as the stereogenic center, such as in DIPAMP *(25)* and modified DIPAMP; these complexes have been characterized spectroscopically. X-Ray structures were obtained for (S,S)-DIOP and (S,S)-CHIRAPHOS. These preformed catalysts P*P-Ru(2-methylallyl)$_2$ **6** are useful precursors, after protonation with HX (X=Cl,Br,I), of the corresponding chiral dihalides X$_2$Ru(P*P) **7** *(26)*. In addition we found that it was possible to prepare dihalide ruthenium catalysts *in situ* from (COD) Ru (2-methylallyl)$_2$ at room temperature by adding HX in the presence of the appropriate chiral ligand *(27)*. A wide range of chiral ruthenium (II) catalysts can be prepared very easily (scheme 4). These new catalysts are efficient in the asymmetric hydrogenation of a wide range of prochiral substrates.

**scheme 4.** *In situ* preparation of dihalide ruthenium catalysts

$$\text{5} \xrightarrow[\text{30 min, R.T.}]{\text{HX, P*P}} \text{8}$$
$$\text{acetone}$$
$$\text{in situ}$$
$$\text{HX = HCl, HBr, HI, ...}$$

Heiser *(28)* has also used (COD) Ru (2-methylallyl)$_2$ **5** as a precursor for the preparation of Ru(II) biacetate and bis trifluroacetate complexes **9** and **10** containing atropisomeric ligands BINAP, BIPHEMP and MeO-BIPHEP (scheme 5). Brown described displacement of cyclooctadiene from its ruthenium 2-propenyl hexafluoropentanedional complex by diphosphines (DIOP, BPPM, BINAP) affording the corresponding complexes which are precursors of active hydrogenation catalysts *(29)*.

**scheme 5.** Synthesis of chiral Ru(II) containing atropisomeric ligands

$$\text{5} \xrightarrow[\text{(ii) P*P}]{\text{(i) CF}_3\text{COOH}} \text{Ru(OCOCF}_3)_2 \xrightarrow[\text{MeOH}]{\text{NaOAc}} \text{Ru(OCOCH}_3)_2$$
$$\text{9} \quad\quad\quad\quad \text{10}$$

P*P = BINAP, BIPHEMP, MeO-BIPHEP

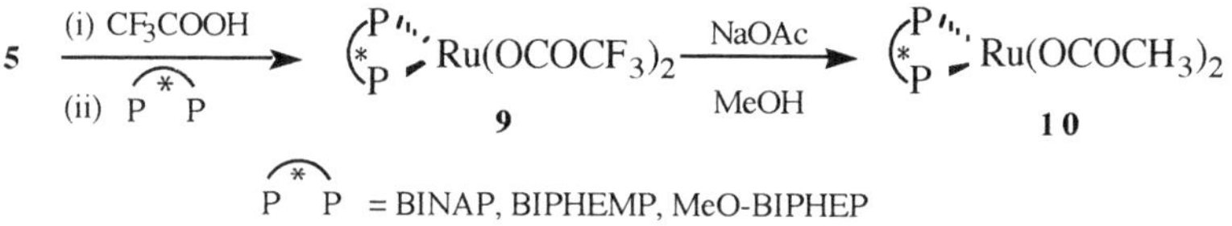

Noyori has reported a novel *in situ* preparation of cationic [(arene) Ru(X) BINAP]Cl **12** by ligand exchange between [RuCl$_2$(benzene)]$_2$ **11** and BINAP *(30,31)* (scheme 6). Subsequently, the same methodology was applied to the synthesis of sulfonated BINAP ruthenium (II) *(32)*.

**scheme 6.** Synthesis of BINAPRu(arene) catalysts

$$[\text{RuCl}_2\text{C}_6\text{H}_6)]_2 + \text{BINAP} \xrightarrow[\text{EtOH/hexane}]{50\text{-}55°\text{C}} [\text{BINAP-Ru C}_6\text{H}_6(\text{Cl})]\text{Cl}$$
$$\text{11} \quad\quad\quad\quad\quad\quad\quad\quad\quad \text{12}$$

Due to the success of BINAP-ruthenium complexes in asymmetric hydrogenation, several groups developed other procedures for the synthesis of such catalysts. Taber found a simplified procedure for the preparation of the Ikariya and Saburi complex **2** *(33)*. In the presence of trace amounts of strong acid, [Ru$_2$Cl$_4$(BINAP$_2$)]NEt$_3$ **2** was also found to be remarkably efficient *(34)*. (S) - BINAP Ru(acac)$_2$ was generated by hydrogenation of a mixture of BINAP and Ru(acac)$_3$ in methanol *(35)*. Some other preparations of chiral complexes of ruthenium containing chelating phosphines have been developed (*36-44*). A new synthetic improvement of Noyori's catalyst **3** has been recently reported (*45*). Three new arene ruthenium(II) catalysts containing atropisomeric ligands such as BICHEP (*46*), FUPMOP (*47*), BIMOP (*48*) have been prepared using an analogous procedure decribed by Noyori (*30*).

**Ru - catalyzed asymmetric hydrogenation.** One of the most remarkable advances in asymmetric hydrogenation in the last ten years is the development of the extremely effective BINAP chiral ruthenium (II) catalysts. The outstanding performance of these complexes has been reported by Noyori. Several reviews dealing with these BINAP-Ru catalysts have appeared in recent years (*49-53*). However, new chiral ruthenium(II) catalysts **8** prepared *in situ* containing a wide range of chiral diphosphines ligands have also been reported and are active catalysts in asymmetric hydrogenation (*54-55*).

**Asymmetric hydrogenation of C=C bonds**
  **Allylic alcohols.** Hydrogenation of olefins is one of the most useful synthetic operations. The ability to do this asymmetrically provides a versatile route to stereodefined tertiary centers of chiral molecules. Chemoselective hydrogenation of the C(2)-C(3) double bond of geraniol and nerol with BINAP Ru(OAc)$_2$ gave citronellol with 92 % ee (*56-57*). Noyori observed a specific relationship between starting material geometry (Z or E), BINAP chirality (S,R) and product configuration (S,R). However the two-carbon-homologue of geraniol is inert under the same conditions.
This procedure offers a practical method for the synthesis of enantiomerically pure terpenes. The same BINAP catalyst was used in the synthesis of α-tocopherol side chain **13**. Heiser and coworkers have also applied the (S)-BIPHEMP and (S)-MeO-BIPHEPRu(O$_2$CCF$_3$)$_2$ catalysts to this asymmetric hydrogenation. Thus, (2E,7R)-tetrahydrofarnesol **12** gave α- tocopherol side chain **13** in 100 % yield with an enantioselectivity of 98.7 % (*28*) (scheme 7).

**scheme 7.** Synthesis of (3R,7R)-3,7,11-trimethyldodecanol

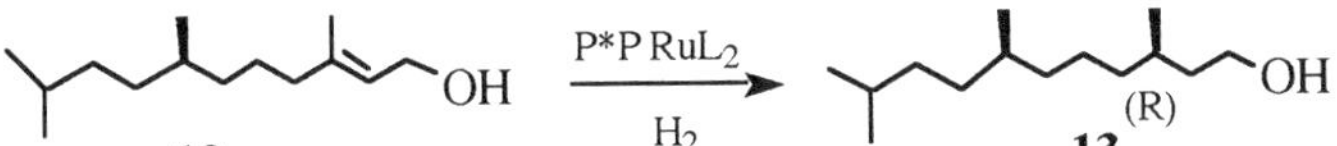

$$\underset{\textbf{12}}{\text{structure}} \xrightarrow[\text{catalyst}]{\text{P*P RuL}_2 \atop \text{H}_2} \underset{\textbf{13}}{\text{structure}}^{(R)}$$

(S)-BINAPRu(OAc)$_2$     ee = 98%
(S)-MeO-BIPHEPRu(O$_2$CCF$_3$)$_2$     ee = 98.7%

Polyprenols gave dolicol, a metabolite in glycoprotein biosynthesis (*58*). Noyori found that the kinetic resolution of racemic allylic alcohols can be accomplished by the BINAP-ruthenium catalysts with high enantio-recognitions. This method provides a practical way to prepare protected (R)-4-hydroxy-2-cyclopentenone **14**, an intermediate in prostaglandin synthesis (*59*) (scheme 8).

**scheme 8.** Synthesis of 4(R)-hydroxy cyclopentenone

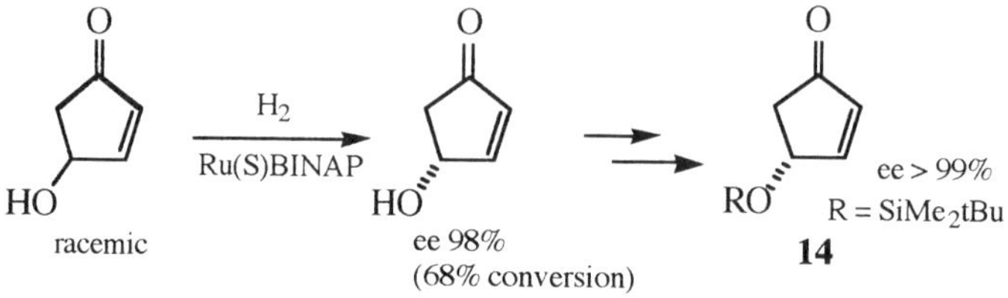

[Structure: racemic HO-cyclopentenone] →(H₂, Ru(S)BINAP)→ [HO-cyclopentenone, ee 98%, (68% conversion)] → [RO-cyclopentenone, **14**, R = SiMe₂tBu, ee > 99%]

Noyori established that a chiral unit present in the olefin affects the stereochemical outcome of the hydrogenation. Thus, hydrogenation of the enantiomerically pure azetidinone **15** in the presence of (S)-tolylBINAPRu(OAc)$_2$, gave a mixture of the β-diastereomers **16** and **17** in 99.9 : 0.1 ratio (60).
We also found that preformed (R)-BINAP-Ru(2-methylallyl)$_2$ and easily available (R)-MeO-BIPHEPRuBr$_2$ catalyts are efficient catalysts giving the 1β-methyl carbapenem intermediate **16** with 93-99% diastereoselectivity *(54,61)*. This high diastereoselectivity is a result of the combined effect of chirality transfer from (R)-BINAP or (R)-MeO-BIPHEP Ru catalysts to the olefinic face and the asymmetric induction caused by the preexisting stereogenic center of the azetidinone **15** (scheme 9).

**scheme 9.** Synthesis of 1β-methyl carbapenem intermediate

[Structures of azetidinones **15**, **16**, **17** with OTBDMS groups]

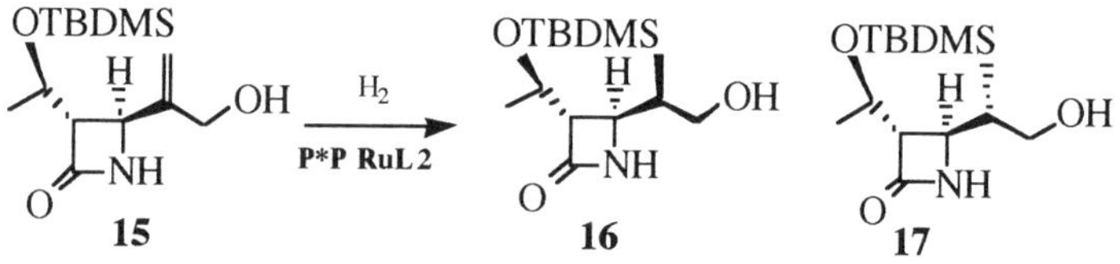

| catalyst | 16 | 17 |
|---|---|---|
| (S)-tolyl-BINAPRu(OAc)$_2$ | 22 | 78 |
| (R)-tolyl-BINAPRu(OAc)$_2$ | 99.9 | 0.1 |
| (S)-BIPHEMPRu(2-methylallyl)$_2$ | 22 | 78 |
| (R)-BINAPRu(2-methylallyl)$_2$ | 97.5 | 2.5 |
| (R)-MeO-BIPHEPRuBr$_2$ *(in situ)* | >99 | |

**α, β-unsaturated acids.** Ru complexes are particularly efficient catalysts for the hydrogenation of olefins. The presence of α-heteroatoms on the olefins facilitates ligand exchange. Thus, unsaturated carboxylic acids undergoing ligand exchange with BINAPRu(OAc)$_2$ are good substrates. Noyori *(49-53)* used this catalyst under mild conditions (4 atm of H$_2$, RT in methanol) for the hydrogenation of a wide range of α, β-unsaturated carboxylic acids *(22,62)* with high enantioselectivites (97 % ee). The full set of P*PRu(2-methylallyl)$_2$ catalysts **6** was also investigated in the hydrogenation of tiglic acid **17a** as model substrate *(24)*. These catalysts are very active (3 atm of hydrogen at RT). The CBD and DIOP ligands coordinated to ruthenium gave moderate enantioselectivity (70% ee). Interestingly, (R,R)Me-DUPHOSRu(2-methylallyl)$_2$ displayed good efficiency, giving the saturated acid **18a** with 80% enantioselectivity (not optimized). This study revealed that ruthenium having an allyl ligand is highly efficient *(54)* under our standard conditions ; tiglic acid was hydrogenated to **18a** with 90% ee using (S)-BIPHEMPRu(2-methylallyl)$_2$ as catalyst.*(54)* (scheme 10).

**scheme 10.** Asymmetric hydrogenation of disubstituted acrylic acids

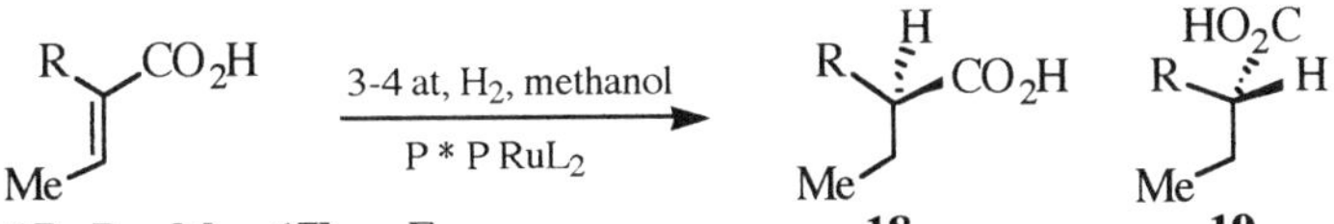

| substrate | Catalyst | ee % | Product |
|---|---|---|---|
| 17a | R-BINAP Ru(OAc)$_2$ | 91 | 18a |
| | R-R (CBD)Ru ($\rangle\!\!-\!\!)_2$ | 70 | 19a |
| | R-R (DUPHOS)Ru ($\rangle\!\!-\!\!)_2$ | 80 | 18a |
| | S-BIPHEMP Ru ($\rangle\!\!-\!\!)_2$ | 90 | 18a |
| 17b | Ru$_2$Cl$_4$BINAP$_2$NEt$_3$ | 90 | 18b |

a:R=Me; b:R=F

(Ru$_2$Cl$_4$(BINAP)$_2$ NEt$_3$ and others [(RuX(BINAP)(arene)] Y (X = halogen, Y = BF$_4$) also serve as highly efficient catalysts. The α-fluorocarboxylic acid **17b** can be hydrogenated to **18b** with high enantiomeric excess using BINAP-Ru catalysis *(63)*. An important application of this technique is the enantioselective synthesis of chiral arylacetic acids (naproxen, ibuprofen). These antiinflammatory agents are obtained in high enantiomeric excesses. Noyori observed that hydrogenation of **20** under high pressure (135 atm of H$_2$, 12h,RT) (S)-BINAPRu(OAc)$_2$ afforded naproxen with 95 % ee *(62)*. (S)-BINAPRu(2-methylallyl)$_2$ was also effective at 70 atm of hydrogen, (S) naproxen being obtained with 85-90 % ee *(54)*. Davis recently designed a sulfonated BINAP-Ru supported catalyst which has been shown to be an excellent catalyst (96% ee) *(64)*. This technology offers new opportunities in ruthenium chemistry. Improved conditions (137 atm of H$_2$, -7°C) using the Ikariya-Saburi catalyst **2** were reported *(65)* for the asymmetric hydrogenation of 2-(6'methoxy-2'naphthyl) acrylic acid **20** to (S)-naproxen **21** with 98.7% ee (scheme 11).

**scheme 11.** Enantioselective synthesis of antiinflammatory arylacetic acids

- Naproxen

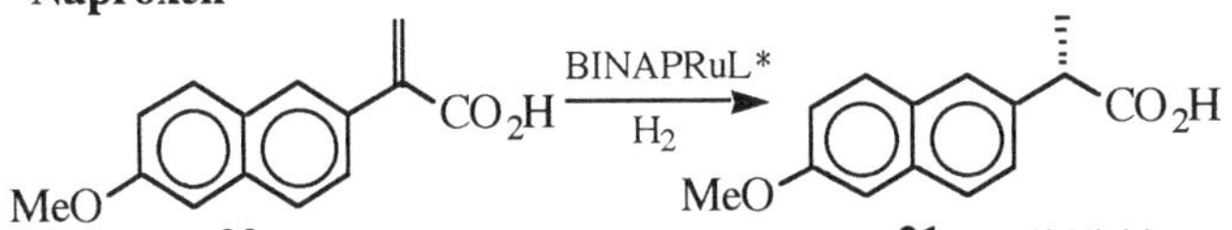

| Catalyst | Pressure(atm) | ee % |
|---|---|---|
| (S)-BINAP Ru(OAc)$_2$ | 135 | 97 |
| (S)-BINAP Ru ($\rangle\!\!-\!\!)_2$ | 70 | 87-90 |
| (S)-BINAP Ru Sulfonated supported | 30-70 | 96 |
| Ru$_2$Cl$_4$(S)-(BINAP)$_2$,NEt$_3$ | 137(-7°C) | 98.7 |

The (S)-BINAPRu(acac)$_2$ complex (70 atm of H$_2$, 24°C, 22h) catalyzes the asymmetric hydrogenation of 2-(4 isobutylphenyl)propenoic acid **22** to give (S)-ibuprofen (**23**) with about 90% ee *(35)*. Use of a modified (S)-BINAPRu(OAc)$_2$ produces **23** with 97%ee *(66)* as shown in scheme 12.

**Scheme 12.** Synthesis of (S)-ibuprofen
- **Ibuprofen**

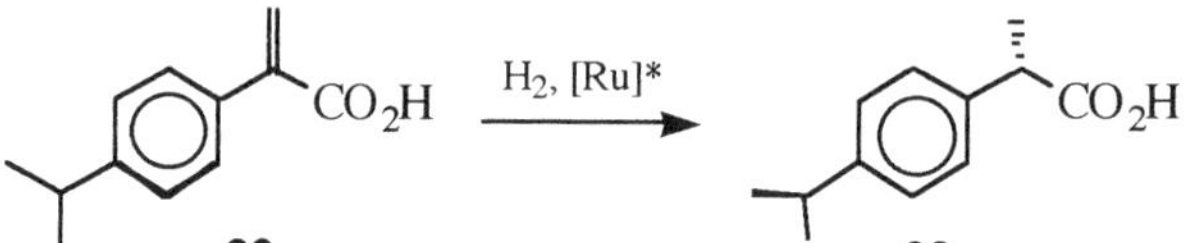

| Catalyst [Ru]* | ee % |
|---|---|
| Ru(acac)$_3$ + (S)-BINAP | 90 |
| Ru(OAc)$_2$ (S)-H$_8$BINAP | 97 |

Geranic and homogeranic acids having two olefinic double bonds are hydrogenated proximal to the carboxyl group with 87 % ee *(62)*. The utility of this general reaction was shown in the synthesis of δ or γ lactones using BINAP-Ru(OAc)$_2$ with high enantiomeric excesses *(62)*. 1,3-Butadiene 2,3-dicarboxylic acid was hydrogenated with high diastereoselectivity and enantioselectivity *(67)*.

**Dehydroamino acids and related substrates.** The hydrogenation of dehydroamino acids catalyzed by ruthenium (II) catalysts yields amino acids with enantioselectivities varying from 70-92 % ee. Interestingly, the chirality induced in the product by the use of various ruthenium catalysts with BINAP *(68,69)* is opposite to that obtained with cationic rhodium complexes. A water-soluble chiral ruthenium (II) sulfonated BINAP complex in methanol at room temperature has been shown to be an excellent catalyst in asymmetric hydrogenation of dehydroamino acids in methanolic solvents with 90% ee *(70)*. BINAP(RuOAc)$_2$ served as a very good catalyst for the hydrogenation of (E)-β-acylaminoacrylic acids giving the corresponding (E) β- substituted β-acylamino acids with high enantioselectivity (96%)*(71)*. The same catalyst was extremely efficient in the asymmetric hydrogenation of *N*-acyl-1-alkylidene tetrahydroisoquinoline **24** to isoquinoline alkaloid precursors **25** *(22)*. The reactions are conducted at RT, 100 atm of H$_2$ for 45 h. A variety of optically active benzomorphans and morphans are obtainable by using this Ru-chemistry *(71-73)*. Heiser and co-workers further applied this catalytic process to the asymmetric hydrogenation of enamides **26** using (S)-BIPHEMP-Ru(OAc)$_2$ as catalyst (25°C and 60 atm) for the synthesis of the corresponding (S)-*N*-formyl octahydroisoquinoline **27** with 96 % ee *(28)* (scheme13).

**scheme 13.** Synthesis of indoline alkaloids

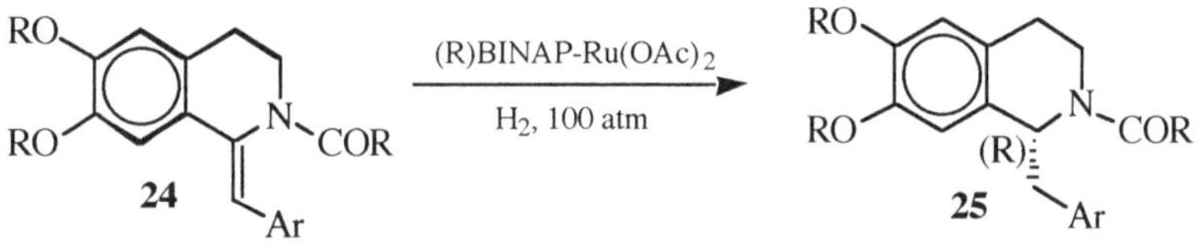

**scheme 13.** Continued.

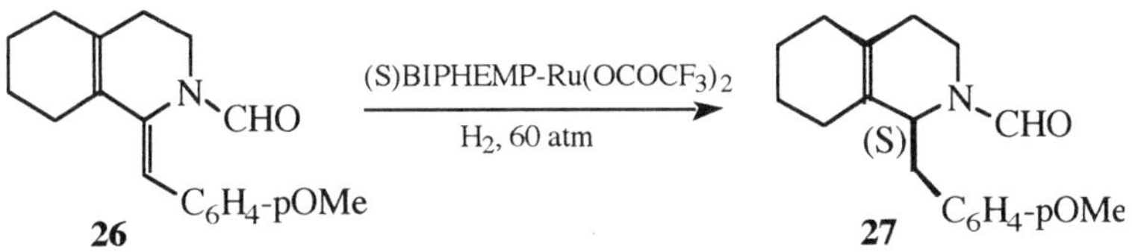

26 → (S)BIPHEMP-Ru(OCOCF₃)₂, H₂, 60 atm → 27

The availability of both enantiomers of (R) and (S) BINAP *(74-75)*, BIPHEMP *(76)* and MeO-BIPHEP*(77)* offers practical and flexible routes to a variety of optically-active, clinically-effective morphine-based analgesics.

**Alkylidene lactones.** The hydrogenation of exo double bonds in several lactones also provided good results. The Ru-BINAP hydrogenation of diketene *(78)* or γ-lactone *(79)* afforded saturated ketones with enantioselectivities up to 98%. The homologues as well as cyclic lactones with endocyclic double bonds were hydrogenated with low enantioselectivities *(80)*.

**Hydrogenation of Carbonyl Groups.** The chiral diphosphines Ru(II) diacetate **3, 8** and diallyl complexes **6** are generally ineffective for the hydrogenation of β-keto esters, even at high hydrogen pressure and long reaction times. The carboxylate and the allyl ligands may be exchanged by other anions, such as Cl⁻, I⁻, Br⁻ etc, to give the corresponding dihalide Ru(II) complexes. Thus, methyl acetoacetate was smoothly and quantitatively reduced by these complexes at high pressure (25-100 atm of $H_2$) at room temperature with ratio S/C over 1000 affording 3-hydroxybutyrate with 99 % ee *(81)*. Methanol or ethanol are the solvents of choice, but aprotic solvents such as dichloromethane can also be used. At higher temperature (50-100°C), the hydrogenation proceeds smoothly at 3.5 to 4 atm of hydrogen using [BINAPRu(benzene)Cl]Cl complexes **12** *(30)*. However, no satisfactory results were obtained at atmospheric pressure, conversion being slow even at 100°C.
It is interesting to note that the chiral ruthenium dihalide catalysts **8** prepared *in situ* *(27)* are particularly efficient. We found that it was possible to carry out the asymmetric hydrogenation of methyl acetoacetate at 50°C (2 mol.% of catalyst) with excellent enantiomeric excess up to 99%*(82)*. Thus, as exemplified in table 1, a wide range of β-keto esters **28** were hydrogenated to **29** in nearly quantitative yields and with high enantioselectivities*(82)*.

The interesting feature of this method resides in the easy accessibility of our Ru(II) catalysts. Moreover, our technique avoids the usual limitations of special apparatus such as stainless steel autoclaves. This procedure should be useful for synthetic purposes on a laboratory scale as shown in table 1.

**Screening of chiral diphosphines and catalysts.** This asymmetric hydrogenation technique finds wide generality in organic synthesis. One of the most popular ligands is the biphenyl oriented BINAP. The asymmetric hydrogenation of β-keto esters **30** to **31 or 32** using BINAP and modified BINAPRu(II) catalysts have been reported by Noyori *(49-53)* and others *(34,83)*.(scheme 14)

**table 1.** Asymmetric hydrogenation at atmospheric pressure of β-keto esters with P*PRuBr$_2$ ruthenium(II) catalysts **8**

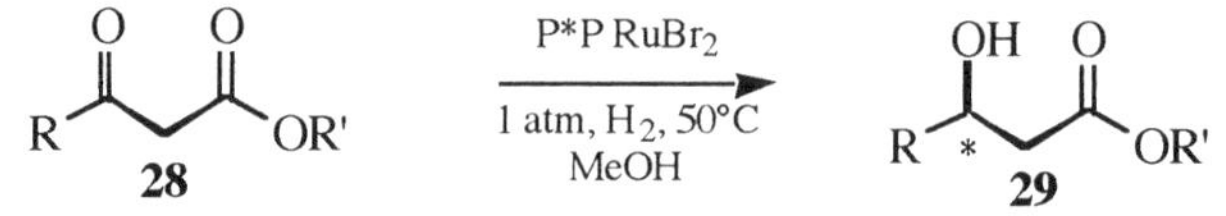

$$R\underset{28}{\overset{O\quad O}{\underset{\|\quad\|}{C-CH_2-C}}}OR' \xrightarrow[\text{MeOH}]{P^*P\,RuBr_2,\ 1\,atm,\,H_2,\,50°C} R\underset{29}{\overset{OH\quad O}{\underset{|\quad\|}{*CH-CH_2-C}}}OR'$$

| R | R' | Catalyst | % Yield | % ee | confign |
|---|---|---|---|---|---|
| Me | Me | (S)-BINAPRuBr$_2$ | 100 | 97 | S |
| Et | Me | (S)-BINAPRuBr$_2$ | 100 | 99 | S |
| Et | Me | (R)-MeO-BIPHEMPRuBr$_2$ | 100 | 99 | R |
| iPr | Me | (S)-BINAPRuBr$_2$ | 100 | 97 | S |
| C$_5$H$_{11}$ | Me | (S)-MeO-BIPHEMPRuBr$_2$ | 100 | 97 | S |
| C$_{15}$H$_{31}$ | Me | (S)-BINAPRuBr$_2$ | 100 | 96 | S |
| Ph | Et | (R)-MeO-BIPHEMPRuBr$_2$ | 82 | 96 | S |
| Ph | Et | (R,R)-Me-DUPHOSRuBr$_2$ | 73 | 78 | S |
| Thienyl | Me | (S)-BINAPRuBr$_2$ | 100 | 87 | R |

**scheme 14.** Asymmetric hydrogenation of β-keto esters with chiral Ru(II) catalysts

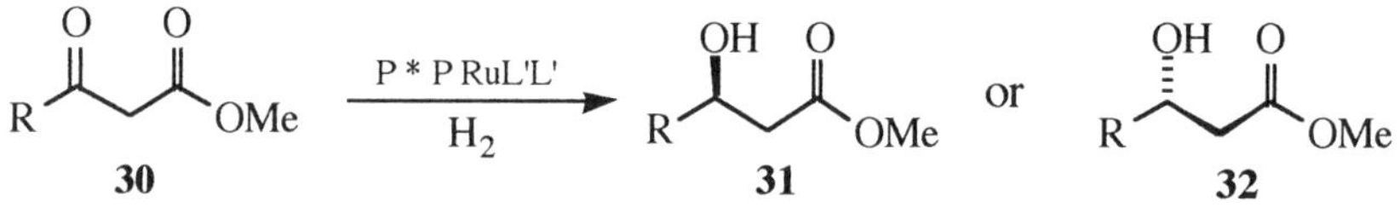

$$R\underset{30}{\overset{O\quad O}{\|\quad\|}}OMe \xrightarrow[H_2]{P^*P\,RuL'L'} R\underset{31}{\overset{OH\quad O}{|\quad\|}}OMe \quad\text{or}\quad R\underset{32}{\overset{OH\quad O}{|\quad\|}}OMe$$

Our *in situ* procedure for the preparation of chiral dihalide Ru catalysts allows screening of a wide range of chiral chelating phosphines including atropisomeric ligands.
The chiral Ru(II) catalysts containing DIOP, CBD, BDPP and the phosphinite ligand GLUCOPHOS (scheme 1) under standard conditions (5-10 atm of H$_2$, 50°C) gave poor enantioselectivities in the range of 5-22% *(54)*. A higher enantioselectivity was observed with PROPHOS (56%) and CHIRAPHOS (60%). We found that the crude dibromo-ruthenium(II) bearing C$_2$ symmetric bisphospholane (R,R)-MeDUPHOS as chelating phosphine afforded (R)-3-hydroxybutyrate in quantitative yield approaching 90% ee *(54)*. Later, Burk observed using (R,R)iPr-BPE-RuBr$_2$ catalyst, that a complete conversion of methyl-3-oxobutyrate was achieved in 4 h, with 4 atm of H$_2$ to yield (S)-methyl 3-hydroxybutyrate with an extremely high enantioselectivity (99.3%) *(84)*.

Thus, this new class of chelating diphosphines *(85)* emerges as efficient ligands in ruthenium catalysis. The results of hydrogenation of several β-keto esters with several chelating biphosphine ligands including atropisomeric ligands are shown in table 2.

## 2. GENET  Chiral Ruthenium (II) Catalysts for Asymmetric Hydrogenation

**table 2.** Screening of P*P Ru(II) catalysts for hydrogenation of β-keto esters

| Catalyst | substrate | product | % ee | ref. |
|---|---|---|---|---|
| (S,S)CHIRAPHOSRuBr$_2$ | Me | 32 | 60 | 54 |
| (S)BIPHEPRuBr$_2$ | Me | 32 | 99 | 54 |
| (R-CpRu(S)BINAPCl | Me | 32 | 77 | 80 |
| (COD)$_2$Ru$_2$Cl$_4$(CH$_3$CN)Me + (R)-BIPHEMP | C$_{11}$H$_{23}$ | 31 | 97.3 | 28 |
| Ru$_2$Cl$_4$[(S)-BINAP)]$_2$RuBr$_2$ | C$_4$H$_9$ | 32 | 90 | 30 |
| (R)-BINAPRuBr$_2$ | Me | 31 | | 81 |
| (R)-MeO-BIPHEMPRuBr$_2$ in situ | Me | 31 | 99 | 54 |
| (R,R)-Me-DUPHOSRuBr$_2$ | Me | 31 | 87 | 54 |
| [(R,R)-iPr-BPE]RuBr$_2$ | Me | 31 | 99.3 | 84 |
| 1/2 Ru$_2$I$_4$ p.cymene + (R)-BIMOP | Me | 31 | 99 | 48 |

**Generality.** A wide range of functionalized ketones are converted to the corresponding secondary alcohols with high enantioselectivity. β-keto esters are the substrate of choice. The chemoselective asymmetric hydrogenation of β-keto esters having an unsaturated chain has been realised *(54)*. Using *in situ* prepared catalysts **8**, the keto group was more rapidly reduced than the olefin moiety. For example, under optimized conditions hydrogenation of methyl 3-oxo-5-(E)-octen-5-oate with 0.2-0.3% of (S)BIPHEMPRuBr$_2$ or (S)BINAPRuBr$_2$ at 6 atm of hydrogen, 80°C for 5 min. gave methyl (3)-(S)-hydroxy-5-(E)-octen-5-oate in 95% yield and 99% ee. This ketone reduction was possible without alkene hydrogenation using crude Ru$_2$Cl$_4$BINAP$_2$NEt$_3$ 3.5 atm of H$_2$, 80°C, 30 h *(33)*.
Hydrogenations of diketones with Ru-BINAP are also satisfactorily hydrogenated. For example, 2,4 pentadione gave *anti* 2,4 pentane diol selectively. In the first step, asymmetric hydrogenation is effected with high enantioselectivity and the second hydrogenation step is assisted by the hydroxy group giving *anti* 1,3 diol *(86)*.
α-keto esters or amides are also hydrogenated in high chemical yields and with excellent enantioselectivities (up to 99%) using ruthenium catalysts containing atropisomeric ligands*(46,87)*. In our screening of new catalysts we found that C$_2$ symmetric bis phospholanes are efficient ligands for the asymmetric hydrogenation of α-keto esters. For instance, methyl phenyl glyoxylate was hydrogenated at 20 atm of H$_2$,50°C with *in situ* R,R,MeDUPHOSRuBr$_2$ to give (R) methyl mandelate with 80% enantioselectivity (not optimized) *(54)*. Some enantiomerically enriched alcohols are shown in scheme 15.

**scheme 15.** Synthesis of enantiomerically enriched alcohols by Ru-catalyzed asymmetric hydrogenation of functionalized ketones

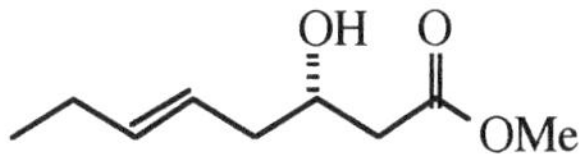

|  | ee | Ref. |  | ee | Ref. |
|---|---|---|---|---|---|
| OH O<br>⋮<br>\~/\~/\~OMe<br>(S)BIPHEMP RuBr$_2$<br>in situ | 99 | (54) | 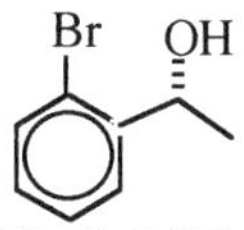 Br OH<br>⋮<br>C$_6$H$_4$-CH(OH)CH$_3$<br>1/2(RuCl$_2$C$_6$H$_6$)$_2$<br>(S)BINAP | 96 | (86) |

Continued on next page.

**scheme 15.** Continued.

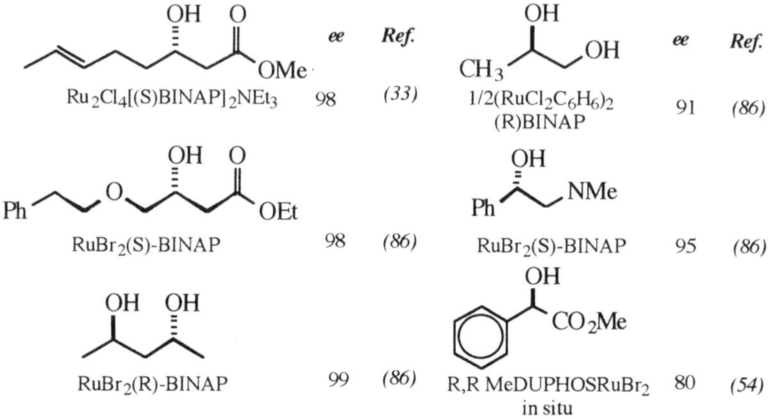

| Structure | Catalyst | ee | Ref. |
|---|---|---|---|
| CH₂=CH-CH₂-CH₂-CH(OH)-CH₂-C(O)-OMe | Ru₂Cl₄[(S)BINAP]₂NEt₃ | 98 | (33) |
| Ph-CH₂-O-CH₂-CH(OH)-CH₂-C(O)-OEt | RuBr₂(S)-BINAP | 98 | (86) |
| CH₃-CH(OH)-CH₂-CH(OH)-CH₃ | RuBr₂(R)-BINAP | 99 | (86) |
| CH₃-CH(OH)-CH₂-OH | 1/2(RuCl₂C₆H₆)₂ (R)BINAP | 91 | (86) |
| Ph-CH₂-CH(OH)-NMe | RuBr₂(S)-BINAP | 95 | (86) |
| Ph-CH(OH)-CO₂Me | R,R MeDUPHOSRuBr₂ in situ | 80 | (54) |

**Sense of the enantioselectivity.** The scope of this asymmetric hydrogenation is wide. Noyori, using BINAPRu catalysts, has demonstrated that substituted ketones bearing a heteroatom at the $\alpha, \beta$ and $\gamma$ position are smoothly and quantitatively reduced with high enantioselectivity, suggesting that the key intermediate might be the 5 to 7 membered intermediate **33** in which the Ru(II) atom interacts with the carbonyl group and at the coordinating heteroatom A *(49-53)*. This stereochemical influence through interaction with ruthenium-centered ligands such as BINAP, BIPHEMP, MeO-BIPHEP *(27,28,54,86)* allows a general prediction of the absolute configuration of the alcohol by the appropriate choice of the axial biphenyl ligands. A specific relationship between chiral atropisomeric ligands e.g : BIPHEMP, MeO-BINAP, BINAP, (R,S) and alcohol configuration (R,S) is shown in scheme 16.

**scheme 16.** General sense of hydrogenation of functionalized β-ketones with atropisomeric ligands

$L^{**}$ = (S) configuration of the biaryl ligand

$L^{*}$ = (R) configuration of the biaryl ligand

A=heteroatom
C=Sp³carbon, n=0,1,2

**33**

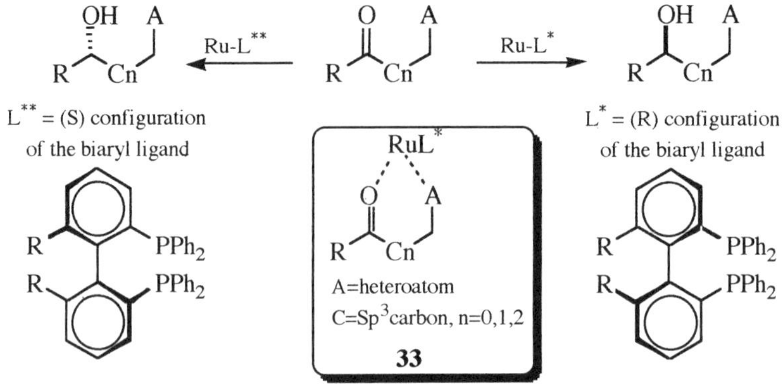

**Dynamic kinetic resolution: α-substituted β-keto esters.** The asymmetric hydrogenation of racemic α- substituted β-keto esters **34** should in principle provide four stereomers **35, 36, 37, 38** as shown in scheme 17.

## 2. GENET   *Chiral Ruthenium (II) Catalysts for Asymmetric Hydrogenation*   43

**scheme 17**

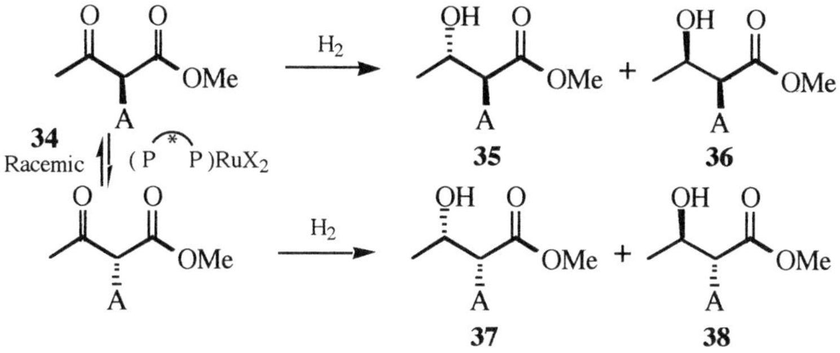

A = Alkyl, Aryl, Hetero atom (halogen, nitrogen)

However, under certain conditions a racemic substrate may be converted to a major stereomer. Noyori *(68,89)* and our group *(90,26)* successfully and simultaneously discovered the first example of stereo- and enantioselective hydrogenation through dynamic kinetic resolution of racemic α-acetamido-β-ketoesters (A = NHCOR).
The ruthenium dihalide catalysts **7** containing chelating diphosphine such as CHIRAPHOS catalyzed the hydrogenation of methyl-2-acetamido-3-oxobutanoate affording the corresponding *syn* β-hydroxy-α-amino esters with high diastereoselectivity (up to 97%) and enantioselectivity approaching 90% *(26,90)*.
Noyori and coworkers found that using BINAP Ru the hydrogenation in dichloromethane proceeds with higher ee than in methanol and with *syn* selectivity. He also established the generality of this so-called dynamic kinetic resolution *(89,91)*. The production of hydroxy esters from α-methylated substrates is not diastereoselective but highly enantioselective (*syn:anti* 1:1). A better diastereoselectivity is observed with cyclic substrates. For example, asymmetric hydrogenation of carboalkoxy cyclopentanone gives the *trans* β-hydroxy ester with 92% ee and 98% diastereoselectivity. The efficiency of this powerful dynamic kinetic resolution was profoundly affected by the nature of the carbon skeleton, as diastereoselectivity is decreased by increasing the ring size *(91)*.
We found that with racemic β-ketoesters having a tetralone structure a very high diastereoselectivity was obtained using (R)-MeO-BIPHEP as well as (R)-BINAP RuBr$_2$ prepared *in situ*. Under standard conditions (10 atm of H$_2$, 80°C, 48 h, in methanol) the *trans*-product (1S,2R) was formed almost exclusively (97% de) with high enantioselectivity (95% ee) *(92)*.
It is interesting to note that acyclic substrates bearing an amide or carbamate group exhibit remarkable *syn* selectivity to yield β-hydroxy-α-amino acids with excellent ee. This procedure offers an efficient route for the synthesis of *threonine (27,89,90)* and threo-3,4-dihydroxy phenyl serine *(89)*, an *anti*-Parkinsonian agent. A mathematical treatment of this unique dynamic kinetic resolution has been reported by Noyori *(93)*. The BINAP-Ru catalyzed hydrogenation utilizing this technology has been extended to α-bromo-β-keto phosphonates *(94)*. Thus, in the presence of RuCl$_2$(S)BINAP(dmf)$_n$ (4 atm of H$_2$, 25°) a 90:10 *syn/anti* selectivity is observed. This asymmetric reaction offers a practical route to fosfomycin, a clinically useful

antibiotic *(94)* and β-hydroxy phosphonic esters *(95)*. A new $C_2$ symmetric ligand (BITAP) has been devised using this aymmetric catalysis *(96)*.

Interestingly, we recently successfully realized the dynamic kinetic resolution of racemic open chain α-chloro-β-keto esters. We reported a very simple catalytic system consisting of a mixture of COD $Ru_2$(2-methylallyl)$_2$ and (R) or (S) MeO-BIPHEP. Under our standard conditions (80°C, 80 atm of $H_2$, 20 h in dichloromethane), racemic α-chloro β-ketoesters were smoothly reduced to the corresponding α-chloro β-hydroxyesters with excellent *anti* selectivity and high enantioselectivity. For example racemic methyl or ethyl 2-chloro-3-oxo butyrates were hydrogenated using (S)-MeO-BIPHEP and (R)BINAP to the corresponding enantiomerically enriched anti α-chloro-β-hydroxy esters. The scope of this reaction is wide *(97)*. The useful synthetic examples cited above are shown in scheme 18.

**scheme 18.** Synthesis of enantiomerically enriched alcohols containing an adjacent stereogenic center via dynamic kinetic resolution *(98)*

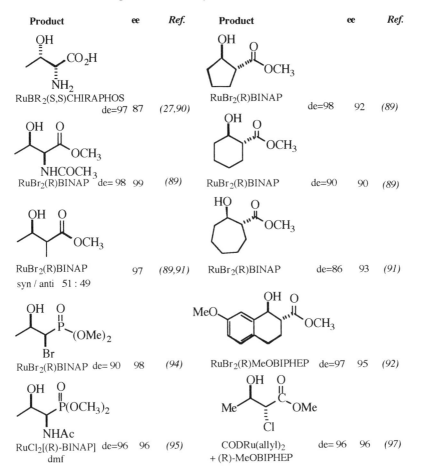

**scheme 18.** Continued.

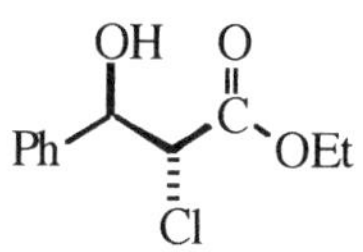

Ph—CH(OH)—CH(Cl)—C(=O)—OEt
CODRu(allyl)₂ + (R)-BINAP   de=98   99   *(97)*

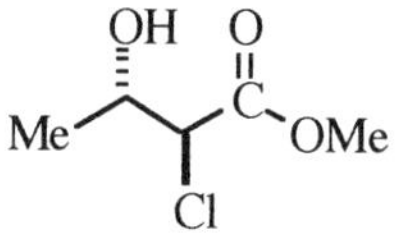

Me—CH(OH)—CH(Cl)—C(=O)—OMe
CODRu(allyl)₂ + (S)-MeOBIPHEP   de=96   96   *(97)*

**Hydrogenation of non-chelating C=O groups.** The hydrogenation of non-chelating aromatic ketones with hydrogen and BINAP proceeds generally with low enantioselectivity. In 1993, we reported that hydrogen transfer conditions (Meerwein-Pondorf-Verley-reduction) of aryl methyl ketones catalyzed by our set of chiral-ruthenium(II) catalysts **6** and **7** proceeds smoothly over a short reaction time (2-10 min) at reflux in propanol containing KOH. Under these reactions conditions, racemisation of the product was minimized. Thus, acetophenone was reduced to (S)-1-phenylethanol with 62% ee and in 98% yield*(99)*.

Very recently, Noyori reported a highly efficient system consisting of $RuCl_2$-[S-BINAP](dmf) and (S,S)diphenylethylenediamine. For example, the hydrogenation of acetonaphtnone in 2-propanol, in the presence of KOH, (4 atm of $H_2$, 28°C, 6 h) proceeded with complete conversion yields to give (R)-1-(1-napthtnyl)ethanol in 97% ee*(100)*. He also found that hydrogen transfer conditions were particularly efficient. Thus, the reaction in isopropanol with KOH in the presence of (S,S)-TsDPEN and $[RuCl_2(mesitylene)]_2$ or $[RuCl_2(\eta^6\text{-}C_6Me_6)_2]$ and chiral aminoalcohols at low concentration of substrate (0.1M) yields to (S)-phenyl ethanol and derivatives with high enantioselectivities (up to 99%) *(101)*.

**Hydrogenation of carbon-nitrogen double-bond.** In contrast to rhodium chemistry the asymmetric hydrogenation of a carbon-nitrogen double bond by chiral ruthenium catalysts is less documented. The hydrogenation of most imines is much more difficult than ketones and has been reported to be poorly selective (25% ee) *(8)*. To our knowledge, there is only one successful example of asymmetric hydrogenation using **2** as catalyst. Thus, the cyclic sulfonamide **39** was hydrogenated in 84% yield to sultam **40** with high ee (99%) after recrystallization*(102)* (scheme 19)

**scheme 19.** Synthesis of optically pure sultam

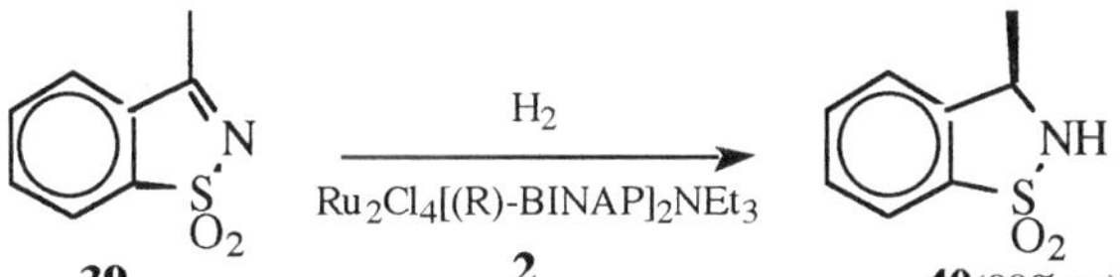

**39** →[$H_2$, $Ru_2Cl_4[(R)\text{-BINAP}]_2NEt_3$ / **2**]→ **40** (99% ee)

**Synthetic applications.** In addition to the examples given above, this asymmetric catalysis allows enantioselective synthesis of various compounds of practical significance, for example glycerol building blocks containing two different or only one protecting groups at the hydroxyls *(103-105)* as well several pharmaceutical agents and their intermediates, such as L-DOPS *(89)*, Mevinolin *(106)* an inhibitor of hydroxylmethyl glutaryl coenzyme reductase, carnitine *(107,54)*, optically pure 1,3 diols(*108*). We recently developed an efficient synthesis of enantiomerically pure *anti* N-Boc-α-hydrazino-β-hydroxyesters **41, 42** by sequential ruthenium catalyzed hydrogenation and electrophilic amination *(109)* as shown in scheme 20.

**scheme 20.** Synthesis of *anti* β-hydroxy α-hydrazinoesters

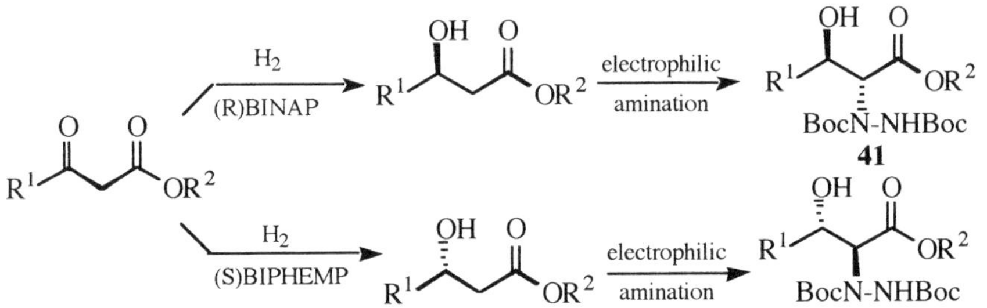

This methodology offers a new and powerful route to *anti* β-hydroxy α-amino acids. Moreover, reduction of racemic 2-chloro-3-keto esters under optimal conditions *(97)* offers a new route to optically active glycidate key intermediates **43** and **45** useful for synthesis of the C-13 side chain of taxol, taxotere **46** and diltiazem **44** (a potent channel blocker used for the treatment of hypertension) (scheme 21).

**scheme 21.** Synthesis of optically active phenyl glycidates

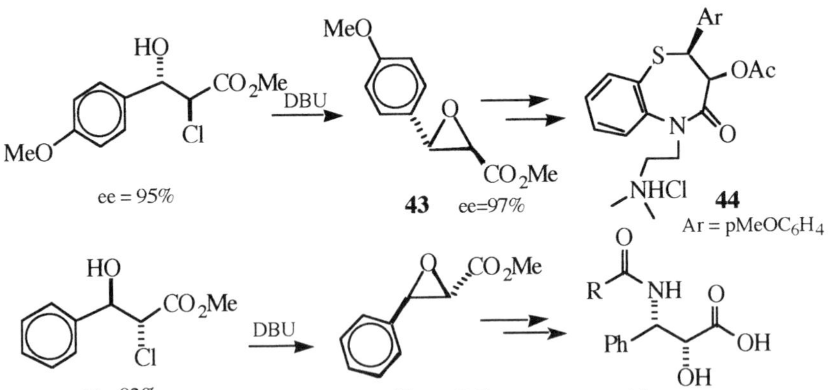

R = Ph (taxol C-13 side chain)
R = OtBu (taxotere C-13 side chain)

In addition to the examples given above, this Ru-P*P technology has been widely used for the synthesis of a natural products, such as FK-506 (a T-cell activation gene inhibitor *(110)* (+) Brefeldin(A) *(111)*, Biphenomycine A *(112)*, HMG-CoA synthase inhibitor 1233A *(113)*(-)Carbacephem *(114)*- Roxaticin *(115)* and Tetrahydrolistatin **47** *(116)* (Scheme 22).

**scheme 22.** Synthesis of tetrahydrolistatin

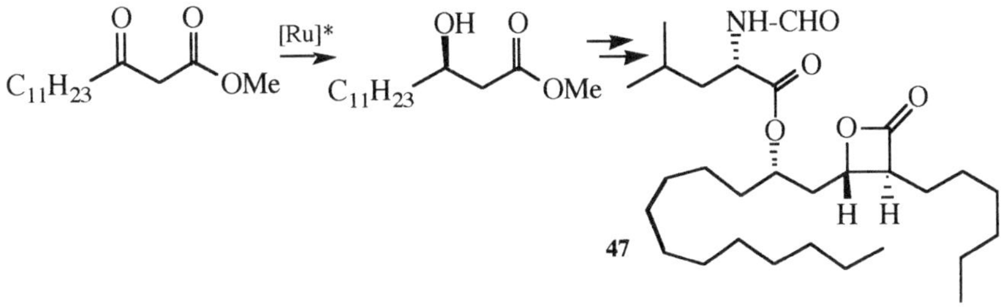

[Ru]*=[(R)-BINAP][*p*-cymene]RuCl$_2$ ,H$_2$(40-60 psi),cat HCl, MeOH,40°C,24 h, 72%

A number of heterocycles of biological interest and their intermediates have been prepared using Ru-based asymmetric hydrogenation of β-keto esters, for instance, Indolizidine 223AB **48** *(117)*, (3S,4S)-4-hydroxy-2,3,4,5 tetrahydropyridazine-3-carboxylic acid **49** (an unusual amino acid constituent of luzopeptin A) *(118)*, and both enantiomers of *trans* 3-hydroxypipecolic acid **50** *(119)*. The synthesis of (-)-Ptilomycalin, a complex guanidinium alkaloid, was accomplished in a convergent fashion *(120)* from enantioenriched secondary alcohols. One of these was prepared using ruthenium technology.

**scheme 23.** Synthesis of heterocycles

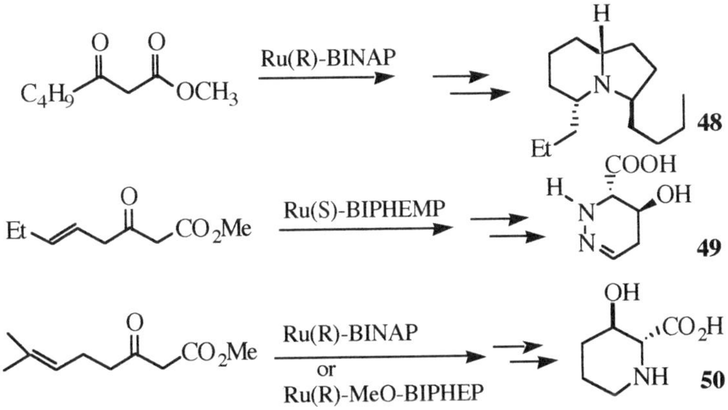

The enantiomerically pure alcohol **16** is a very useful intermediate for the preparation of 1β-methyl carbapenem analogues **51** via π-allyl palladium cyclization *(61,121)* (scheme 24).

**scheme 24.** Synthesis of bicyclic carbapenem analogues

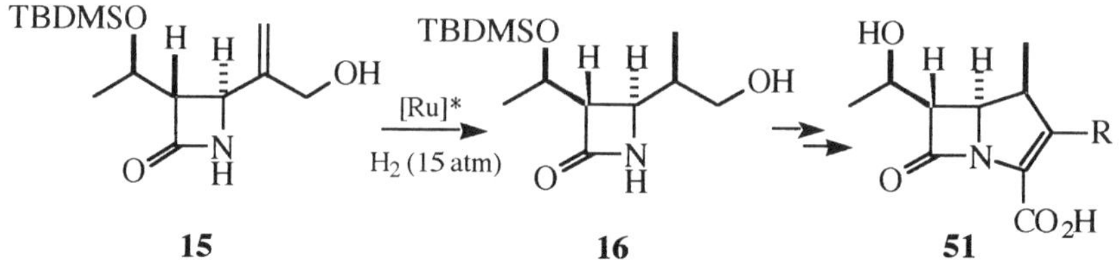

[Structures: 15 → 16 (via [Ru]*, $H_2$ (15 atm)) → 51]

Acetoxyazetidinone **54**, a key intermediate in carbapenem and penem antibiotic synthesis, is produced (150 t/year) by Takasago from racemic methyl-2-benzoylaminomethyl 3-keto butanoate **52**. The desired *syn* diastereomeric product **53** is prepared with 98% ee (*syn : anti* 96 : 4) using the dynamic kinetic resolution technique *(122)* (scheme 25).

**scheme 25.** Practical synthesis of acetoxyazetidinone **54**

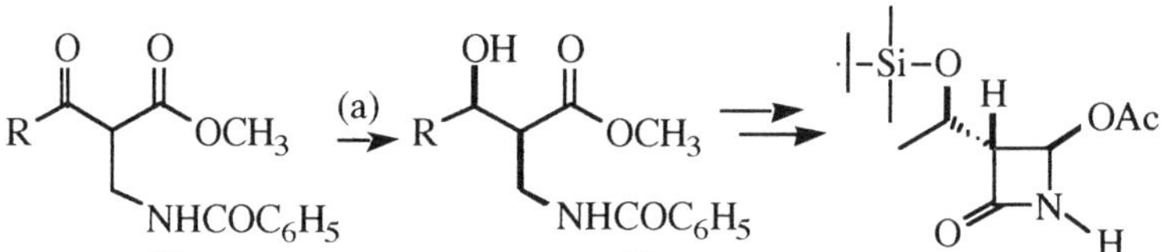

[Structures: 52 → 53 (a) → 54]

(a) $(RuCl_2,(R)\text{-BINAP})NEt_3, CH_2Cl_2$

**Acknowledgments** : I would like to express my heartfelt thanks to my co-workers who are mentioned individually in the references, as well as to Dr. R. Schmidt (Hoffman La Roche) for a generous gift of (R) and (S)-MeO-BIPHEP and BIPHEMP.

References

1. Kagan,H.B. *Comprehensive Organomet.Chem*.Ed.Wilkinson,G.Perg.Press **1982**,8,463
2. Kagan,H.B. *Bull. Soc. Chim. Fr.* **1988**,846
3. Consiglio,G.,Waymouth,R.M. *Chem. Rev.* **1989**,89,257
4. ApSimon,J.W.;Collier,T. *Tetrahedron* **1986**,42,5157
5. Blystone, S.L. *Chem. Rev.* **1989**,89,1664
6. Ojima,I.;Clos,N.;Bastos,C. *Tetrahedron* **1989**,45,6901
7. Whitesell,J.K. *Chem. Rev.* **1989**,89,1581
8. James,B.;Joshi,A.M.;Kvintovics,P.;Morris,R.M.;Thorburn Blackburn,D. W Ed. *Catalysis of Organic Reactions* Marcel Dekker. N.Y.**1990**,11
9. Noyori,R. *Tetrahedron* **1994**, 50, 4259
10. Collins,A.N.;Sheldrake,G.N.;Crosby,J. Eds. *Chirality in Industry*, J. Wiley,N.Y. **1992**
11. Noyori,R. *Asymmetric Catalysis*, J. Wiley ,N.Y. **1994**

12. Akutagawa,S. *Applied Catalysis A : general* **1995**,128,171
13. Koenig,K.E. *Asymmetric Catalysis*, Morrisson J.E., Ed. Academic Press, New York **1985**,5,71
14. Dunina,V.V.;Beletskaya,I.P. *Zhurnal Organicheskoi Khimii* **1992**,28,1929
15. Brown,J.M. *Angew. Chem. Int. Ed. Eng.* **1987**,26,190
16. Inoguchi,K.;Sakuraba,S.;Achiwa,K.*Synlett* **1992**,169
17. Kagan,H.B.*Asymmetric Synthesis*,Morrison,J.D.,Ed. Academic Press **1985**,5
18. Brunner,H.*Topics in Stereochemistry*, Eliel, E.L. and Wiley, S.H. Eds. **1988**,18,541-63
19. James,B.R.;Wang,D.;Voigt,R.F.*J.Chem.Soc.Chem.Commun.* **1975**,574
20. Ikariya,T.;Ischii,Y.;Kawano,H.;Arai,T.;Saburi,M.;Yoshikawa,S.;Akutagawa,S. *J. Chem. Soc. Chem. Commun.* **1985**,922
21. Kitamura,M.;Tokunaga,M.;Noyori,R. *J. Org. Chem.* **1992**,57,4053
22. Noyori,R.;Ohta,M.;Hsiao,Y.;Kitamura,M.;Ohta,T.;Takaya,H. *J. Am. Chem. Soc.* **1986**,108,7117
23. Ohta,T.;Takaya,H.;Noyori,R. *Inorg. Chem.* **1988**,27,566
24. Genet,J.P.;Mallart,S.;Pinel,C.;Jugé,S. *Tetrahedron : Asymmetry* **1991**,2,1,43
25. Genet,J.P.;Mallart,S.;Pinel,C.;Jugé,S.;Laffitte,J.A. *Tetrahedron Lett.* **1992**,33,5343
26. Genet,J.P.;Mallart,S.;Pinel,C.;Thorimbert,S;Jugé,S.;Laffitte,J.A. *Tetrahedron : Asymmetry* **1991**,2,555
27. Genet,J.P.;Pinel,C.;Ratovelomanana-Vidal,V.;Pfister,X.;Caño de Andrade,M.C.; Laffitte,J.A. *Tetrahedron Asymmetry* **1994**,5,665
28. Heiser,B.;Broger,E.A.;Crameri,Y. *Tetrahedron : Asymmetry* **1991**,2,51
29. Alcock,N.W.;Brown,J.M.;Rose,M.;Wienand,A. *Tetrahedron : Asymmetry* **1991**,2,47
30. Kitamura,N.;Iokunaga,M.;Ohkuna,T.;Noyori,R. *Tetrahedron Lett.* **1991**,32,4163
31. Mashima,K.;Kusano,K.H.;Noyori,R;Takaya,H. *Chem. Comm.* **1989**,1208
32. Wam,K.;Davis,M.E. *Tetrahedron : Asymmetry* **1993**,4,2461
33. Taber,D.F.;Silverberg,L.J. *Tetrahedron Lett.* **1991**,32,4227.
34. King,S.A.;Thompson,A.S.;King,A.O.;Verhoeven,T.R. *J. Org. Chem.* **1992**,57,6689
35. Fronczek,F.R.;Watkins,S.E.;Stahly,G. *Organometallics* **1993**,12,1467
36. Mezzetti,A.;Consiglio,G. *J. Chem. Soc. Chem. Commun.* **1991**,1675
37. Mezzetti,A.;Costella,L.;Del Zotto,A.;Rigo,P.;Consiglio, G. *Gazetta Chim. It.* **1993**,123,155
38. Kawano,H.;Ishii,Y.;Kodama,T.;Saburi,M.;Uchida,Y. *Chem. Lett.* **1987**,1311
39. Tsukahara,T.;Kawano,H.;Ishii,Y.;Akutagawa,S.;Takahashi,T.;Uchida,Y.;Saburi,M. *Chem. Lett.* **1988**,2055
40. Saburi,M.;Takeuchi,H.;Osagawara,M.;Tsukahara,T.;Ishii,Y.;Takahashi,T.;Uchida,Y.;Ika riya,T. *J.Organomet. Chem.* **1992**,428,155
41. Bottechi,C.;Gladiali,S.;Matteoli,U.;Frediani,P. *J. Organomet. Chem.* **1977**,140,221
42. Matteoli,U.;Menchi,G.;Frediani,P.;Bianchi,M.;Piacenti,F. *J. Organomet. Chem.* **1985**,285,281
43. Joshi,A.M.;Thorburn,I.S.;Rettig,S.J.;James,B.R. *Inorganica Chimica Acta* **1992**, 283
44. James,B.R.;Pacheco,A.;Rettig,S.J.;Thorburn,I.S. *J. Mol. Cat.* **1987**,47,147
45. Chan,S.C.;Laneman,S. *Inorganica Chimica Acta* **1994**,223,165
46. Chiba,T.;Myashita,A.;Nohira, H.;Takaya, H. *Tetrahedron Lett.* **1993**,34,2351
47. Murata,M.;Morimoto,T.;Achiwa,K. *Synlett* **1991**,827
48. Yamamoto,N.;Murata,M.;Morimoto,T.;Achiwa,K. *Chem. Pharm. Bull.* **1991**,39,1085
49. Noyori,R.;Kitamura,M. *Modern Synthetic Methods* Ed. Scheffold, R. Spinger Verlag **1989**, 128
50. Noyori,R. *Chem. Soc. Rev.* **1989**,18,187
51. Noyori,R.;Takaya,H. *Acc.Chem.Res.* **1990**,23,345
52. Noyori,R.*Science*, **1990**,248,1194
53. Noyori,R. *Organic Synthesis in Japan Past, Present and Future*, Noyori,R. Ed. Tokyo Kagaku Dozin **1992**,301.
54. Genet,J.P.;Ratovelomanana-Vidal,V. ; Pfister,X. ;Caño.de.Andrade, M.C.;Laffitte,J.A.;Darses,S.;Pinel,C.;Bischoff,L.;Galopin.C. *Tetrahedron : Asymmetry* **1994**,5,675
55. Genet,J.P. *Acros Organics Acta* **1995**,1,4

56. Takaya,H.;Kasahara,I.;Ohta,T.;Sayo,N.;Kumobayashi,H.;Akutagawa,S.;Inoue,S.I.;Kasahara,I.;Noyori,R. *J. Am. Chem. Soc.* **1987**,109,5,1596
57. Takaya,H.; Ohta,T.;Inoue,S.I.;Tokunaga,M.;Kitamura,M. and Noyori,R. *Organic Syntheses* **1993**,72,74
58. Imperiali,B.;Zimmerman,J.N. *Tetrahedron Lett.* **1988**,29,5343
59. Kitamura,M.;Kasahara,I.;Manabe,K.;Noyori,R.;Takaya,H. *J. Org. Chem.* **1988**,53,708
60. Kitamura,M.;Nagai,K.;Hsiao,Y. and Noyori,R. *Tetrahedron Lett.* **1990**,31,549
61. Roland,S. Thesis University Pierre and Marie Curie **1995**
62. Ohta,T.;Takaya,H.;Kitamura,M.;Nagai,K. and Noyori,R *J. Org. Chem.* **1987**,52,3174
63. Saburi,M.;Shao,L.;Sakurai,T.;Uchida,Y. *Tetrahedron Lett.* **1992**,33,51,7877
64. Kam T. Wan;Davis,M.E. *Nature* **1994**,370,449
65. Chan,A.S.C.;Laneman,S.A.;Miller,R.E.;Wagenknecht,J.H.;Coleman,J.P. *Chem.Ind.* **1994**,53,49
66. Zhang,X.;Uemura,T.;Matsumura,K.;Sayo,N.;Kumobayashi,H.;Takaya,H. *Synlett* **1994**,501
67. Muramatsu,H.;Kawano,H.;Ishii,Y;Saburi,M.;Uchida,Y. *J. Chem. Soc. Chem. Commun.* **1989**,769
68. Noyori,R.;Ikeda,T.;Ohkuma,T. et al *J. Am. Chem. Soc.* **1989**,111,9134
69. Kawano,H.;Ikariya,T.;Ishii,Y.;Saburi,M. *Chem. Commun. Perkin I.* **1989**,1571
70. Kam to Wan,Davis,M.E. *Tetrahedron : Asymmetry* **1993**,4,2461
71. Lubell,W.O.; Kitamura,M.;Noyori,R.*Tetrahedron : Asymmetry* **1991**, 2,7,543
72. Kitamura,M.;Hsiao,Y.;Noyori,R.;Takaya,H. *Tetrahedron Lett.* **1987**,28,4829
73. Kitamura,M.;Hsiao,Y.;Ohta,M.;Takaya,H.;Noyori,R. *J. Org. Chem.* **1994**,59,297
74. Miyachita,A.;Takaya,H.,Souchi,T.;Noyori,R. *Tetrahedron* **1984**,40,1245
75. Miyachita,A.;Yasuda,A.;Takaya,H.;Souchi,T.;Noyori,R. *J. Amer. Chem. Soc.* **1980**,102,7932
76. Knierzinger,A.;Schönholzer,P. *Helv. Chim. Acta.* **1992**,75,1211
77. Schmidt,R.;Foricher,J.;Cereghetti,M. and Schönholzer,P. *Helv.Chim.Acta* **1991**,74,370
78. Ohta,T.;Miyake,N.;Takaya,H. *J. Chem. Soc. Commun.* **1992**,1725
79. Ohta,T.;Miyake,N.;Akutagawa,S.;Seido,N.;Kumobayashi,H.;Takaya,H. *J. Org. Chem.* **1995**,60,357
80. Ohta,T.;Miyake,N.;Seido,N.;Takaya,H. *Tetrahedron Lett.* **1992**,33,635
81. Noyori,R.;Ohkuma,T.;Kitamura,M.;Takaya,H.;Sayo,N.;Kumobayashi,H.;Akutagawa,S. *J. Am. Chem. Soc.* **1987**,109,5856
82. Genet,J.P.;Ratovelomanana-Vidal,V.;Pfister,X.;Caño de Andrade,M.C.; Guerreiro,P. ; Lenoir,J.Y. *Tetrahedron Lett.* **1995**,36,4801
83. Hoke,J.B.;Hollis,L.S.;Stern,E.W. *J. Organomet. Chem.* **1993**,455,193
84. Burk,M.J.;Harper,T.G.P.;Kalberg,C.S. *J. Am. Chem. Soc.* **1995**,117,4423
85. Burk,MJ.;Feaster,J.E.;Nugent,W.A.;Harlow,R.L. *J.Am. Chem. Soc.* **1993**,115,10125
86. Kitamura,M.;Ohkuma,T.;Takaya,H.;Noyori,R. *J. Am. Chem. Soc.***1988**,110,629
87. Nozaki,K.;Sato,N.;Takaya,H. *Tetrahedron : Asymmetry.* **1993**,4,2179
88. Ohta,T.;Miyake,N.;Takaya,H. *J. Chem. Soc. Commun.***1992**,1725
89. Kitamura,M.;Tokunaga,M.;Ohkuma,T.; Noyori,R. *Organic Syntheses* **1992**,71,1
90. Genet,J.P.;Mallart,S.;Jugé,S.French Patent, 8911159 (Aug:**1989**)
91. Kitamura,M.;Ohkuma,T.;Tokunaga,M.;Noyori,R. *Tetrahedron : Asymmetry* **1990**,1,1
92. Genet,J.P.;Pfister,X.; Ratovelomanana-Vidal,V.;Pinel,C.;Laffite, J.A. *Tetrahedron Lett.* **1994**,35,4559
93. Kitamura,M.;Tokunaga,M.;Noyori,R. *J. Am. Chem. Soc.* **1993**,115,144
94. Kitamura,M.;Tokunaga,M.;Noyori,R. *J. Am. Chem. Soc.* **1995**,117,2931
95. Kitamura,M.;Tokunaga,M.;Pham,T.;Noyori,R. *Tetrahedron Lett.* **1995**,5569
96. Fukuda,N.;Mashima,K.;Matsumura,Y.I.;Takaya,H. *Tetrahedron Lett.* **1990**,7185
97. Genet,J.P.;Caño de Andrade,M.C.;Ratovelomanana-Vidal,V. *Tetrahedron Lett.* **1995**,36,2063
98. Review : Noyori,R.;Tokunaga,M.;Kitamura,M. *Bul. Chem. Soc. Jpn.* **1995**,68,36
99. Genet,J.P.; Ratovelomanana-Vidal,V.;Pinel,C. *Synlett.* **1993**,479
100. Ohkuma,T.;Ikariya,T.;Ooka,H.;Hashiguchi,S.;Ikariya,T.;Noyori,R. *J. Am. Chem. Soc.* **1995**,117, 2675

101.a)Hashiguchi,S.;Fuji,A.;Takahara,J.;Ikariya,T.;Noyori,R. *J.Am. Chem. Soc.* **1995**,117,7562
b)Takehara,J.;Hashiguchi,S.;Fujii,A.;Inoue,S.I.;Ikariya,T.;Noyori,R. *Chem. Commun* **1996**,233
102.Oppolzer,W.;Willis,M.; Bernardinelli,G. *Tetrahedron Lett.* **1990**,31,4117
103.Cesarotti,E.;Mauri,A.;Villa,L.;Pallavicini,M. *Tetrahedron Lett.* **1991**,32,4384
104.Cesarotti,E.;Antognazza,P.;Mauri,A.;Pallavicini,M.;Villa,L. *Helv. Chim. Acta* **1992**,75,2563
105.Buser,H.P.;Spindler,F. *Tetrahedron : Asymmetry* **1993**,4,2451
106.Shao,L.;Seki,T.;Kawano,H.;Saburi,M. *Tetrahedron Lett.* **1991**,32,7699
107.Kitamura,M.;Ohkuma,T.;Noyori,R.; *Tetrahedron Lett.* **1989**,29,1555
108.Rychnovsky,S.D.;Griesgraber,G.;Zeller,S.;Skalitzky,D.J. *J. Org. Chem.* **1991**,56,5161
109.Greck,C.;Bischoff,L.;Pinel,C.;Ferreira,F.;Piveteau,E.;Genet,J.P. *Synlett* **1993**,475
110.Jones,A.B.;Yamaguchi,M.;Patten,A.;Ragan,J.A.;Smith,D.B.;Danishefsky,S.J.;Schreiber, S.L. *J. Org. Chem.* **1989**,54,17
111.Taber,D.F.;Silverberg,L.J.;Robinson,E.D. *J. Am.Chem. Soc.* **1991**,113,6639
112.Schmidt,U.;Leitenberger,V.;Griesser,H. *Synthesis* **1992**,1248
113.Wovkulich,P.M.;Shankaran,K.;Uskokovic,M.R. *J. Org.Chem.* **1993**,58,832
114.Guzo,P.R.; Miller,M.J. *J. Org. Chem.* **1994**,59,4862
115.Rychnovsky,S.D.;Hoye,R.C. *J. Amer. Chem. Soc.* **1994**,116,1753
116.Pommier,A.;Pons,J.M.;Kocienski,P.J.;Wong,L. *Synthesis* **1994**,1294
117.Taber,D.F.;Deker,P.B.;Silverberg,L.J. *J. Org. Chem.* **1992**,57,5990
118.Greck,C.;Bischoff,L.;Genet,J.P. *Tetrahedron : Asymmetry* **1995**,6,1989
119.Greck,C.;Ferreira,F.;Genet,J.P. *Tetrahedron Lett.* **1996**,37,2031
120.Overman,L.E.;Rabinowitz,M.H.;Renhove,P.A. *J. Am. Chem. Soc.* **1995**,117,2657
121.Roland,S.;Durand,J.O.;Savignac,M.;Genet,J.P. *Tetrahedron Lett.* **1995**,36,4801
122.Mashima,K.;Matsumura,Y.I.;Kusano,K.;Kumobayashi,H.;Sayo,N.;Hori,Y.;Ishizaki,T.; Akutagawa,S.;Takaya,H. *Chem. Comm.* **1991**,9,609

# Chapter 3

# Tartrate-Derived Ligands for the Enantioselective LiAlH$_4$ Reduction of Ketones

## A Comparison of TADDOLates and BINOLates

**Albert K. Beck, Robert Dahinden, and Florian N. M. Kühnle**

Laboratorium für Organische Chemie, Eidgenössischen Technischen Hochschule, ETH-Zentrum, Universitätstrasse 16, CH–8092 Zürich, Switzerland

The enantioselectivities of the LAH reductions performed in our group using the tartrate derivatives 1,4-bis(dimethylamino)-(2S,3S)-butane-2,3-diol (**3**, DBD) and α,α,α',α'-tetraaryl-1,3-dioxolane-4,5-dimethanol (**5**, TADDOL) as chiral ligands are reviewed and compared to the results of other groups with biaryl-derived LAH complexes of BINOL, biphenol and biphenanthrol. When (R,R)-TADDOLs are applied in the reaction, the preferentially formed 1-arylalkanols have (S)-configuration, as is observed for the products obtained with the corresponding (P)-BINOL, (P)-biphenol and (P)-biphenanthrol derivatives. A common mechanistic model is discussed.

The synthesis of enantiomerically pure secondary alcohols can be achieved by two different approaches: the reduction of unsymmetrically substituted ketones (*1-5*) and the nucleophilic addition to aldehydes (Scheme 1).

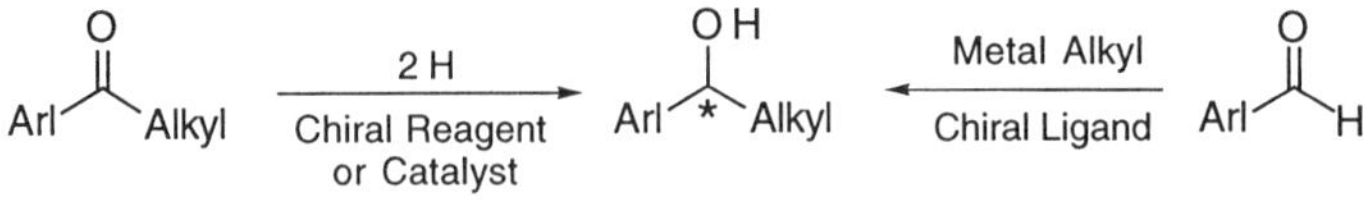

$$\text{Arl}-\underset{\text{Alkyl}}{\overset{\text{O}}{\text{C}}} \xrightarrow[\text{Chiral Reagent or Catalyst}]{2\text{ H}} \text{Arl}-\underset{\text{Alkyl}}{\overset{\text{OH}}{\underset{*}{\text{C}}}} \xleftarrow[\text{Chiral Ligand}]{\text{Metal Alkyl}} \text{Arl}-\underset{\text{H}}{\overset{\text{O}}{\text{C}}}$$

**Scheme 1**

For both routes, highly efficient *catalytic* methods now exist. Thus, the reduction of ketones can be carried out with H$_2$ and a transition metal complex following Noyori's procedures (*6-8*), with BH$_3$ and an oxazaborolidine (*9*) as in the work of Itsuno (*10*) and Corey (*11,12*) or with NaBH$_4$ and a Co complex as shown by Mukaiyama (*13*). In the most successful realization of enantioselective group transfer to aldehydes, organozinc or titanium reagents are employed in the presence of chiral aminoalcohols (Noyori (8,14)) or Ti complexes of N,N'-1,2-cyclohexane-diyl-bistrifluoromethanesulfonamide (CYDIS) (Ohno (15,16), Knochel (17)) or of α,α,α',α'-tetraaryl-1,3-dioxolane-4,5-dimethanols (TADDOLs) (our group (18-20)).

One of the oldest approaches to enantioselection is the use of stoichiometric amounts of chirally modified lithium aluminum hydride (LAH). The idea of exchanging hydrides with chiral alcohols in the coordination sphere of aluminum was born in 1951 by Bothner-By (21) when he used a complex obtained from the reaction of camphor and LAH to reduce aliphatic ketones. Later it was shown that the optical activity of the product was due to contamination by the chiral reagent (22,23). In the 1960s, Červinka and Landor changed the reducing system from monodentate (24,25) to bidenate chiral ligands, which were derived from sugars or alkaloids (23,26) and obtained enantioselectivities in the range of 50 to 75 % ee.

Considering our long-standing relationship with tartaric acid used as a chiral building block in the preparation of chiral auxiliaries in both enantiomeric forms (EPC syntheses (27)) and as chiral solvents (28-34), it seemed an obvious starting point for the modification of LAH which results in the formation of a chiral reducing agent. This goal was achieved in the mid-seventies with 1,4-bis(dimethylamino)-(2S,3S)-butane-2,3-diol (3, DBD (35)), a bidentate ligand that is easily accessible, in both enantiomeric forms, in two steps from diethyl tartrate (1) *via* the diamide 2 (Scheme 2).

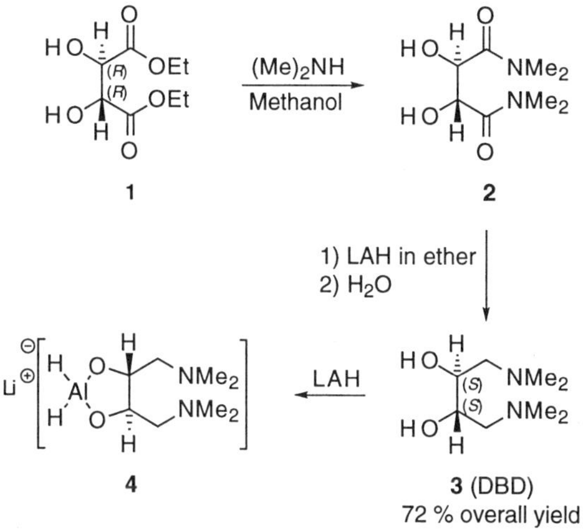

**Scheme 2**

The selectivities obtained in the reduction of aryl alkyl ketones with the LAH-DBD complex 4 are displayed in Table I (35). Even though these selectivities may appear rather unsatisfactory from today's viewpoint, they were among the best enantiomer ratios achieved at that time and the method was successfully applied in natural product synthesis (36,37).

**Table I.** Enantiomer Ratios (er) for the Reduction of Aryl Alkyl Ketones with the Al-DBD Complex **4** (1 equiv., rt, 4 h, yields >80%).

| Entry | Ketone | sec. Alcohol er (S)/(R) |
|---|---|---|
| 1 | Acetophenone | 71:29 |
| 2 | Propiophenone | 72:28 |
| 3 | Valerophenone | 74:26 |
| 4 | Isobutyrophenone | 64:36 |
| 5 | Pivalophenone | 61:39 |
| 6 | Cyclohexyl phenyl ketone | 77:23 |
| 7 | 2,4,6-Trimethylacetophenone | 88:12 |
| 8 | 1-Acetylnaphtalene | 66:33 |

A new impulse to this research effort in our group was triggered by the synthesis of TADDOL **5** from tartaric acid in three steps (*38-42*). This new class of ligand turned out to be very useful for the synthesis of enantiomerically pure compounds (*18-20,43-46*). (For a practically complete list of TADDOLs with references for their syntheses and crystal structures, see Table I in a recent full paper (*44*)). During the investigation of the potential of these new ligands in the early eighties, we attempted to enantioselectively reduce propiophenone by employing the chiral LAH derivative **6** (*47*). These reactions were carried out under conditions which had been optimized by our group for reductions with **4** (*35*). Using one equivalent of the chiral LAH-complex **6** and one equivalent of propiophenone, the reaction in diethyl ether at room temperature produced the corresponding *sec.* alcohol with only 6 % ee. Other ketones gave higher selectivities: 14 % ee with 2-hexanone and up to 40 % ee with acetophenone depending on the reaction time (*47*). Considering these low selectivities, we did not carry out further experiments at that time.

A major step forward in the pursuit of high enantioselectivities in the asymmetric LAH reduction was achieved by Noyori (*48-54*). He and his co-workers replaced one of the two remaining hydrides on the LAH-BINOL complex **7** by an alkoxide moiety such that only one hydride remains on the metal (*cf.* **8**, **9**) (*49,54*). This involved rediscovering a principle first found by Landor in 1967, which led, in some cases, to a dramatic increase in the asymmetric induction (*55*). Unfortunately, in the case of DBD, no improvement in the selectivity was achieved by this method (*35*) but, in contrast, Noyori's BINAL-H complexes **8** or **9** and Suda's biphenol complex **10** (*56*) are striking examples of this effect. The $C_2$-symmetric complex **7** or **11** led only to the disappointing enantiomer ratio (er) of 51:49 in the reduction of acetophenone, whereas the $C_1$-symmetric complexes **8** and **10** reached a ratio of over 96:4 (*cf.* Table II and III, entry 1 each) (*49,54,56*).

Excellent results were obtained in the reduction of aryl alkyl ketones and of α-alkynyl ketones (*cf.* Table II; for a more recent result, see (*57*)). However, the enantioselectivities drop severely when purely aliphatic ketones are employed as substrates. For example, the treatment of the unbranched 2-octanone with **8** gave the corresponding alcohol only with an er of 62:38 (Table II, entry 11). The 1-phenyl-2-propanol from the β-aryl substituted benzyl methyl ketone was isolated with the even lower er of 56:44 (Table II, entry 10). Comparable results were obtained by Yamamoto using the biphenanthrol complex **12** (Table III, entry 3 and 4) (*58*).

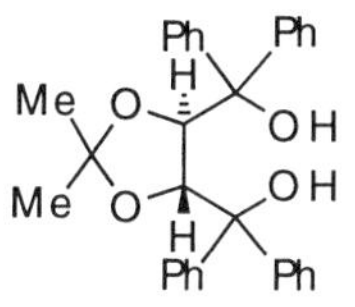

(*R,R*)-**5**

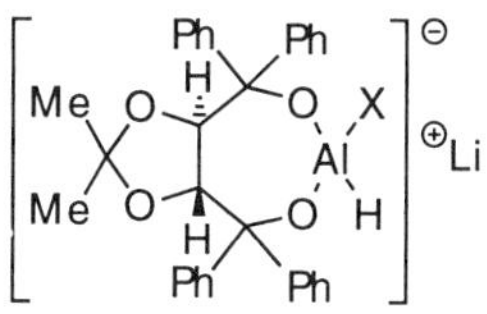

(*R,R*)-**6**  X = H
(*R,R*)-**15** X = OEt

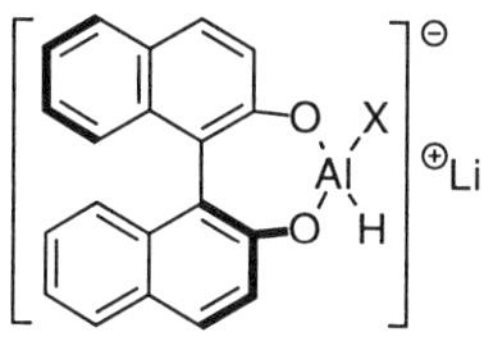

(*P*)-**7**  X = H
(*P*)-**8**  X = OEt
(*P*)-**9**  X = OMe

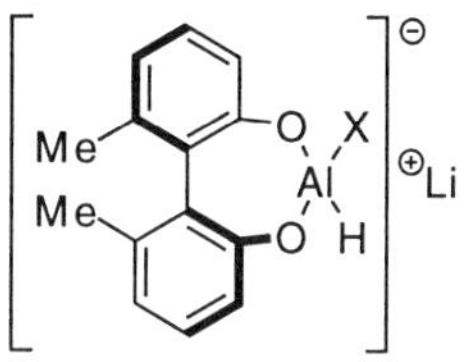

(*P*)-**10** X = OEt
(*P*)-**11** X = H

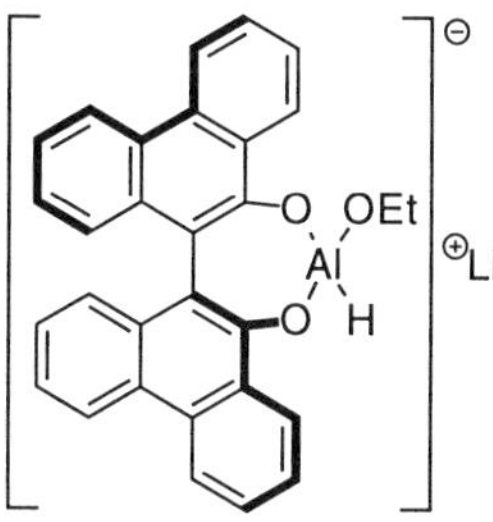

(*P*)-**12**

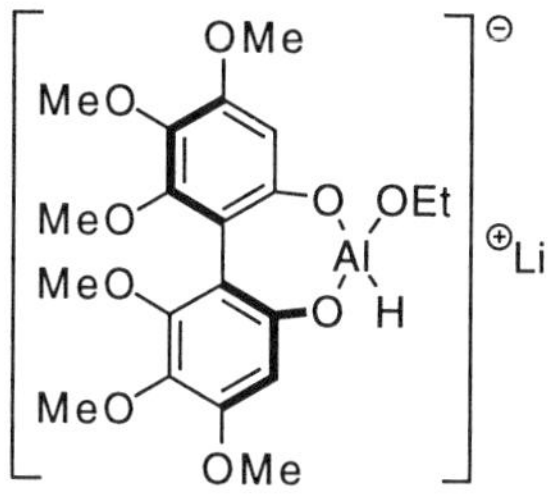

(*P*)-**13**

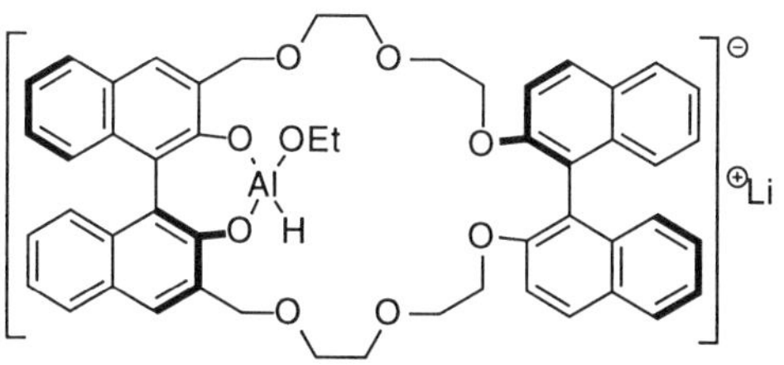

(*P,P*)-**14**

This substrate problem was later solved by two groups. Meyers and co-workers used the hexamethoxybiphenyl ligand **13** to reduce 2-octanone with an er of 88:12 and benzyl ethyl ketone with an er of 97:3 (Table III, entry 8 and 9) (*59*). These selectivities are in the range of those found for aryl alkyl ketones using Noyori's method. Another approach was undertaken by Yamamoto's group who introduced the crown 2,2'-dihydroxy-1,1'-binaphthyl complex **14** and obtained reasonable selectivities for aliphatic ketones (*cf.* Table III, entries 10-12) (*60*).

**Table II.** Enantiomer Ratios (er) and Reaction Conditions for the Reduction of Different Ketones with 3 Equivalents of the BINAL-H Complexes **8** and **9** to the Corresponding *sec*. Alcohols

| Entry | BINAL-H | Ketone | Reaction Conditions (°C) (time (h)) | *sec*. Alcohol Yield (%) | er (S)/(R) |
|---|---|---|---|---|---|
| 1 | (*M*)-**8** | Acetophenone | -100 (3), -78 (16) | 61[a] | 3:97 |
| 2 | (*P*)-**9** | Propiophenone | -100 (3), -78 (16) | 62[a] | 99:1 |
| 3 | (*P*)-**8** | Butyrophenone | -100 (3), -78 (16) | 78 | >99:1 |
| 4 | (*P*)-**8** | Isobutyrophenone | -100 (3), -78 (16) | 68[a] | 86:14 |
| 5 | (*M*)-**8** | Pivalophenone | -100 → rt (10) | 80 | 28:72 |
| 6 | (*P*)-**8** | 1-Octyn-3-one | -100 (1), -78 (2) | 71 | 94:6 |
| 7 | (*P*)-**8** | 1-Undecyn-3-one | -100 (1), -78 (2) | 74 | 95:5 |
| 8 | (*P*)-**9** | 1-Undecyn-3-one | -100 (1), -78 (2) | 80 | 98:2 |
| 9 | (*P*)-**9** | 1-Tetradecyn-3-one | -100 (1), -78 (2) | 90 | 96:4 |
| 10 | (*P*)-**8** | Benzyl methyl ketone | -100 (2), -78 (16) | 74[a] | 56:44 |
| 11 | (*P*)-**8** | 2-Octanone | -100 (2), -78 (16) | 67[a] | 62:38 |

[a] Determined by GC analysis.


**Table III.** Reduction of Different Ketones with Suda's 6,6'-Dimethylbiphenyl Complex **10**, Yamamoto's Biphenanthryl Complex **12** or Crown Binaphthyl Complex **14** and Meyers Hexamethoxybiphenyl Complex **13**

| Entry | LAH-complex | Ketone | Reaction Conditions (°C) (time (h)) | *sec*. Alcohol Yield (%) | er (S)/(R) |
|---|---|---|---|---|---|
| 1 | 10 | Acetophenone | -100 (3), -78 (16) | 52 | 96:4 |
| 2 | 10 | Butyrophenone | -100 (4), -78 (16) | 60 | 99:1 |
| 3 | 12 | Acetophenone | -5 (1) | 75 | 99:1 |
| 4 | 12 | Propiophenone | -5 (1) | 78 | 99:1 |
| 5 | 13 | Acetophenone | -100 (2), -78 (12) | 93 | 98:2 |
| 6 | 13 | Valerophenone | -100 (2), -78 (12) | 84 | 97:3 |
| 7 | 13 | 1-Acetyl-1-cyclohexene | -100 (2), -78 (12) | 68 | 92:8 |
| 8 | 13 | 2-Octanone | -100 (2), -78 (12) | 76 | 88:12 |
| 9 | 13 | Benzyl ethyl ketone | -100 (2), -78 (12) | 72 | 97:3 |
| 10 | 14 | Acetophenone | 0 (24) | 76 | 95:5 |
| 11 | 14 | Benzyl methyl ketone | 0 (24) | 75 | 94:6 |
| 12 | 14 | 2-Hexanone | 0 (24) | 78 | 93:7 |

In summary, it can be stated that Noyori's BINAL-H complexes **8** and **9** and the related chiral biaryl LAH complexes **13** and **14** are powerful reagents for the reduction of several ketone classes and therefore it is not much of a surprise that numerous applications have been published in recent years, such as the reduction of acylstannanes (Scheme 3, (a) and (b) (*61-65*)), trifluoromethyl ketones (Scheme 3, (c) (*66*)), and phosphinyl imines (*67*), reductive ring opening of lactones (Scheme 3, (d) (*68,69*)), and *meso*-1,2-dicarboxylic anhydrides (Scheme 3, (e) (*70,71*)) (for a wide variety of natural product syntheses see (*72-84*)).

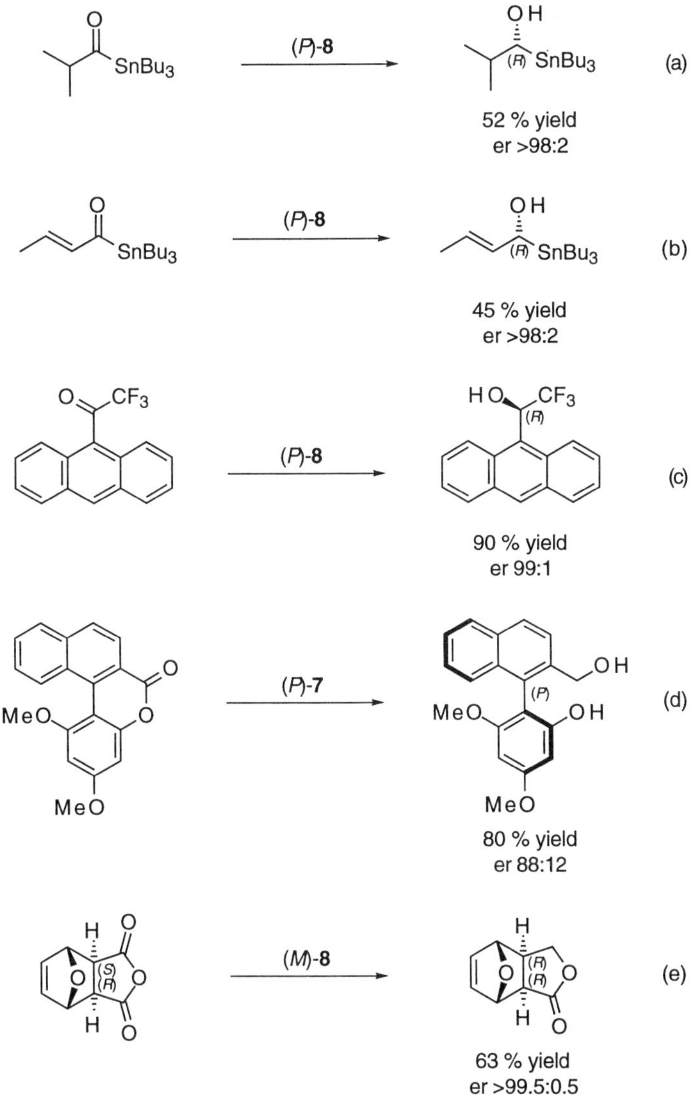

**Scheme 3**

However, the major disadvantages of these methods are the large excess of the expensive or not commercially available chiral reagents (up to four equivalents) and the very low reaction temperature of -100°C. The method of Yamamoto is the only exception with a reaction temperature of either -5 or 0°C. Having considered this situation, we returned to our efforts with the readily available TADDOL ligands. However, only monohydride LAH complexes were now used. We tested the TADDOL complex **15** under conditions similar to those reported for the BINOL ligand. The first experiment, the reduction of acetophenone, gave the corresponding *sec.* alcohol in 45 % yield and with a selectivity of 94:6. This result prompted us to begin a systematic investigation using acetophenone as the standard substrate.

## LAH-TADDOLates as Enantioselective Reducing Agents

**Preparative Results.** In order to avoid aging effects (*85,86*), which are known to decrease the enantioselectivity, we used only purified LAH always handling it in a glove box. The chiral reducing agent **15** was prepared by the reaction of LAH with one equivalent of ethanol and one equivalent of TADDOL **5** in THF (Scheme 4). If a considerable amount of precipitate separates, the corresponding batch has to be discarded (*47,49*). The resulting complex **15** was not isolated and the ketone (0.5 equivalent) was added as a THF solution *via* syringe (Scheme 4a).

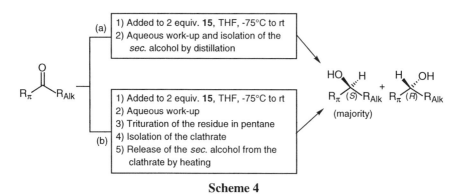

Scheme 4

The best results with the TADDOL system in our group were obtained when the ketone was added to the solution of two equivalents of the reducing agent at -78°C and the mixture was then allowed to warm up to room temperature over four hours. Lower temperatures and larger excesses of the reducing agent (up to 4:1) did not increase the selectivity of the reaction. Thus, the LAH-TADDOLates are not suffering from the major disadvantages of the biaryl derived systems. Variation of the alkoxy group from ethoxide to methoxide, isopropoxide, *t*- and *n*-butoxide decreased the selectivity but did not affect the yield (*49,87*). Unlike most of the TADDOL mediated reactions performed by our group, the highest selectivities in LAH reductions were observed with the first TADDOL ever synthesized (**5**, Table IV, entry 1). All other TADDOLs except for the 2-phenyl-α,α,α',α'-tetraphenyl-1,3-dioxolane-4,5-dimethanol displayed a lower asymmetric induction in the reaction. Scheme 4a summarizes the points mentioned above and shows the optimized reaction conditions for TADDOL **5** (*47*).

**Table IV.** Enantiomer Ratios (er) and Reaction Conditions (see Scheme 4a) for the Reduction of Different Ketones with the Al-TADDOLate **15** to the Corresponding *sec*. Alcohols

| Entry | Ketone | Reaction Conditions (°C) (time (h)) | *sec*. Alcohol Yield (%) | er (*S*)/(*R*) |
|---|---|---|---|---|
| 1 | Acetophenone | -74 → rt (4) | 95 | 95:5 |
| 2 | Propiophenone | -82 → rt (4) | 89 | 94:6 |
| 3 | Isobutyrophenone | -79 → rt (4) | 90 | 93:7 |
| 4 | Pivalophenone | -72 → rt (5) | 89 | 77:23 |
| 5 | 2-Bromoacetophenone | -74 → rt (5) | 76 | 95:5 |
| 6 | 4-Bromoacetophenone | -80 → rt (5) | 62 | 96:4 |
| 7 | 2-Fluoroacetophenone | -94 → rt (6) | 77 | 74:26 |
| 8 | 4-Methoxyacetophenone | -71 → rt (5) | 87 | 86:14 |
| 9 | 2-Furyl methyl ketone | -76 → rt (6) | 66 | 89:11 |
| 10 | 1-Acetylcyclohexene | -63 → rt (5) | 72 | 82:18 |
| 11 | 1-Octyn-3-one | -78 → rt (4) | 84 | 76:24 |
| 12 | Benzyl methyl ketone | -74 → rt (5) | 82 | 21:79 |
| 13 | 2-Octanone | -76 → rt (5) | 83 | 61:39 |
| 14 | 3-Nonanone | -76 → rt (4) | 82 | 50:50 |

**Reduction of Different Ketones.** As the intention was to develop a general method for the use of TADDOL ligands in the asymmetric reduction of ketones, over a dozen different substrates were reduced under our optimized conditions. First, the effects of varying the alkyl groups were examined in the reduction of aryl alkyl ketones. This showed that the selectivity drops with increasing size of the alkyl group. From methyl to ethyl and isopropyl, the changes are almost negligible (Table IV, entries 1-3) but for *t*-butyl (Table IV, entry 4), the tendency becomes obvious (enantiomer ratio 77:23). Next, the substitution pattern on the aromatic ring was varied. *Ortho*- and *para*-bromo substitution had no effect on the reaction (≥95:5, entries 5 and 6 in Table IV) but the more strongly electron withdrawing *ortho*-fluorine caused a drop in selectivity (Table IV, entry 7). The electron donating *para*-methoxy group also had an unfavorable influence (Table IV, entry 8). With the heteroaromatic furyl substituent (Table IV, entry 9) or with acetylcyclohexene (Table IV, entry 10), the enantioselectivities are moderate. With 1-octyn-3-one, the er is 76:24 (Table IV, entry 11). However, our selectivities were less than those reported by the Noyori group for this substrate type (*48*). In the case of purely aliphatic ketones, rather poor selectivities were observed (Table IV, entries 13, 14). For example, the reduction of 2-octanone leads to the corresponding alcohol with a disappointing er of 61:39 and 3-nonanone gives a racemic product, whereas Meyers (*59*) and Yamamoto (*60*) report considerably higher selectivities for these substrates (see Table III). Strikingly, with benzyl methyl ketone, a reversal of the selectivity was in fact found (er = 21:79, Table IV, entry 12) when using the TADDOL complex **15**. For a very similar substrate, benzyl ethyl ketone (Table III, entry 9), an excellent 97:3 selectivity with no reversal of configuration was observed by Meyers (see also mechanistic part).

**Enhancement of Enantiopurity by Clathrate Formation in the Crude Product Mixture.** As it is well known in the literature (*88-93*) that TADDOLs enantioselectively form inclusion compounds with chiral alcohols, we attempted to use this phenomenon to enhance the enantiopurity of our products by a special work-up procedure: after aqueous work-up, drying of the organic phase and evaporation of the solvents, pentane was added to the resulting oily residue. This led to an immediate precipitation of a white powder which was stirred with the supernatant liquid for 24 hours. The solid was isolated by filtration and heated in a bulb-to-bulb distillation apparatus to yield the entrapped *sec*. alcohol (Scheme 4b). By applying this new work-up procedure for the reduction of acetophenone, 1-phenylethanol was isolated in 85 % yield and with a ratio of (*S*)- and (*R*)-enantiomers of 97.5:2.5 (Table V, entry 1).

Table V. Enantiomer Ratios and Reaction Conditions (see Scheme 4b) for the Reduction of Different Substrates with Complex **15** and with Enantiomer Enrichment by Clathrate Formation of the Crude Product

| Entry | Ketone | Reaction Conditions (°C) (time (h)) | *sec*. Alcohol Yield (%) | er (*S*)/(*R*) |
|---|---|---|---|---|
| 1 | Acetophenone | -78 → rt (4) | 85 | 97.5:2.5 |
| 2 | Acetophenone | -76 → rt (5) | 73 | 98.5:1.5[a] |
| 3 | Propiophenone | -65 → rt (4) | 43 | 98.5:1.5 |
| 4 | 2-Fluoroacetophenone | -78 → rt (4) | 49 | 99:1 |
| 5 | 1-Acetylcyclohexene | -78 → rt (4) | 38 | 98.5:1.5 |
| 6 | 2-Furyl methyl ketone | -78 → rt (4) | 46 | 95.5:4.5 |
| 7 | Benzyl methyl ketone | -78 → rt (4) | 25 | 11.5:88.5 |

[a] Repeat of the trituration and isolation steps 3, 4 and 5 in Scheme 4b.

By repeating the trituration and isolation steps 3, 4 and 5 in Scheme 4b, the enantiomer ratio was increased to 98.5:1.5 and the overall yield remained at 73 % (Table V, entry 2). The same enrichment procedure was applied to several product alcohols and the enantiopurities achieved are shown in Table V. From these results, it can be concluded that under these conditions a branched substituent on the alcohol or a long aliphatic chain prevents clathrate formation. This simple modification of the work-up procedure is suitable for increasing the enantiopurity of the alcohols formed in the LiAl-TADDOLate hydride reductions considerably.

## Comparison of the Different Mechanistic Proposals

The different mechanisms proposed in the literature (47,49,59,94) assume that only one chiral biaryl ligand is coordinated to the Al atom and that the hydride must be close to the carbonyl C-atom in the transition state. Therefore, Noyori (49) suggests a "*quasi*-aromatic" chair-like six-membered ring transition state, in which the lithium ion acts as a Lewis acid activating the carbonyl group (Figure 1). The oxygen of the

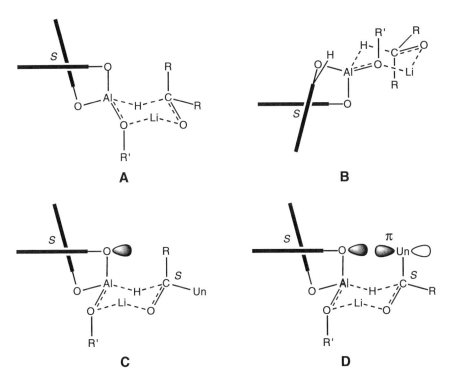

**Figure 1.** Possible transition state arrangements for the reduction of ketones with LAH modified BINOL proposed by Noyori.

achiral alkoxy group is more likely to act as bridging atom because it displays a much higher basicity than the binaphthol derived oxygens. The O-atoms of the chelating binaphthol occupy axial and equatorial positions on the ring. This leads to a pair of diastereoisomeric transition state arrangements (Figure 1, **A** and **B**) of which one suffers from steric repulsion between the binaphthyl and the alkoxy moiety (Figure 1, **B**). For the other arrangement (Figure 1, **A**), two different orientations (Figure 1, **C** and **D**) of the ketone are still possible. As the best results were obtained with unsaturated substrates, Noyori restricts the explanation to ketones with a π-system in the α-position. The stereoselectivity is controlled by the interactions of the axial substituents, where a destabilizing n/π type electronic repulsion in **D** in Figure 1 is assumed to be the dominating effect and the reaction therefore proceeds *via* the arrangement **C** in Figure 1, where no such electronic effect exists. The 1,3-diaxial type steric repulsion is contrary to this effect and increases with the bulkiness of the alkyl sidechain resulting in lower selectivities but cannot overcome the electronic interaction. The same mechanistic proposal was used by Meyers to rationalize his results (*59*).

For us, the first puzzling question was whether the reactive species is really a mononuclear complex or whether polynuclear aggregates, which would not be unusual for lithium and aluminum derivatives are involved (*95-97*). In order to verify Noyori's assumption that the reactive species is a mononuclear complex, we checked whether the enantiopurity of the ligand was linearly correlated to the enantiomer ratio of the product (*98,99*). It can be seen from Figure 2 that this is actually the case.

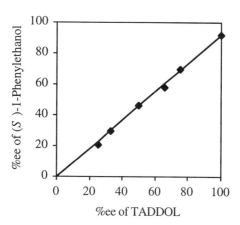

**Figure 2.** Linear relationship between enantiopurity of the ligand **5** and the reduction product (*S*)-1-phenylethanol.

A linear correlation has also been observed for several metal-TADDOLate mediated reactions (*43,100*) with the Ti-TADDOLate catalyzed Diels-Alder reaction being an exception (*44*). In order to gain information about the structure of the chiral

LAH-complex (see formula **15**) a deutero-THF solution of the reducing agent was examined by $^1$H NMR spectroscopy. In comparison to diisopropoxy-Ti-TADDOLate (*101*), the methine protons of the TADDOL ligand in **15** are shifted to higher field and split into two coupled doublets indicating that the complex is no longer $C_2$-symmetric and the integral ratios suggest a TADDOL to ethanol ratio of 1:1 on the aluminum. These results lead to a proposed structure of the reducing complex as a seven-membered O-Al-O containing ring. If we assume - as Noyori, Meyers and their respective co-workers do - that the hydride is transferred in a bimetallic six-membered ring chair-like transition state, the product complex, which is initially formed, should be as pictured in Figure 3. The hydride is added to the carbonyl carbon from the *Re* face (*102,103*) with the unsaturated, usually larger $R_\pi$ group in an equatorial position on the six-membered ring. In this arrangement, $R_\pi$ is located parallel to a *quasi*-axial phenyl group and the saturated $R_{Alk}$ points towards a *quasi*-equatorial phenyl ring on the other side of the TADDOL moiety.

The different selectivities obtained with different ketones are compatible with this model: the ideal substrate for our system is an aryl alkyl ketone with an unbranched alkyl substituent of moderate chain length. The arrangement in Figure 3 might be stabilized by a π-π-stacking effect of the ketone aryl substituent ($R_\pi$) with one of the *quasi*-axial phenyl groups ($Ph_{ax}$) of the TADDOL ligand (*cf.* the higher enantioselectivities of reductions of unsaturated ketones, such as 1-acetylcyclohexene, in comparison to dialkyl ketones which give almost racemic products, see Table IV). Highly branched alkyl groups such as *t*-butyl or long chain unbranched alkyl substituents are subject to steric interactions with the *quasi*-equatorial TADDOLate phenyl group ($Ph_{eq}$). The reaction of benzyl methyl ketone was found to proceed with a topicity opposite to that found in all the other cases (*Si* instead of *Re* hydride transfer). Consideration of Figure 3 shows that, for this latter substrate, π-π interaction between its aromatic ring and the *quasi*-axial phenyl group ($Ph_{ax}$) of the TADDOL in the Al complex might not be possible.

We notice that hydride transfer from the *Re* face is observed not only with (*R,R*)-TADDOLate modified LAH but also with the corresponding LAH derivatives of (*P*)-BINOL (*48,49*), (*P*)-biphenol derivatives (*56,59*) and biphenanthrol (*58*). Thus, the relative topicity of the process is *like* (*lk*) with TADDOLs and *unlike* (*ul*) (*102,103*) with the axially chiral phenol derivatives. Do these ligands have any common structural features?

Inspection of superpositions of 29 X-ray crystal structures of the (*R,R*)-TADDOLate moieties (Figure 4a), of 8 known (*R,R*)-Ti-TADDOLates (Figure 4b), of 9 (*P*)-BINOLs/Ti-BINOLate derivatives (Figure 4c), of 14 (*P*)-biphenols/metal-biphenolates (Figure 4d), and of 7 (*P*)-biphenanthrols (Figure 4e) reveals that they can all be assigned λ arrangement (*44*) of the aromatic groups, demonstrating the great difference in available space in the upper left and lower right compared to the lower left and upper right hand side of the center of these structures. Thus, the preference for the $R_\pi$ group to occupy an equatorial position in a six-membered cyclic transition state appears to be the common feature of these reductions.

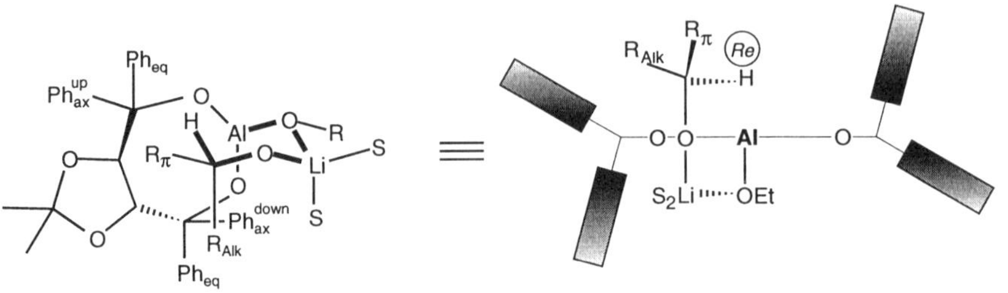

**Figure 3.** Schematic picture of the aluminum complex with the *sec.* alcohol after hydride transfer to the carbonyl C-atom. The unsaturated $R_\pi$-group is parallel to the *quasi*-axial Ph group of the ligand. The alkyl residue $R_{Alk}$ is pointing towards the *quasi*-equatorial Ph group. S represents neutral solvent molecules. The shaded bars indicate the Ph groups of the ligand.

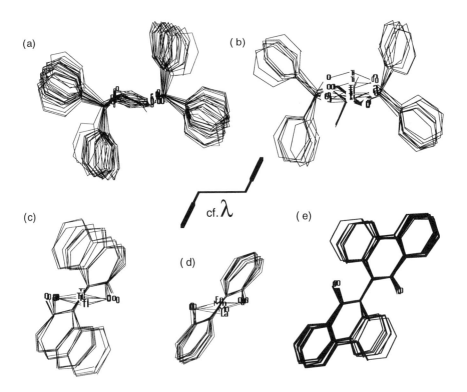

**Figure 4.** The superpositions of a) 29 structures of TADDOLs and TADDOL analogues, b) the 8 known Ti-TADDOLates all in the $(R,R)$-configuration, c) 9 BINOLates, d) 14 biphenol derivatives, and e) 7 biphenanthrols in the $(P)$-configuration are shown. Some of these structures had to be inverted for this purpose. The arrangement of the *quasi*-axial aryl groups is designated $\lambda$.

## Summary

In conclusion, we present a comparison of the application of LAH complexes modified by tartrate derived bidentate ligands (DBD, TADDOL) and the corresponding biaryl complexes (BINOL, biphenol and biphenanthrol) in the enantioselective reduction of ketones. All of the ligands give excellent selectivities with aryl alkyl ketones, but alkyl alkyl ketones turn out to be less suitable substrates. However, two biaryl complexes (**13** and **14**) lead to reasonable selectivities in most cases. The advantages of a readily accessible TADDOL ligand (no resolution step in the synthesis), of temperatures in the range of dry ice and of only a two-fold excess of the chiral reducing agent are obvious. We also propose a simple, common stereochemical model for BINOLate, biphenolate, biphenanthrolate and TADDOLate mediated LAH reductions of aryl alkyl or alkenyl alkyl ketones to the corresponding secondary alcohols.

## Acknowledgments

We are indebted to our great mentor, Prof. Dr. D. Seebach, for all his help and encouragement during the preparation of this manuscript. We are grateful to Dr. J. Matthews for the careful proof-reading of the final version of the manuscript. We thank the *Stiftung Stipendien-Fonds der Chemischen Industrie*, Germany, for a scholarship granted to F.N.M.K. The generous financial support of SANDOZ AG Basel and the Swiss National Science Foundation (SNF-2127-044199.94/1) is gratefully acknowledged.

## Literature Cited

1.  Nógrádi, M. *Stereoselective Synthesis: A Practical Approach;* Second ed.; VCH: Weinheim, 1995.
2.  Singh, V. K. *Synthesis* **1992**, 605 - 617.
3.  Brown, H. C.; Park, W. S.; Cho, B. T.; Ramachandran, P. V. *J. Org. Chem.* **1987**, *52*, 5406 - 5412; Erratum: *J. Org. Chem.* **1988**, *53*, 3396.
4.  Haubenstock, H. *Topics in Stereochemistry* **1983**, *14*, 231 - 300.
5.  Morrison, J. D.; Mosher, H. S. *Asymmetric Organic Reactions*; Prentice-Hall: Englewood Cliffs, 1970, pp 160 - 280.
6.  Ohkuma, T.; Ooka, H.; Hashiguchi, S.; Ikariya, T.; Noyori, R. *J. Am. Chem. Soc.* **1995**, *117*, 2675 - 2676.
7.  Noyori, R.; Kitamura, M. In *Modern Synthetic Methods*; R. Scheffold, Ed.; Springer: Berlin, Heidelberg, 1989; Vol. 5; pp 115 - 198.
8.  Noyori, R. *Asymmetric Catalysis in Organic Synthesis*; Wiley: New York, 1994.
9.  Wallbaum, S.; Martens, J. *Tetrahedron Asymm.* **1992**, *3*, 1475 - 1504.

10. Hirao, A.; Itsuno, S.; Nakahama, S.; Yamazaki, N. *J. Chem. Soc., Chem. Commun.* **1981**, 315 - 317.
11. Corey, E. J.; Bakshi, R. K.; Shibata, S. *J. Am. Chem. Soc.* **1987**, *109*, 5551 - 5552.
12. Corey, E. J.; Helal, C. J. *Tetrahedron Lett.* **1993**, *34*, 5227 - 5230.
13. Nagata, T.; Yorozu, K.; Yamada, T.; Mukaiyama, T. *Angew. Chem.* **1995**, *107*, 2309 - 2311; *Angew. Chem., Int. Ed. Engl.* **1995**, *34*, 2145 - 2147.
14. Noyori, R.; Kitamura, M. *Angew. Chem.* **1991**, *103*, 34 - 55; *Angew. Chem., Int. Ed. Engl.* **1991**, *30*, 49 - 69.
15. Yoshioka, M.; Kawakita, T.; Ohno, M. *Tetrahedron Lett.* **1989**, *30*, 1657 - 1660.
16. Takahashi, H.; Kawakita, T.; Ohno, M.; Yoshioka, M.; Kobayashi, S. *Tetrahedron* **1992**, *48*, 5691 - 5700.
17. Knochel, P. *Synlett* **1995**, 393 - 403.
18. Minireview covering the literature up to 1994: Dahinden, R.; Beck, A. K.; Seebach, D. In *Encyclopedia of Reagents for Organic Synthesis*; L. A. Paquette, Ed.; Wiley: Chichester, GB, 1995; pp 2167 - 2170.
19. Ito, Y. N.; Ariza, X.; Beck, A. K.; Bohác, A.; Ganter, C.; Gawley, R. E.; Kühnle, F. N. M.; Tuleja, J.; Wang, Y. M.; Seebach, D. *Helv. Chim. Acta* **1994**, *77*, 2071 - 2110.
20. Schäfer, H.; Seebach, D. *Tetrahedron* **1995**, *51*, 2305 - 2324.
21. Bothner-By, A. A. *J. Am. Chem. Soc.* **1951**, *73*, 846.
22. Portoghese, P. S. *J. Org. Chem.* **1962**, *27*, 3359 - 3360.
23. Landor, S. R.; Miller, B. J.; Tatchell, A. R. *J. Chem. Soc. C* **1966**, 1822 - 1825.
24. Červinka, O. *Chimia* **1959**, *13*, 332.
25. Červinka, O. *Coll. Czech. Chem. Commun.* **1961**, *26*, 673 - 680.
26. Červinka, O.; Belovsky, O. *Coll. Czech. Chem. Commun.* **1967**, *32*, 3897 - 3908.
27. Seebach, D.; Hungerbühler, E. In *Modern Synthetic Methods*; R. Scheffold, Ed.; Salle + Sauerländer: Frankfurt/Aarau, 1980; Vol. 2; pp 91 - 173.
28. Seebach, D.; Dörr, H.; Bastani, B.; Ehrig, V. *Angew. Chem.* **1969**, *81*, 1002 - 1003; *Angew. Chem., Int. Ed. Engl.* **1969**, *8*, 982 - 983.
29. Seebach, D.; Kalinowski, H.-O.; Bastani, B.; Crass, G.; Daum, H.; Dörr, H.; Du Preez, N. P.; Ehrig, V.; Langer, W.; Nüssler, C.; Oei, H.-A.; Schmidt, M. *Helv. Chim. Acta* **1977**, *60*, 301 - 325.
30. Seebach, D.; Oei, H.-A.; Daum, H. *Chem. Ber.* **1977**, *110*, 2316 - 2333.
31. Seebach, D.; Langer, W. *Helv. Chim. Acta* **1979**, *62*, 1701 - 1709.
32. Langer, W.; Seebach, D. *Helv. Chim. Acta* **1979**, *62*, 1710 - 1722.
33. Seebach, D.; Crass, G.; Wilka, E.-M.; Hilvert, D.; Brunner, E. *Helv. Chim. Acta* **1979**, *62*, 2695 - 2698.
34. Seebach, D.; Kalinowski, H.-O.; Langer, W.; Crass, G.; Wilka, E.-M. *Organic Syntheses* **1983**, *61*, 24 - 34; *Organic Syntheses, Collective Volume VII* **1990**, 41 - 50.
35. Seebach, D.; Daum, H. *Chem. Ber.* **1974**, *107*, 1748 - 1763.

36. Seebach, D.; Meyer, H. *Angew. Chem.* **1974**, *86*, 40 - 41; *Angew. Chem., Int. Ed. Engl.* **1974**, *13*, 77 - 78.
37. Meyer, H.; Seebach, D. *Liebigs Ann. Chem.* **1975**, 2261 - 2278.
38. Seebach, D.; Beck, A. K.; Schiess, M.; Widler, L.; Wonnacott, A. *Pure Appl. Chem.* **1983**, *55*, 1807 - 1822.
39. Seebach, D.; Weidmann, B.; Widler, L. In *Modern Synthetic Methods*; R. Scheffold, Ed.; Sauerländer: Aarau, 1983; Vol. 3; pp 217 - 353.
40. Seebach, D.; Beck, A. K.; Imwinkelried, R.; Roggo, S.; Wonnacott, A. *Helv. Chim. Acta* **1987**, *70*, 954 - 974.
41. Beck, A. K.; Bastani, B.; Plattner, D. A.; Petter, W.; Seebach, D.; Braunschweiger, H.; Gysi, P.; La Veccia, L. *Chimia* **1991**, *45*, 238 - 244.
42. Beck, A. K.; Gysi, P.; La Vecchia, L.; Seebach, D. *Org. Syntheses* in preparation.
43. Weber, B.; Seebach, D. *Tetrahedron* **1994**, *50*, 6117 - 6128; [Tetrahedron Symposia-in-Print No. 55 on "Mechanistic Aspects of Polar Organometallic Chemistry"].
44. Seebach, D.; Dahinden, R.; Marti, R. E.; Beck, A. K.; Plattner, D. A.; Kühnle, F. N. M. *J. Org. Chem.* **1995**, *60*, 1788 - 1799.
45. Seebach, D.; Devaquet, E.; Ernst, A.; Hayakawa, M.; Kühnle, F. N. M.; Schweizer, W. B.; Weber, B. *Helv. Chim. Acta* **1995**, *78*, 1636 - 1650.
46. Seebach, D.; Jaeschke, G.; Wang, Y. M. *Angew. Chem.* **1995**, *107*, 2605 - 2606; *Angew. Chem., Int. Ed. Engl.* **1995**, *34*, 2395 - 2396.
47. Seebach, D.; Beck, A. K.; Dahinden, R.; Hoffmann, M.; Kühnle, F. N. M. *Croat. Chem. Acta* **1996**, *69*, in press.
48. Noyori, R.; Tomino, I.; Yamada, M.; Nishizawa, M. *J. Am. Chem. Soc.* **1984**, *106*, 6717 - 6725.
49. Noyori, R.; Tomino, I.; Tanimoto, Y.; Nishizawa, M. *J. Am. Chem. Soc.* **1984**, *106*, 6709 - 6716.
50. Noyori, R. *Pure Appl. Chem.* **1981**, *53*, 2315 - 2322.
51. Nishizawa, M.; Yamada, M.; Noyori, R. *Tetrahedron Lett.* **1981**, *22*, 247 - 250.
52. Nishizawa, M.; Noyori, R. *Tetrahedron Lett.* **1980**, *21*, 2821 - 2824.
53. Noyori, R.; Tomino, I.; Nishizawa, M. *J. Am. Chem. Soc.* **1979**, *101*, 5843 - 5844.
54. Noyori, R.; Tomino, I.; Tanimoto, Y. *J. Am. Chem. Soc.* **1979**, *101*, 3129 - 3131.
55. Landor, S. R.; Miller, B. J.; Tatchell, A. R. *J. Chem. Soc. C* **1967**, 197 - 201.
56. Suda, H.; Kanoh, S.; Umeda, N.; Ikka, M.; Motoi, M. *Chem. Lett.* **1984**, 899 - 902.
57. Bringmann, G.; Gassen, M.; Lardy, R. *Tetrahedron* **1994**, *50*, 10245 - 10252.
58. Yamamoto, K.; Fukushima, H.; Nakazaki, M. *J. Chem. Soc., Chem. Commun.* **1984**, 1490 - 1491.
59. Rawson, D.; Meyers, A. I. *J. Chem. Soc., Chem. Commun.* **1992**, 494 - 496.
60. Yamamoto, K.; Ueno, K.; Naemura, K. *J. Chem. Soc., Perkin Trans. 1* **1991**, 2607 - 2608.

61. Chan, P. C.-M.; Chong, J. M. *J. Org. Chem.* **1988**, *53*, 5584 - 5586.
62. Chong, J. M.; Park, S. B. *J. Org. Chem.* **1992**, *57*, 2220 - 2222.
63. Marshall, J. A.; Luke, G. P. *J. Org. Chem.* **1991**, *56*, 483 - 485.
64. Marshall, J. A.; Welmaker, G. S.; Gung, B. W. *J. Am. Chem. Soc.* **1991**, *113*, 647 - 656.
65. Marshall, J. A.; Luke, G. P. *J. Org. Chem.* **1993**, *58*, 6229 - 6234.
66. Chong, J. M.; Mar, E. K. *J. Org. Chem.* **1991**, *56*, 893 - 896.
67. Hutchins, R. O.; Abdel-Magid, A.; Stercho, Y. P.; Wambsgans, A. *J. Org. Chem.* **1987**, *52*, 702 - 704.
68. Bringmann, G.; Hartung, T. *Synthesis* **1992**, 433 - 434.
69. Bringmann, G.; Hartung, T. *Tetrahedron* **1993**, *49*, 7891 - 7902.
70. Matsuki, K.; Inoue, H.; Ishida, A.; Takeda, M. *Heterocycles* **1993**, *36*, 937 - 940.
71. Matsuki, K.; Inoue, H.; Takeda, M. *Tetrahedron Lett.* **1993**, *34*, 1167 - 1170.
72. Matsuki, K.; Sobukawa, M.; Kawai, A.; Inoue, H.; Takeda, M. *Chem. Pharm. Bull.* **1993**, *41*, 643 - 648.
73. Hwu, J. R.; Robl, J. A.; Gilbert, B. A. *J. Am. Chem. Soc.* **1992**, *114*, 3125 - 3126.
74. Sahoo, S. P.; Graham, D. W.; Acton, J.; Biftu, T.; Bugianesi, R. L.; Girotra, N. N.; Kuo, C.-H.; Ponpipom, M. M.; Doebber, T. W.; Wu, M. S.; Hwang, S.-B.; Lam, M.-H.; MacIntyre, D. E.; Bach, T. J.; Luell, S.; Meurer, R.; Davies, P.; Alberts, A. W.; Chabala, J. C. *Bioorg. Med. Chem. Lett.* **1991**, *1*, 327 - 332.
75. Karibe, N.; Rosser, R. M.; Ueda, M.; Sugimoto, H. *J. Labelled Compd. Radiopharm.* **1990**, *28*, 823 - 830.
76. Kurek-Tyrlik, A.; Wicha, J.; Zarecki, A. *J. Org. Chem.* **1990**, *55*, 3484 - 3492.
77. Fried, J.; John, V.; Szwedo Jr., M. J.; Chen, C.-K.; O'Yang, C. *J. Am. Chem. Soc.* **1989**, *111*, 4510 - 4511.
78. Burke, S. D.; Takeuchi, K.; Murtiashaw, C. W.; Liang, D. W. M. *Tetrahedron Lett.* **1989**, *30*, 6299 - 6302.
79. Ireland, R. E.; Thaisrivongs, S.; Dussault, P. H. *J. Am. Chem. Soc.* **1988**, *110*, 5768 - 5779.
80. Nakamura, T.; Namiki, M.; Ono, K. *Chem. Pharm. Bull.* **1987**, *35*, 2635 - 2645.
81. Suemune, H.; Akashi, A.; Sakai, K. *Chem. Pharm. Bull.* **1985**, *33*, 1055 - 1061.
82. Tanaka, T.; Okamura, N.; Bannai, K.; Hazato, A.; Sugiura, S.; Manabe, K.; Kamimoto, F.; Kurozumi, S. *Chem. Pharm. Bull.* **1985**, *33*, 2359 - 2385.
83. Baeckström, P.; Björkling, F.; Högberg, H.-E.; Norin, T. *Acta Chem. Scand. Ser. B* **1983**, *37*, 1 - 5.
84. Ishiguro, M.; Koizumi, N.; Yasuda, M.; Ikekawa, N. *J. Chem. Soc., Chem. Commun.* **1981**, 115 - 117.
85. Yamaguchi, S.; Mosher, H. S.; Pohland, A. *J. Am. Chem. Soc.* **1972**, *94*, 9254 - 9255.
86. Yamaguchi, S.; Mosher, H. S. *J. Org. Chem.* **1973**, *38*, 1870 - 1877.
87. Vigneron, J. P.; Jacquet, I. *Tetrahedron* **1976**, *32*, 939 - 944.

88. Von dem Bussche-Hünnefeld, C.; Beck, A. K.; Lengweiler, U.; Seebach, D. *Helv. Chim. Acta* **1992**, *75*, 438 - 441.
89. Weber, E.; Dörpinghaus, N.; Wimmer, C.; Stein, Z.; Krupitsky, H.; Goldberg, I. *J. Org. Chem.* **1992**, *57*, 6825 - 6833.
90. Minireview: Kaupp, G. *Angew. Chem.* **1994**, *106*, 768 - 770; *Angew. Chem., Int. Ed. Engl.* **1994**, *33*, 728 - 729.
91. Toda, F. *J. Synth. Org. Chem. Jpn.* **1994**, *52*, 923 - 934.
92. Toda, F.; Tanaka, K.; Okada, T. *J. Chem. Soc., Chem. Commun.* **1995**, 639 - 640.
93. Toda, F. *Acc. Chem. Res.* **1995**, *28*, 480 - 486.
94. Pavlov, V. A.; Klabunovskii, E. I.; Struchkov, Y. T.; Voloboev, A. A.; Yanovsky, A. I. *J. Mol. Catal.* **1988**, *44*, 217 - 243.
95. Seebach, D. *Angew. Chem.* **1988**, *100*, 1685 - 1715; *Angew. Chem., Int. Ed. Engl.* **1988**, *27*, 1624 - 1654.
96. Weiss, E. *Angew. Chem.* **1993**, *105*, 1565 - 1587; *Angew. Chem., Int. Ed. Engl.* **1993**, *32*, 1501 - 1523.
97. Mole, T.; Jeffery, E. A. *Organoaluminium Compounds*; Elsevier Publishing Company: Amsterdam, 1972.
98. Puchot, C.; Samuel, O.; Duñach, E.; Zhao, S.; Agami, C.; Kagan, H. B. *J. Am. Chem. Soc.* **1986**, *108*, 2353 - 2357.
99. Guillaneux, D.; Zhao, S.-H.; Samuel, O.; Rainford, D.; Kagan, H. B. *J. Am. Chem. Soc.* **1994**, *116*, 9430 - 9439.
100. Schmidt, B.; Seebach, D. *Angew. Chem.* **1991**, *103*, 1383 - 1385; *Angew. Chem., Int. Ed. Engl.* **1991**, *30*, 1321 - 1323.
101. Seebach, D.; Plattner, D. A.; Beck, A. K.; Wang, Y. M.; Hunziker, D. *Helv. Chim. Acta* **1992**, *75*, 2171 - 2209.
102. Prelog, V.; Helmchen, G. *Angew. Chem.* **1982**, *94*, 614 - 631; *Angew. Chem., Int. Ed. Engl.* **1982**, *21*, 567 - 583.
103. Seebach, D.; Prelog, V. *Angew. Chem.* **1982**, *94*, 696-702; *Angew. Chem., Int. Ed. Engl.* **1982**, *21*, 654 - 660.

# Chapter 4

# Electrophilic Assistance in the Reduction of Six-Membered Cyclic Ketones by Alumino- and Borohydrides

### Jacqueline Seyden-Penne[1]

### Laboratoire des Carbocycles, Université Paris Sud, 91405 Orsay, France

The nature of the cation associated with the tetracoordinated alumino- and borohydrides as well as the solvent are of prime importance in performing selective reductions of ketones. Electrophilic assistance takes place, either when the associated cation acts as a Lewis acid (complexation control) or when the reaction is run in a protic solvent. It induces 1,2 reduction of α-enones which is favored over 1,4 reduction; saturated ketones can be reduced, leaving α-enones unchanged; the stereoselectivity of the reduction of cyclohexanones increases towards axial attack.

Electrophilic catalysis in the reduction of ketones by alumino- or borohydrides either by the associated cation or by a protic solvent is well documented (*1-3*). For instance, reduction of ketones by $LiAlH_4$ in THF is faster than reduction by $NaAlH_4$ or $n$-$Bu_4NAlH_4$ (*4, 5*). Pierre and Handel (*6*) have shown that, in the presence of [2.1.1] cryptand, the reduction of some ketones by $LiAlH_4$ is inhibited. Later on, it was evidenced in our group (*7, 8*) that the reduction of arylmethylketones, 2-cyclohexenone and cyclohexanone by $LiAlH_4$ and $LiBH_4$ in THF or $Et_2O$ is strongly slowed down in the presence of [2.1.1] cryptand. Wiegers and Smith (*5*) also noticed a decrease in the rate of the reduction of camphor by $LiAlH_4$ in THF when a crown-ether was added to the reaction medium. These results are explained by electrophilic assistance resulting from coordination between the cation associated with the hydride reagent and the lone pairs of electrons on the carbonyl oxygen. The stronger the coordination, the faster is the reduction For example, sodium cation would have a weaker coordination and consequently a slower reduction rate than a stronger Lewis acid such as lithium cation.

[1]Current address: Le Vallet de Vermenoux, 84220 Goult, France

In the presence of [2.1.1] cryptand, the coordination of the lithium cation does not take place.

The decrease in the reduction rate due to absence of electrophilic assistance may also be accompanied by changes in selectivity as seen in the following :
- 1,2 vs 1,4 regioselectivity in the reductions of α-enones.
- selective reductions of saturated vs α,β-unsaturated ketones.
- stereoselectivity of the reductions of cyclic ketones.

The influence of electrophilic assistance on these three domains of selective reductions by tetracoordinated alumino- and borohydrides will be discussed in this chapter.

## 1. Regioselectivity in the Reduction of α-Enones

Reductions by $LiAlH_4$ and $LiBH_4$ will be discussed first (8). The reduction of ketones by these reagents in THF is first order in each of them provided that their concentration is low (4, 5). It seems likely that such is also the case with α-enones. In order to eliminate the eventual problem of s-cis / s-trans conformational equilibrium in α-enones, 2-cyclohexenones **1a-e** are selected as models. The 1,2 reduction of the enones **1a-e** leads to the formation of the allylic alcohols **2a-e** while the 1,4-reduction gives the saturated ketones **3a-e**. These ketones may be further reduced to the saturated alcohols **4a-e**. The allylic alcohols **2** are not reduced to the saturated alcohols **4** under the reaction conditions. The results of the reduction of the cyclohexenones **1a-e** by $LiAlH_4$ in THF at 25°C are listed in Table I. The concentration of each reagent was 0.08 M and the total yields were over 90% based on recovered starting materials.

**a** : R = H, **b** : R = 3-Me, **c** : R = 5,5-Me$_2$, **d** : R = 3,5,5-Me$_3$, **e** : R = 2-Me

The general trends which appear from Table I are the following :
- The presence of [2.1.1]cryptand in the reaction medium causes both a decrease in the reduction rate and a change in regioselectivity compared to the reaction without any addend (exp.1, 4, 5, 6, 8, 9, 10-18).
- The presence of 12-crown-4 induces similar trends but to a far lesser extent (exp.1-3, 6, 7).

### Table I : LiAlH$_4$ reduction of 2-cyclohexenones

| Exp. | α-Enone | Addend (1.2 eq.) | Reaction time | Yield % | C$_2$ vs C$_4$ attack |
|---|---|---|---|---|---|
| 1 | 1a | none | 1mn | >98 | 86-14 |
| 2 | 1a | 12-crown-4 | 1mn | >98 | 75-25 |
| 3 | 1a | 12-crown-4* | 1mn | >98 | 62-38 |
| 4 | 1a | [2.1.1] | 1mn | 85 | 14-86 |
| 5 | 1a | [2.1.1] | 15mn | >98 | 14-86 |
| 6 | 1b | none | 1mn | >98 | 95- 5 |
| 7 | 1b | 12-crown-4* | 1mn | >98 | 88-12 |
| 8 | 1b | [2.1.1] | 15mn | 50 | 24-76 |
| 9 | 1b | [2.1.1] | 2hrs | 95 | 26-74 |
| 10 | 1c | none | 1mn | 90 | 58-42 |
| 11 | 1c | [2.1.1] | 2hrs | 35 | 22-78 |
| 12 | 1c | [2.1.1] | 8hrs | 90 | 20-80 |
| 13 | 1d | none | 1hr | >98 | 84-16 |
| 14 | 1d | [2.1.1] | 2hrs | 30 | 26-74 |
| 15 | 1d | [2.1.1] | 8hrs | 95 | 26-74 |
| 16 | 1e | none | 1mn | >98 | 95- 5 |
| 17 | 1e | [2.1.1] | 5mn | 80 | 42-58 |
| 18 | 1e | [2.1.1] | 15mn | >98 | 42-58 |

*2.5 eq. 12-crown-4 instead of 1.2 eq.

Similar rate decreases and changes in regioselectivity were observed in the reduction of 1a, 1b, and 1d with LiBH$_4$ in THF, with and without added [2.1.1] cryptand. The regioselectivity of the reduction of 1a and 1d by $n$-Bu$_4$NBH$_4$ in THF (9) or of 1d by NaBH$_4$ in aprotic phase transfer conditions (9) or in the presence of polyethyleneglycol (10) are similar to those observed with LiBH$_4$ in the presence of [2.1.1] cryptand (85-95% 1,4 reduction). However, alkali borohydrides in alcohols (1, 2, 9) or in water (11) gave substantial amounts of allylic alcohols resulting from 1,2 reduction. In addition, the reduction of enones with LiAlH$_4$ or LiBH$_4$ consistently gives more 1,2-reductions in diethyl ether than in THF (7, 8).

These results are explained based on the assumption that these reductions take place mainly under Frontier control (3, 7-9, 12, 13) and that the α-enone LUMO-cation is the predominant interaction. The lower the LUMO level of the α-enone, according to the strength of the interaction with the cation, the faster is the reduction. The regioselectivity depends upon the relative magnitude of the coefficients $c$ of the carbonyl carbon C-2 and of the C-4 double bond carbon in the LUMO of the free or of the complexed α-enone (*12*). Indeed, in free enones, $c_4 > c_2$ while in Li cation-complexed ones, $c_2 > c_4$. Moreover, in the later case, the positive charge of the carbonyl carbon is higher than when no complexation occurs, so that charges interaction also favors C-2 attack.

$$\underset{4\qquad\quad 2}{\overset{\displaystyle\phantom{|}}{\text{C}}=\overset{|}{\text{C}}-\overset{|}{\text{C}}=\text{O}}$$

• In the presence of [2.1.1] cryptand, LiAlH$_4$ exists as cryptand separated ion-pairs so that the cation does not coordinate to the carbonyl of the α-enone. As in the α-enone LUMO, the C-4 coefficient is larger than the C-2 one (*3, 12, 13*), 1,4 reduction is favored. On the other hand, steric repulsions can explain the differences observed with **1 a-e** according to the ring substituents.

• In THF, in the absence of any additive, LiAlH$_4$ exists mostly as solvent-separated ion-pairs (*4*). The lone pairs of the carbonyl group compete with the solvent in coordinating the Li cation, this coordination induces a lowering of the LUMO energy of the α-enone so that the reduction is faster. As the C-2 coefficient is larger in this later case (*3, 12*) as well as C-2 positive charge , 1,2 reduction predominates. The reaction is said to be under complexation control.

• In the presence of 12-crown-4 which is less efficient as Li$^+$ complexing agent than [2.1.1] cryptand, the complexation of the cation to the carbonyl oxygen lone pairs can still take place. It is however not as strong as in pure THF so that 1,2 reduction is less favored.

The trends observed in diethylether (*8*) follow the same lines although the structure of LiAlH$_4$ in this solvent is ill-defined. The cation-carbonyl interaction is stronger than in THF, resulting in more 1,2 reduction (up to 98% with **1a**).

The reduction of the same α-enone with LiBH$_4$ , NaBH$_4$ or *n*-Bu$_4$NBH$_4$ in THF or in protic solvents can be explained in the same fashion. Electrophilic assistance does not take place in the presence of cryptand (*8*), with *n*-Bu$_4$NBH$_4$ in THF (*9*) or when the sodium cation of NaBH$_4$ is coordinated to PEG (*10*), causing predominant 1,4 reduction. Electrophilic assistance occurs with LiBH$_4$ in THF (*8*) as well as protic solvents (*9, 11*) mainly due to hydrogen bonding (*1-3*). Therefore 1,2 reduction predominates under these conditions, due to complexation control.

The results obtained for the reductions by NaBH$_4$ under phase transfer conditions (*9*) are also in agreement with this explanation. In liquid-liquid conditions, under which some water is transferred into the organic phase, 1,2 and 1,4 reductions occur concomitantly. In solid-liquid conditions, using [2.1.1] cryptand as phase transfer catalyst, 1,4 reduction is highly favored (95%).

A similar explanation may apply to interpret the 1,2 reductions of α-enones by the Luche reagent (NaBH$_4$, CeCl$_3$ in MeOH) (*14*), diisopropoxytitanium tetrahydroborate in CH$_2$Cl$_2$ (*15*), lithium pyrrolidinoborohydride in THF (*16*), and DIBAH-*n*-BuLi ate-complex in hexane (*17*). Ranu (*18*) has shown that zinc borohydride in DME leaves α-enones unchanged. When this reagent was supported on $SiO_2$, which can act as a Lewis acid promoting electrophilic assistance, 1,2 reduction did occur (*18*).

Similar trends were observed with bicyclic or steroidal α-enones although the regioselectivities were often dependent upon the structure and the substitution in these substrates (*1, 14, 15, 16, 19-21*). For instance the selective reduction of steroidal 4-en-3-ones **5** either by Luche reagent (*14*) or by diisopropoxytitanium tetrahydroborate (*15*) gave the corresponding equatorial allyl alcohol **6** with a high regioselectivity under complexation control.

R = $C_8H_{17}$, OH, COOMe

Reduction of steroidal α-enone **7** (*20*) by $NaCNBH_3$ in THF in the presence of HCl or by Zn borohydride in $Et_2O$ is highly regioselective towards allyl alcohol **8**, also due to complexation control.

The regioselectivity of the reduction of cyclic α-enones by more bulky reagents such as LTBA (*23*) or trialkylborohydrides ($LiEt_3BH$, Li *s*-$Bu_3BH$, K *s*-$Bu_3BH$) (*1*) is less clear as steric hindrance intervenes so that mixtures of allylic alcohols and saturated ketones were often obtained. This is indeed the case in reductions of 2-cyclohexenone **1a** by LTBA or $LiEt_3BH$ in THF or of that of 3-methylcyclohexen-2-one **1b** by Li or K *s*-$Bu_3BH$. However, the reaction of **1a** with Li *s*-$Bu_3BH$ in THF, followed by trapping of the 1,4 adduct by PhN(SO$_2$CF$_3$)$_2$ lead to the enol triflate **9** in a high yield.

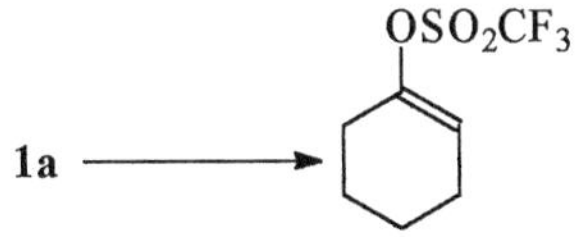

**1a** ⟶ [structure of **9**: cyclohexene with OSO$_2$CF$_3$ substituent]

In contrast, reduction of steroidal α-enone **10** by Li s-Bu$_3$BH (*21*) gave the axial allyl alcohol **11** in a high selectivity, due to the fact that the sterically hindered carbon-carbon double bond is not reduced.

A similar regioselectivity was observed by You and Koreeda (*24*) with other Δ-5 steroidal enones.

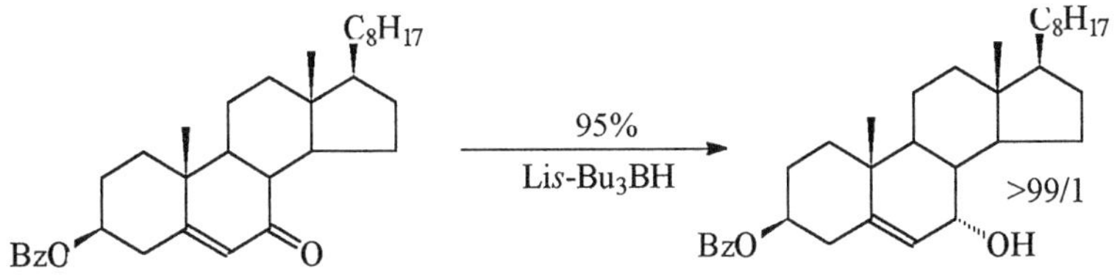

**10** —(95% Li s-Bu$_3$BH)→ **11** (>99/1)

## 2. Competition between Saturated and α,β-Unsaturated Ketones.

According to Frontier orbital theory, the LUMO level of α-enones is lying lower in energy than that of saturated ketones (*3,12*). If frontier control takes place in reductions by tetracoordinated alumino- and borohydrides, α-enones should react faster than saturated ketones having similar steric environment, provided that no electrophilic assistance takes place. This is indeed the case as it has been shown in our group (*8*) that the reduction of 2-cyclohexenone **1a** by LiAlH$_4$ in THF in the presence of [2.1.1] cryptand is faster than that of cyclohexanone **3a** (0.02M solutions, ketone consumption after 5mn : **1a** 70%, **3a** 19%). When electrophilic assistance by a cation or a protic solvent takes place, the reverse can be observed, as the more basic saturated carbonyl group is preferently coordinated, so that its selective reduction might be performed (*1, 3*). Indeed, Ranu and coworkers (*18*) have shown that while α-enones are left intact in the presence of zinc borohydride in DME, at -15°C, saturated ketones are reduced under these conditions. Such is also the case in competitive reductions of α-enones and saturated ketones by LTBA in THF (*23*) or by NaBH₄ in MeOH-CH₂Cl₂ (*1, 25*), as well as by Zn(BH₄)₂.1.5 DMF in suspension in acetonitrile (*1, 26*).

It seemed interesting to examine the possibility of selective reductions of diketones in which an α-enone residue as well as a saturated ketone are present in the same molecule. The selected examples are Wieland-Mischler ketone **12**, androst-4-en-3,17-dione **13**, and progesterone **14**.

**12**  **13**  **14**

The reduction of these three diketones by LiAlH₄ in ethereal medium was non-selective and lead to a mixture of diols resulting from the reduction of both carbonyl groups (*27, 28*). The same result was obtained in the reduction of **13** and **14** by LiBH₄ in refluxing diethylether (*29*). Highly selective reduction of the saturated ketone to alcohols **15-17** (> 90%) is performed in high yield under the following conditions :
 • NaBH₄ in MeOH-CH₂Cl₂ at -78°C (*25*) (from **12** and **13**), in EtOH at -10°C (*30*) or in water (*31*) (from **12**). A lower selectivity was observed in the reduction of **13** (*29*) and **14** (*32*) at higher temperature. Lower yields were obtained when using solid NaBH₄ (*1*).
 •*n*-Bu₄NBH₄ in MeOH at room temperature (*29*) reduced **13** and **14** to secondary alcohols **16** and **17**, with less than 10% diol formation.
 •Zn(BH₄)₂ in DME at -78°C and the complex Zn(BH₄)₂. 1.5 DMF in MeCN (*26*) reduced **12** to **15**. Nevertheless, the later reagent (*26*) as well as Zn(BH₄)₂ in refluxing diethylether (*29*) gave a mixture of **16**, **17**, **18** and **19** from **13** and **14**.
 •LTBA in THF (*34*) reduced androst-4-en-3,17-dione **13** into testosterone **16** while the ate-complex formed from DIBAH and *n*-BuLi transforms Wieland-Mischler ketone **12** into **15** (*35*).

The reaction of progesterone **14** with Luche reagent (*36*) gave preferentially the equatorial allylic alcohol **20** (48%) in addition to the 3β,20β-diol **19** (20%). This result has been explained by the selective formation of a 20-ketal which protects the saturated ketone, to favor the 1,2 reduction of the α-enone.

The selective reduction of the carbon-carbon double bond of the α-enone occurs under conditions in which electrophilic assistance does not take place. An example is

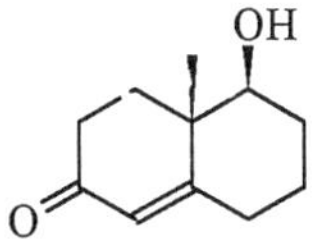

# 4. SEYDEN-PENNE  *Reduction of Six-Membered Cyclic Ketones*  77

**15**

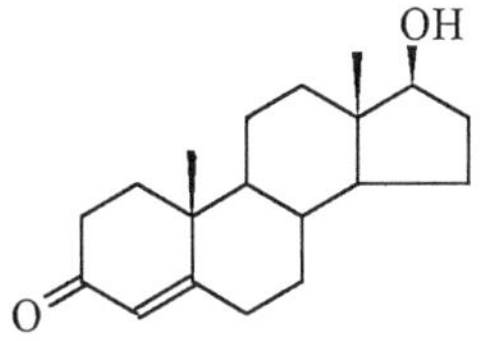

**16**

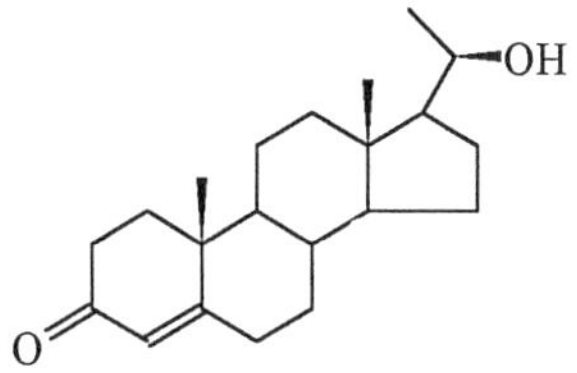

**17**

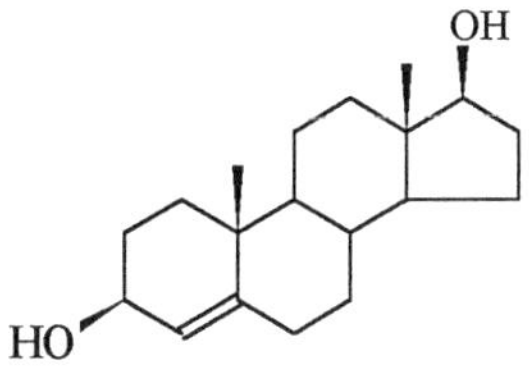

**18**

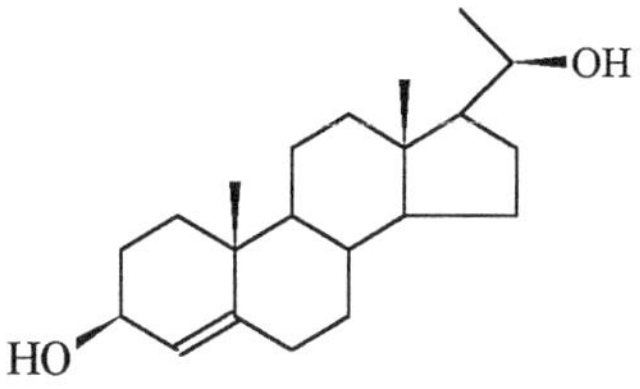

**19**

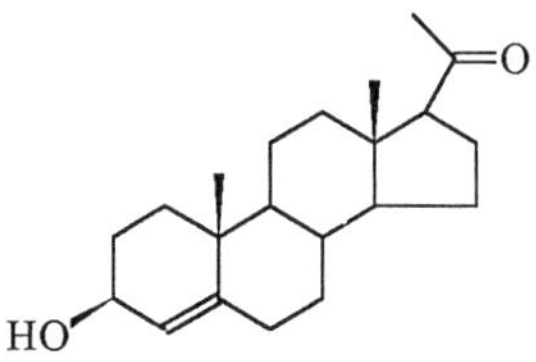

**20**

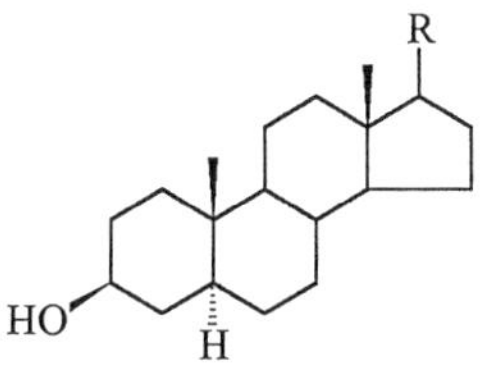

**21**  (R = COCH$_3$)

**25**  (R = =O )

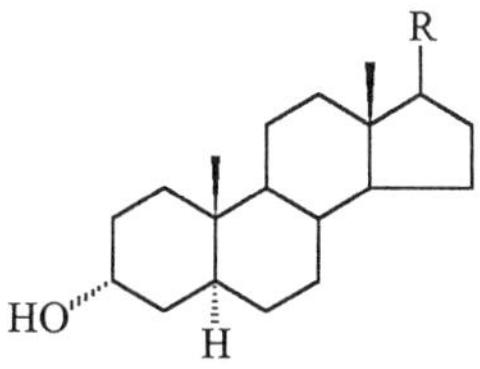

**22**  (R = COCH$_3$)

**26**  (R = =O )

the reduction of progesterone **14** by NaBH₄ in pyridine (*32*) into 5α-20-keto-3-ol **21** in 60% yield, and slight formation of some equatorial allyl alcohol **20**.

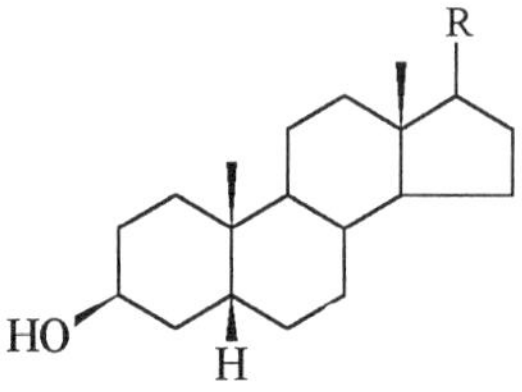

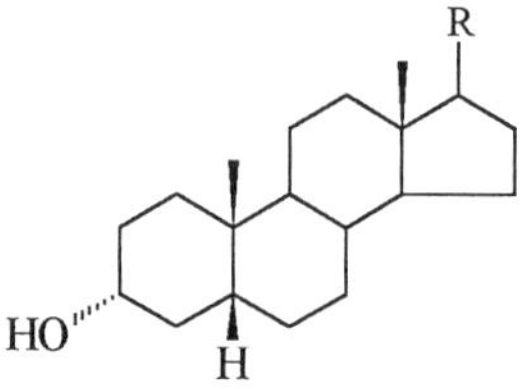

**23** (R = COCH$_3$)

**27** (R = =O )

**24** (R = COCH$_3$)

**28** (R = =O )

The α,β-unsaturated carbonyl functionality of the diketones **13** and **14** is selectively reduced with *n*-Bu₄NBH₄ in THF to the saturated alcohols without affecting the saturated ketone (*29*). Unfortunately, mixtures of stereoisomers at the 3- and 5-positions (**21-28**) were obtained, so that such a methodology does not meet all the selectivity requirements. The stereoselective reduction of the carbon-carbon double bond of Wieland-Mischler ketone **12** to **29** and **30** in a 94/6 ratio and of the steroidal diones **13** and **14** respectively to **31** and **32** has been achieved by complex copper hydrides (*37*) or by sodium dithionite under PTC conditions (*38*).

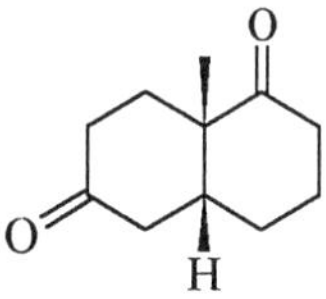

**29**

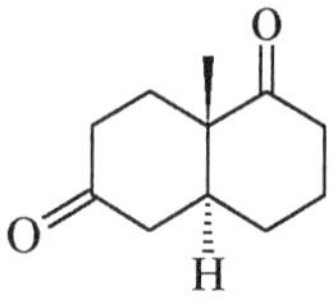

**30**

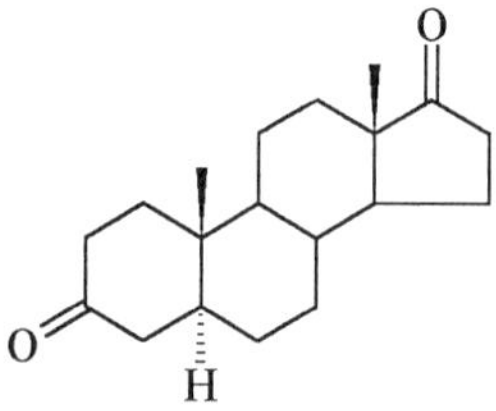

**31**

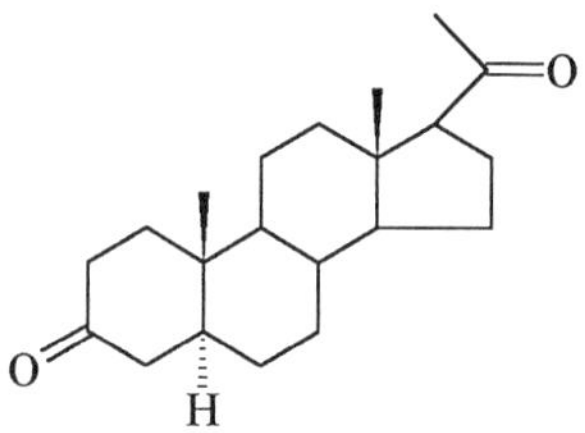

**32**

## 3. Stereoselectivity of Reductions

The addition of nucleophiles, such as hydrides, to carbonyl compounds takes place according to Dunitz-Bürgi approach (*1-3*), i.e. along an obtuse angle close to 109° as indicated in **33**. According to Anh and Eisenstein's calculations (*3*), complexation control induces a decrease of this angle, so that the incoming nucleophile approaches the carbonyl carbon closer to perpendicularity, and one might expect a change in the stereoselectivity of these reductions.

The main factors controlling the reduction of cyclohexanones and cyclohexenones are torsional repulsions, antiperiplanarity with the axial C—H or C—C bonds located on the carbons vicinal to the CO group and eventually, steric interactions (*1-3, 39*). The more flexible and the more flattened (*3, 39, 40*) the ketone at the transition state, the more favorable the geometry to attain antiperiplanarity with vicinal C—H or C—C axial bonds and the higher the amount of axial attack (Felkin-Anh rule) as indicated in **34** and **35**.

Ashby and Boone (*4*) as well as Wiegers and Smith (*5*) reported that electrophilic assistance was accompanied by a change in the stereoselectivity of the reduction of cyclic ketones by alumino- and borohydrides associated to various cations. However, the most significant result was observed in the reduction of 4-*t*-butylcyclohexanone **36** by LiAlH$_4$ in THF with or without added [2.1.1] cryptand (*1, 41*). When no electrophilic assistance was involved, 70% axial attack, leading to **37**, took place while under complexation control the **37/38** ratio was 90/10.

Adsorption of 4-*t*-butylcyclohexanone **36** on Montmorillonite clay, which is a strong Lewis acid, increased the percent of axial attack in its reduction by NaBH$_4$ from 81% in methanol to more than 99% (*1, 42*). Such a high selectivity towards axial attack was also observed when **36** was reduced by Li(*n*-Pr$_2$)NBH$_3$ in THF as 99% equatorial alcohol **37** was formed (*16*) while LiBH$_4$ in THF gave **37** and **38** in a 93/7 ratio. Such differences are indeed significant in terms of activation energies (*43*).

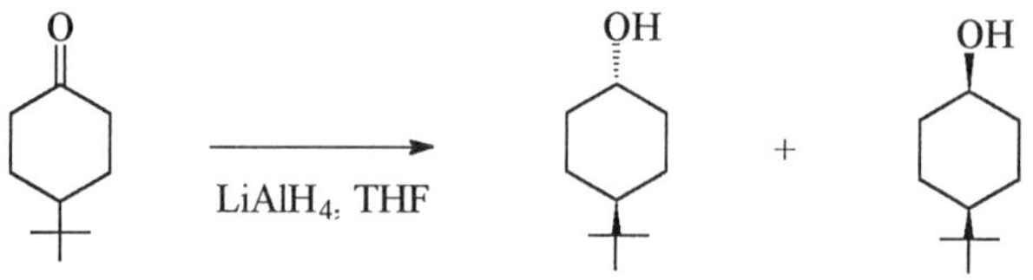

|  | 37 | 38 |
|---|---|---|
| with [2.1.1] | 70 | 30 |
| without [2.1.1] | 90 | 10 |

Similar enhancement of the axial attack was obtained in the reduction of pulegone **39** and of testosterone **16** by NaBH$_4$ in MeOH to give the allylic alcohols **40** and **18** respectively. The ratio of the axial attack increased in the presence of CeCl$_3$ (Luche reagent) (*1, 36*) as a result of a stronger coordination of the Ce(III) cation and the carbonyl group.

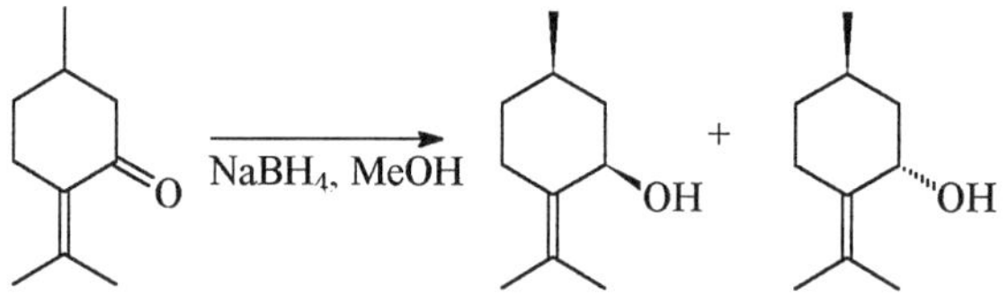

|  | 40 | 41 |
|---|---|---|
|  | 90 | 10 |
| + CeCl$_3$ | 97 | 3 |

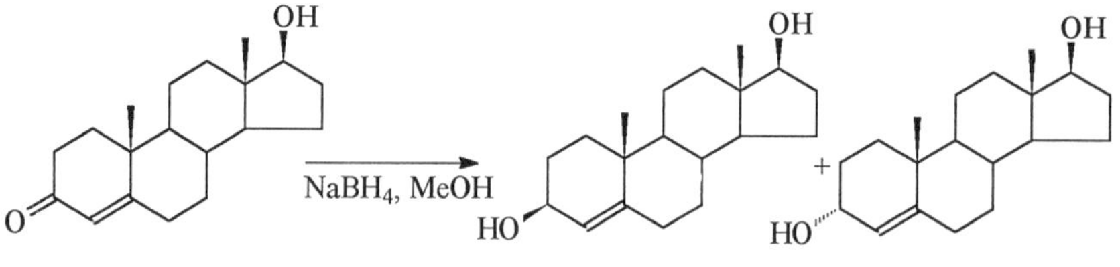

|  | 18 | 42 |
|---|---|---|
|  | 90 | 10 |
| + CeCl$_3$ | 99 | 1 |

A related observation has been made in the reduction of α-enone **10** by NaBH$_4$ in THF-MeOH or by NaCNBH$_3$ in THF-HCl (*1, 20*). However, when the reducing reagent becomes too bulky (Li *s*-Bu$_3$BH, K *s*-Bu$_3$BH or Li(*t*-BuEt$_2$O)$_3$AlH) (*1, 44*) steric hindrance is the predominant interaction so that equatorial attack is favored.

Indeed, the reduction of 4-*t*-butylcyclohexanone by Li(*t*-BuEt$_2$O)$_3$AlH in THF lead to **37** and **38** in a 5/95 ratio (*44*). As shown earlier, the reduction of the steroidal α-enone **10** by Li *s*-Bu$_3$BH (*1, 21*) lead to the axial alcohol **11** (*vide supra*) while Luche reagent which acts as a small hydride donor gave the equatorial diastereoisomer **43**

## Conclusion

Provided that not too bulky hydride donors are used, electrophilic assistance promotes selective reductions of ketones :
- α-enones are reduced to allylic alcohols. Up to date, the most selective reagents are Luche reagent (NaBH$_4$, CeCl$_3$ in MeOH) (*14*), diisopropoxytitanium borohydride in CH$_2$Cl$_2$ (*15*), lithium pyrrolidinoborohydride in THF (*16*) although more examples need to be described with the last two reagents. LiAlH$_4$ in diethylether is also recommended for such a purpose.
- Saturated ketones are reduced to secondary alcohols selectively in the presence of α-enones.

Although the examples discussed in the this chapter deal with six-membered cyclic ketones, these trends are general and can be applied to other cyclic systems or to acyclic ones (*1*). 2-Cyclopentenone, which is prone to 1,4 attack, gives 85% allylic alcohol when reduced by LiAlH$_4$ in diethylether (*23*).

Electrophilic assistance also influences the stereoselectivity of the reduction of cyclohexanones and 2-cyclohexenones, allowing higher selectivity towards axial attack. Again, Luche reagent seems most suitable for the highly stereoselective reduction of 2-cyclohexenones.

## Literature Cited.

1 Seyden-Penne, J. *Reductions by the Alumino- and Borohydrides in Organic Synthesis* ; V. C. H. : New York, N.Y., 1991.

2   Greeves, N. In *Comprehensive Organic Synthesis* ; Trost, B., Fleming, I., Eds. ; Pergamon Press : London, England, 1991, Vol. 8, chap. 1 ; Shambayati, S.; Schreiber, S. Ibid. Vol. 1, chap. 1.10.
3   Anh, N. T. *Topics in Current Chemistry* **1980**, *88*, 146 ; *Orbitales Frontières*, InterEditions-CNRS Editions : Paris, France, 1995.
4   Ashby, E. C. ; Boone, J. R. *J. Am. Chem. Soc.* **1976**, *98*, 5524 ; *J. Org. Chem.* **1976**, *41*, 2890.
5   Wiegers, K. E. ; Smith, S. G. *J. Org. Chem.* **1978**, *43*, 1126.
6   Pierre, J. L. ; Handel, H. *Tetrahedron Lett.* **1974**, 2317 ; Pierre, J. L. ; Handel, H. ; Perraud, R. *Tetrahedron* **1975**, *31*, 2795 ; Handel, H. ; Pierre, J. L. *Tetrahedron* **1975**, *31*, 2799.
7   Loupy, A. ; Seyden-Penne, J. ; Tchoubar, B. *Tetrahedron Lett.* **1976**, 1677.
8   Loupy, A. ; Seyden-Penne, J. *Tetrahedron Lett.* **1978**, 2571 ; *Tetrahedron* **1980**, *36*, 1937.
9   D'Incan, E. ; Loupy, A. *Tetrahedron* **1981**, *37*, 1171.
10  Laxman, M. ; Sharma, M. M. *Synth. Comm.* **1990**, *20*, 111.
11  Geribaldi, S. ; Decouzon, M. ; Boyer, B. ; Moreau, C. *J. Chem. Soc. Perkin Trans. II* **1986**, 1327. These authors used an excess of $NaBH_4$ in dioxane water to avoid the participation of alkoxyborohydrides.
12  Lefour, J. M. ; Loupy, A. *Tetrahedron* **1978**, *34*, 2597.
13  Fleming, I. *Frontier Orbitals and Organic Chemical Reactions*, Wiley Ed. : London, England, 1976.
14  Gemal, A. L. ; Luche, J. L. *J. Am. Chem. Soc* **1981**, *103*, 5454 ; Luche, J. L. *J. Am. Chem. Soc* **1978**, *100*, 2226.
15  Ravikumar, K. S. ; Baskaran, S. ; Chandrasekaran, S. *J. Org. Chem.* **1993**, *58*, 5981.
16  Fuller, J. C. ; Stangeland, E. L. ; Goralski, C. T. ; Singaram, B. *Tetrahedron Lett.* **1993**, *34*, 257 ; Fisher, G. B. ; Fuller, J. C. ; Harrison, J. ; Alvarez, S. G. ; Burkhardt, E. ; Goralski, C. T. ; Singaram, B. *J. Org. Chem.* **1994**, *59*, 6378.
17  Kim, S. ; Ahn, K. H. *J. Org. Chem.* **1984**, *49*, 1717.
18  Ranu, B. C. *Synlett* **1993**, 885 and quoted references.
19  D'Incan, E. ; Loupy, A. ; Maia, A. *Tetrahedron Lett.* **1981**, *22*, 941.
20  Viger, A. ; Marquet, A. ; Barton, D. H. R. ; Motherwell, W. B. ; Zard, S. Z. *J. Chem. Soc. Perkin Trans.I* **1982**, 1937.
21  Kumar, V. ; Amann, A. ; Ourisson, G. ; Luu, B. *Synth. Comm.* **1987**, *17*, 1279.
22  Toromanoff, E. *Topics in Stereochemistry* **1967**, *2*, 158 ; Allinger, N. L. and Eliel, E. L. Eds, New York, N.Y..
23  Malek, J. *Organic Reactions* **1985**, *34*, 1 and quoted references.
24  You, Z. ; Koreeda, M. *Tetrahedron Lett.* **1993**, *34*, 2745 ; Ward, D. E. ; Rhee, C. K. *Can. J. Chem.* **1989**, *67*, 1206.
25  Ward, D. E. ; Rhee, C. K. ; Zoghaib, W. M. *Tetrahedron Lett.* **1988**, *29*, 4977 ; Ward, D. E. ; Rhee, C. K. ; *Can. J. Chem.* **1989**, *67*, 1206.
26  Hussey, B. J. ; Johnstone, R. A. W. ; Boehm, P. ; Entwistle, I. D. *Tetrahedron* **1982**, *38*, 3769.
27  Harada, N. ; Sugioka, T. ; Uda, H. ; Kuruki, T. *Synthesis* **1990**, 53.

28  Sondheimer, F. ; Amendolla, C. ; Rosenkranz, G. *J. Am. Chem. Soc* **1953**, *75*, 5930.
29  D'Incan, E. ; Loupy, A. ; Restelli, A. ; Seyden-Penne, J. ; Viout, P. *Tetrahedron* **1982**, *38*, 1755.
30  Ihara, M. ; Toyota, M. ; Fukumoto, K. ; Kametani, T. *J. Chem. Soc. Perkin Trans.I* **1986**, 2151 ; Tietze, L. ; Utecht, J. *Synthesis* **1993**, 957.
31  Anbrokh, R. V. ; Kamalov, G. L. ; Petrenko, N. F. ; Dimitrieva, T. N. ; Enfiadzhyan, N. O. *Ukr. Khim. Zh.* **1985**, *51*, 319.
32  Kupfer, D. *Tetrahedron* **1961**, *15*, 193.
33  Toda, F. ; Kiyochige, K. ; Yagi, M. *Angew. Chem. Int. Ed. Engl.* **1989**, *28*, 320.
34  Fajkos, J. *Coll. Czech. Chem. Comm.* **1959**, *24*, 2284.
35  Trost, B. M. ; Curran, D. P. *Tetrahedron Lett.* **1981**, *22*, 4929.
36  Gemal, A. L. ; Luche, J. L. *J. Org. Chem.* **1979**, *44*, 4187.
37  Tsuda, T. ; Hayashi, T. ; Satomi, H. ; Kawamoto, T. ; Saegusa, T. *J. Org. Chem.* **1986**, *51*, 537 ; Mahoney, W. S. ; Brestensky, D. M. ; Stryker, J. M. *J. Am. Chem. Soc* **1988**, *110*, 291 ; Lipshutz, B. N. ; Ung, C. S. ; Sengupta, S. *Synlett* **1990**, 64.
38  Akamanchi, K. G. ; Patel, H. C. ; Meenakshi, R. *Synth. Comm.* **1992**, *22*, 1655.
39  Wu, Y. D. ; Houk, K. N ; ; Florez, J. ; Trost, B. M. *J. Org. Chem.* **1991**, *56*, 3656 and quoted references.
40  Calmes, D. ; Gorrichon-Guigon, L. ; Maroni, P. ; Accary, A. ; Barret, R. ; Huet, J. *Tetrahedron* **1981**, *37*, 879 ; Arnaud, C. ;Accary, A. ; Huet, J. *C. R. Acad. Sci.* **1977**, *285*, 325.
41  Agami, C. ; Kazakos, A. ; Levisalles, J. ; Sevin, A. *Tetrahedron* **1980**, *36*, 2977.
42  Sarkar, A. ; Rao, B. R. ; Konar, M. M ; *Synth. Comm.* **1989**, *19*, 2313.
43  Seyden-Penne, J. *Chiral Auxiliaries and Ligands in Asymmetric Synthesis,* Wiley Ed. New York, N.Y., 1995, p.4.
44  Boireau, G. ; Deberly, A. ; Toneva, R. *Synlett* **1993**, 585.

# Chapter 5

# Recent Advances in Asymmetric Reductions with B-Chlorodiisopinocampheylborane

P. Veeraraghavan Ramachandran and Herbert C. Brown

H. C. Brown and R. B. Wetherill Laboratories of Chemistry, Purdue University, West Lafayette, IN 47907–1393

B-Chlorodiisopinocampheylborane (Ipc$_2$BCl) readily prepared from α-pinene, either by the sequential treatment with borane and hydrogen chloride, or by the direct treatment with chloroborane, is an excellent reagent for the asymmetric reduction of aralkyl, α-hindered and α-perfluoroalkyl ketones with predictable stereochemistry. It is used for the kinetic resolution of α-hindered ketones and invariably dictates the diastereomeric outcome of a double asymmetric reduction. A beneficial neighboring group effect allows the synthesis of diols, amino alcohols, hydroxy acids, and lactones in excellent ee via an intramolecular reduction with Ipc$_2$BCl or Ipc$_2$BH. It provides excellent diastereomeric and enantiomeric excess in the reduction of aromatic diketones, providing a route for the synthesis of C$_2$-symmetric bidentate and tridentate ligands for catalytic asymmetric reactions. This reagent has been utilized in the syntheses of several pharmaceuticals and natural products.

Pioneering work in the area of asymmetric reduction via the asymmetric Meerwein-Pondorf-Verley (MPV) (*1-2*) and asymmetric Grignard reductions (*3*) to prepare enantiomerically enriched *sec*-alcohols was reported almost half a century ago. When the hydride reagents, lithium aluminum hydride and sodium borohydride became readily available, chirally modified versions of these reagents were tested for asymmetric reduction (*4*). Although success eluded organic chemists initially, Mosher (*5*) and Noyori (*6*) succeeded in preparing excellent reagents for asymmetric reduction. These reagents have been essentially replaced by borane based asymmetric reducing agents. Although a satisfactory asymmetric reduction of oximes and imines to produce enantiopure amines has not yet been achieved, the phenomenal successes achieved by borane based reagents for the reduction of ketones over the last decade have taken this area to new heights. We have earlier reviewed our own efforts in this area (*7-9*). In this chapter, attention has been focussed on the recent advances in the asymmetric reduction of ketones with B-chlorodiisopinocampheylborane (Ipc$_2$BCl) (*10*).

## $B$-isopinocampheyl-9-borabicyclo[3.3.1]nonane (Alpine-Borane)

A significant breakthrough in asymmetric reduction was achieved when, based on Mikhailov's report on the reactions of trialkylboranes with carbonyl compounds under forcing conditions (150 °C) (*11*), Midland and co-workers established $B$-isopinocampheyl-9-borabicyclo[3.3.1]nonane (Aldrich: Alpine-Borane), as the first successful chiral organoborane reducing agent (*12*) (Figure 1). This reagent is excellent for the reduction of highly reactive carbonyls, such as the deuteroaldehydes, RCDO, $\alpha,\beta$-acetylenic ketones, $\alpha$-keto esters, $\alpha$-halo ketones, and acyl cyanides (*13*). The steric environment of the pinane moiety makes the reactions quite facile at room temperature (rt). The reduction is a MPV type of process, involving a six-membered transition state, and the chiral auxiliary, $\alpha$-pinene, is readily recovered from the reaction mixture. The kinetics of the reaction is bimolecular and the hydride transfer is believed to be the rate limiting step (*14*).

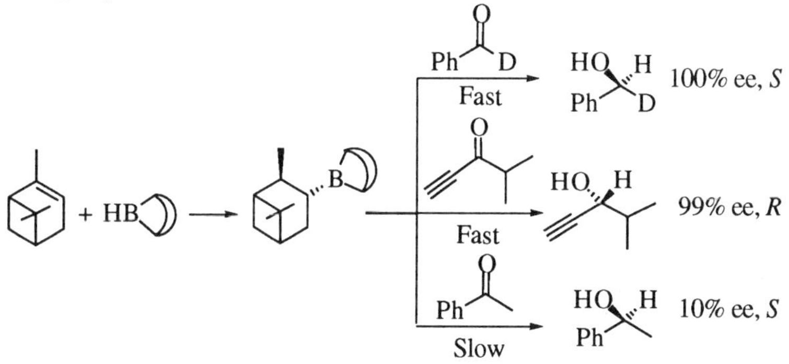

Figure 1. Preparation and reactions of Alpine-Borane

However, Alpine-Borane fails to reduce simple prochiral ketones, such as acetophenone and 3-methyl-2-butanone. The poor selectivity in the reduction of simple ketones with Alpine-Borane is presumed to be due to a concurrent dehydroboration of the reagent in slow reductions followed by an achiral reduction of the carbonyl group by the 9-BBN produced in this stage (Figure 2). This problem can be overcome by minimizing the dissociation either by conducting the reductions in high concentrations (*15*), or at greatly elevated pressures (*16*). These procedures are still impractical for the reduction of unactivated ketones.

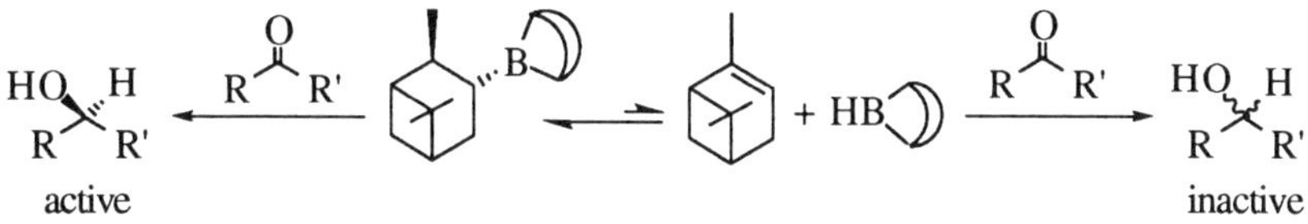

Figure 2. Dehydroboration of Alpine-Borane causes achiral reduction

## $B$-Chlorodiisopinocampheylborane (Ipc$_2$BCl, DIP-Chloride)

We postulated that the rate of reduction of Alpine-Borane might be enhanced by changing the electronic environment of the boron atom. Accordingly, we synthesized $B$-chlorodiisopinocampheylborane (Aldrich: DIP-Chloride) by treating Ipc$_2$BH with hydrogen chloride in ethyl ether (EE) or via the direct hydroboration of $\alpha$-pinene with monochloroborane (Figure 3) (*17*). The former is the preferred procedure since commercial $\alpha$-pinene is enantiomerically enriched during the preparation of Ipc$_2$BH.

**Aralkyl Ketones.** Originally we introduced Ipc$_2$BCl for the reduction of aralkyl ketones. The reagent consistently provides very high ee for the reduction of most types of aralkyl ketones, including heteroaralkyl ketones. The products are obtained with predictable stereochemistry. The (–)-reagent, $^d$Ipc$_2$BCl prepared from (+)-α-pinene, generally provides the $S$-alcohol (Figure 4). The $R$-alcohol produced from the reduction of pivalophenone is attributed to the steric interactions in the transition state (*see below*). Testing the reagent for a series of substituted aralkyl ketones showed that most representative substituents do not affect the chiral outcome. However, in the case of protonic substituents at the *ortho* position in the aromatic ring, a reversal in configuration is observed (*see below*).

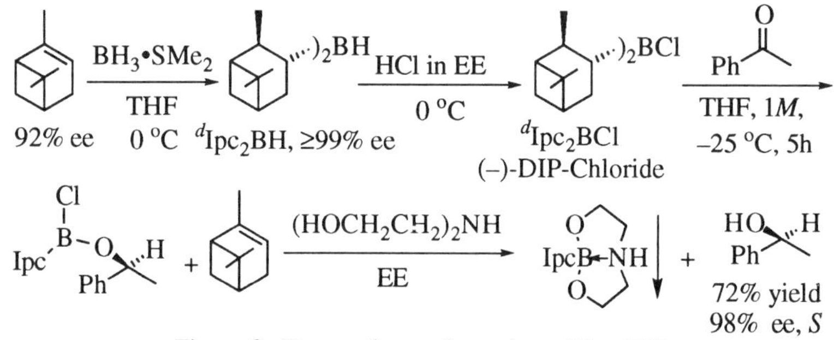

Figure 3. Preparation and reaction of Ipc$_2$BCl

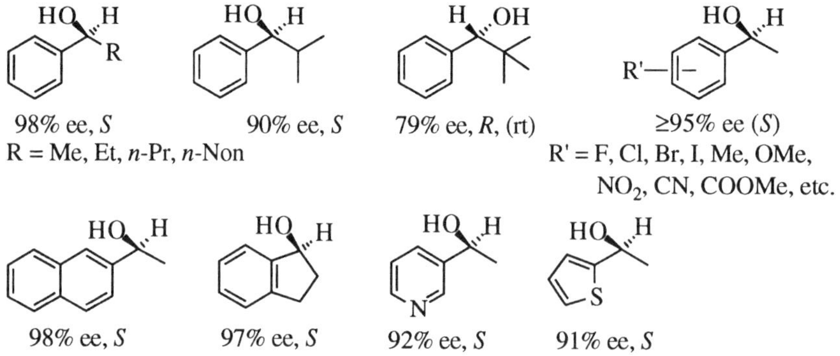

98% ee, $S$
R = Me, Et, $n$-Pr, $n$-Non

90% ee, $S$

79% ee, $R$, (rt)

≥95% ee ($S$)
R' = F, Cl, Br, I, Me, OMe, NO$_2$, CN, COOMe, etc.

98% ee, $S$

97% ee, $S$

92% ee, $S$

91% ee, $S$

Figure 4. Ipc$_2$BCl efficiently reduces aralkyl ketones

**Modified Workup.** The original workup procedure for Ipc$_2$BCl reductions involved a non-oxidative removal of the boron by-product as the diethanolamine complex. We have since developed a considerably simplified workup procedure for the isolation of product alcohols which involves treatment of the reaction mixture following reduction with a molar equiv of acetaldehyde at room temperature (*18*). This achieves the complete recycle of α-pinene (≥99% ee), higher yields of the product alcohols, and the easy disposal of water soluble and environmentally safe side products (Figure 5).

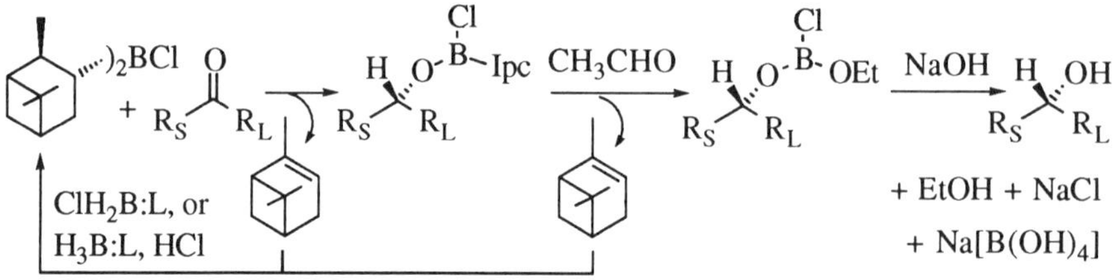

Figure 5. Modified work up procedure for Ipc$_2$BCl reductions

**Mechanism of Reduction.** Although the mechanism of the reaction has been studied in detail using molecular modeling (Rogic, M. M., unpublished data), a simple representation as shown in Figure 6, predicts the configuration of the products obtained from the reductions. The methyl group at the 2-position of α-pinene probably controls the stereochemistry of the product.

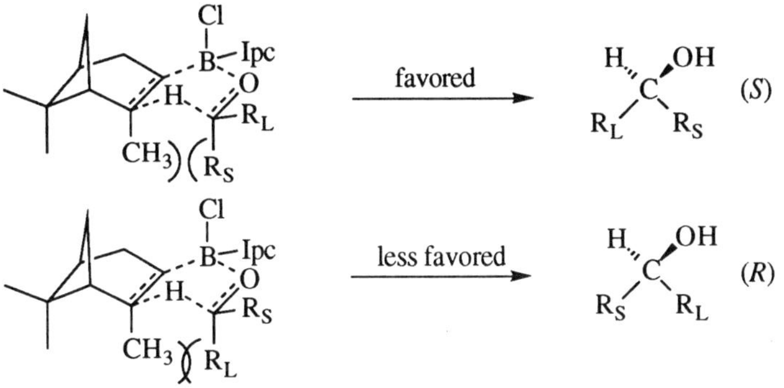

Figure 6. Mechanism of Ipc$_2$BCl reductions

**α-Hindered Ketones.** The steric control of the methyl group at the 2-position of the apopinene structure is clearly presented in Figure 7. This study led to extending the application of Ipc$_2$BCl to α-hindered ketones (19), including hindered acetylenic ketones (18).

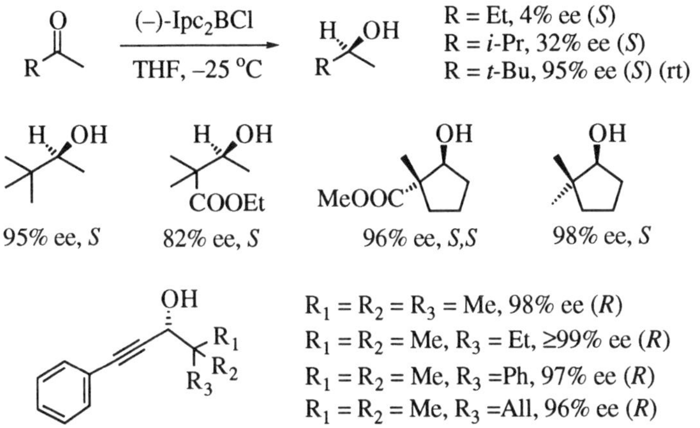

R = Et, 4% ee (S)
R = i-Pr, 32% ee (S)
R = t-Bu, 95% ee (S) (rt)

95% ee, S
82% ee, S
96% ee, S,S
98% ee, S

$R_1 = R_2 = R_3$ = Me, 98% ee (R)
$R_1 = R_2$ = Me, $R_3$ = Et, ≥99% ee (R)
$R_1 = R_2$ = Me, $R_3$ = Ph, 97% ee (R)
$R_1 = R_2$ = Me, $R_3$ = All, 96% ee (R)

Figure 7. Ipc$_2$BCl efficiently reduces α-hindered ketones

**Perfluoroalkyl Ketones.** The importance of chiral fluoro-organic molecules in agricultural, biological, materials, medicinal, and organic chemistry has been reviewed several times (20). The asymmetric reduction of prochiral trifluoromethyl ketones has interested organic chemists in the past, mainly due to the challenging theoretical aspects arising from the influence of the trifluoromethyl group in such reductions (4).

We observed that Ipc$_2$BCl exhibits excellent control in the reduction of α-perfluoroalkyl ketones. The product alcohols are produced in very high ee irrespective of the group flanking the α'-position of the carbonyl moiety (Figure 8), probably due to the interaction of the chlorine atom of the reagent and the fluorine atoms of the ketone. The perfluoroalkyl group acts as the enantiocontrolling group in the transition state (21-23).

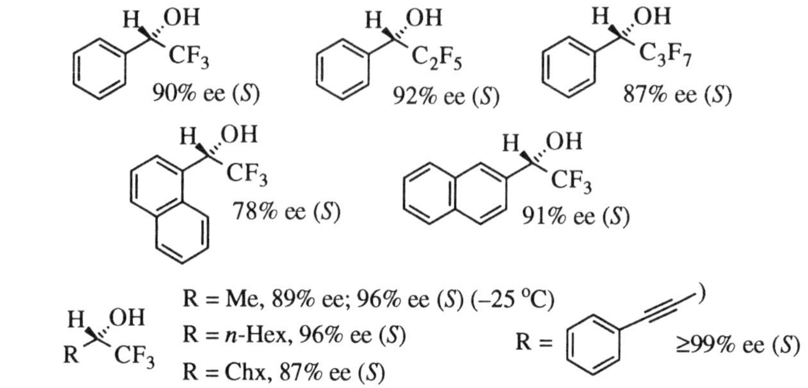

Figure 8. Ipc$_2$BCl efficiently reduces α-perfluoroalkyl ketones

**Perfluoroalkyl Oxiranes.** The consistent and predictable nature of Ipc$_2$BCl in providing α-perfluoroalkyl *sec*-alcohols in very high ee was exploited in achieving a one-pot synthesis of enantiomerically enriched trifluoromethyloxirane (eq 1) (*24*).

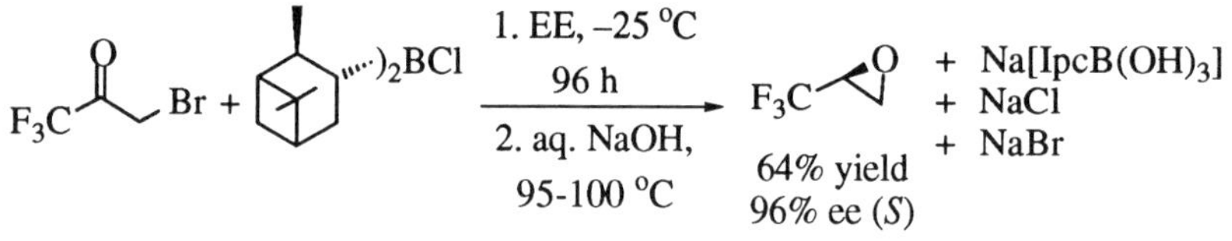

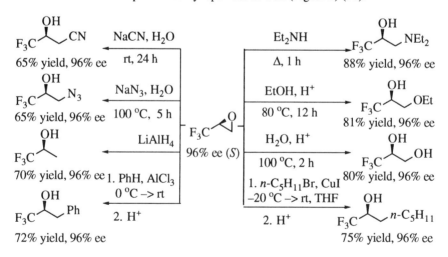

This epoxide serves as an intermediate for the syntheses of a variety of trifluoromethyl alcohols by the appropriate ring opening reactions. This methodology can be extended to other perfluoroalkyl epoxides as well (Figure 9) (*24*).

Figure 9. Ring opening reactions of trifluoromethyloxirane

**Asymmetric Amplification.** Recently, a Merck group discussed an "asymmetric amplification" in chiral reduction, obtaining the product alcohol of higher ee starting with Ipc$_2$BCl prepared from α-pinene of lower enantiomeric purity (70% ee) and chloroborane-methyl sulfide complex, *en route* to the synthesis of an LTD$_4$ antagonist, L-699,392 (Figure 10) (*25*). This procedure avoids the use of enantiomerically pure reagent, though an 0.8 equiv excess of the reagent was used for obtaining optimal ee for the product alcohol.

Kagan studied the non-linear effect between the ee's of Ipc$_2$BCl and the ee's of α-phenethanol produced from the reduction of acetophenone (*26*). He concluded that this effect is a function of the method of preparation of the borane precursor.

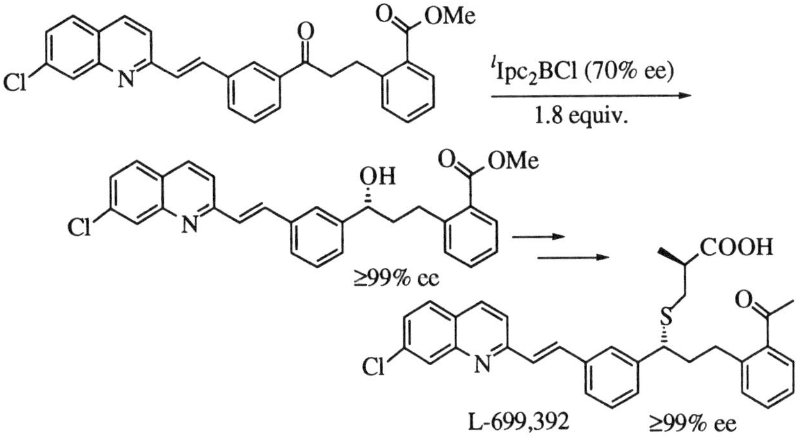

Figure 10. Asymmetric amplification; synthesis of L-699,392

**Kinetic Resolution.** We took advantage of the capability of Ipc$_2$BCl to reduce α-hindered ketones for the kinetic resolution of racemic ketones (Figure 11). The kinetic resolution of a number of bicyclic ketones was also achieved using this methodology (*27*). In all the cases studied, the *R*-isomer of the ketone is recovered when $^d$Ipc$_2$BCl is used for kinetic resolution, while $^l$Ipc$_2$BCl provides the *S*-ketone.

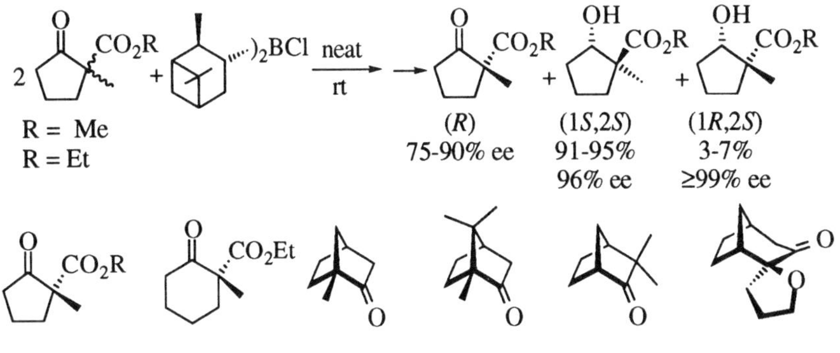

Figure 11. Kinetic resolution of α-hindered ketones

**Double Asymmetric Synthesis.** One of the fascinating areas of research in asymmetric synthesis is the study of double diastereoselection, the match and mismatch between the chirality in the reagent and the substrate. It is interesting to note that pinane-based asymmetric reagents invariably override the influence of the proximal chiral center and control the diastereoselectivity of the reaction (9). This has also been observed in the asymmetric reduction of enantiopure ketones with Ipc$_2$BCl. For example, starting with a particular isomer of the ketone, we obtain the diastereomers of the product alcohol by using the antipodes of the reagent (Figure 12) (28).

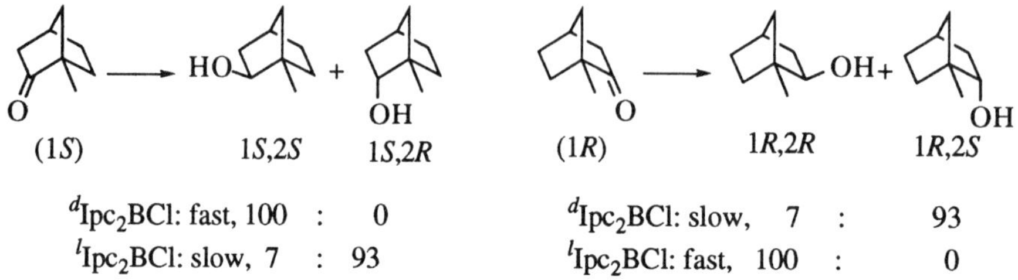

| (1S) | 1S,2S : 1S,2R | (1R) | 1R,2R : 1R,2S |
|---|---|---|---|
| $^d$Ipc$_2$BCl: fast, | 100 : 0 | $^d$Ipc$_2$BCl: slow, | 7 : 93 |
| $^l$Ipc$_2$BCl: slow, | 7 : 93 | $^l$Ipc$_2$BCl: fast, | 100 : 0 |

Figure 12. Double asymmetric reduction of α-chiral ketones

**Reduction of Diketones.** The growing importance of asymmetric syntheses, especially those involving $C_2$-symmetric molecules as chiral directors, provides an impetus for the preparation of such compounds (29). Asymmetric reduction of symmetric diketones offers a promising synthetic route to $C_2$-symmetric diols. Chong and co-workers utilized Ipc$_2$BCl for the asymmetric reduction of 1,4-diphenyl-1,4-butanedione and obtained the product diol in ≥98:≤2, dl:meso, with the dl component showing ≥99% ee (30). They converted the diol to *trans*-2,5-diphenylpyrrolidine (Figure 13).

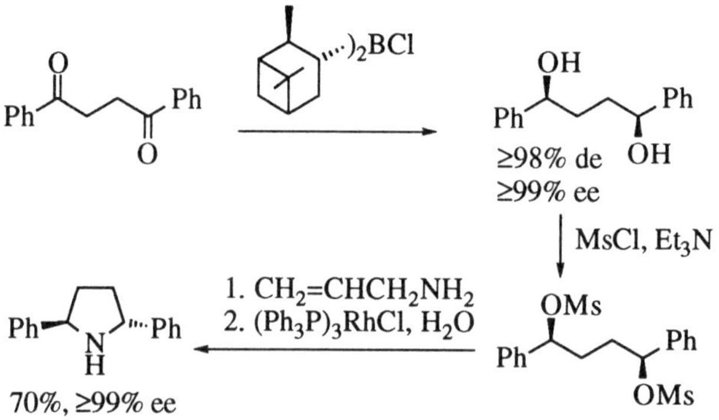

Figure 13. Synthesis of trans-2,5-diphenylpyrrolidine

We (31) and others (32-34) have carried out the reduction of representative diacetylaromatic compounds. In most of the cases tested thus far, the products are obtained in good to excellent yield in essentially enantiomerically pure form (Figure 14).

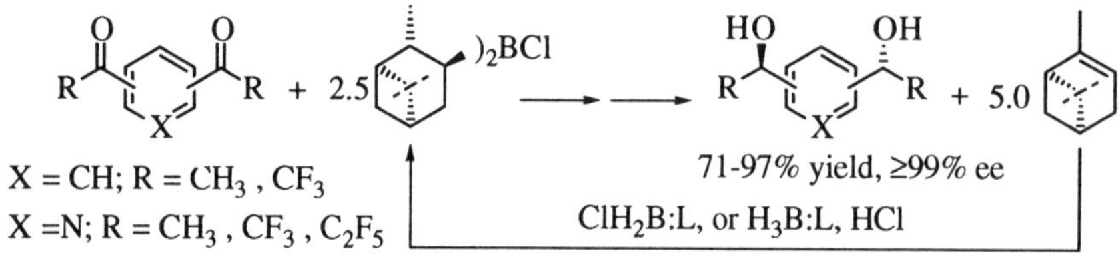

X = CH; R = CH$_3$ , CF$_3$
X = N; R = CH$_3$ , CF$_3$ , C$_2$F$_5$

71-97% yield, ≥99% ee

ClH$_2$B:L, or H$_3$B:L, HCl

Figure 14. Asymmetric reduction of diacylaromatic compounds

The reduction of 2,6-diacylpyridines with differing steric and electronic environments (Figure 15) provides an easy access to enantiomerically pure tridentate ligands for use in transition metal catalyzed reactions (*33*).

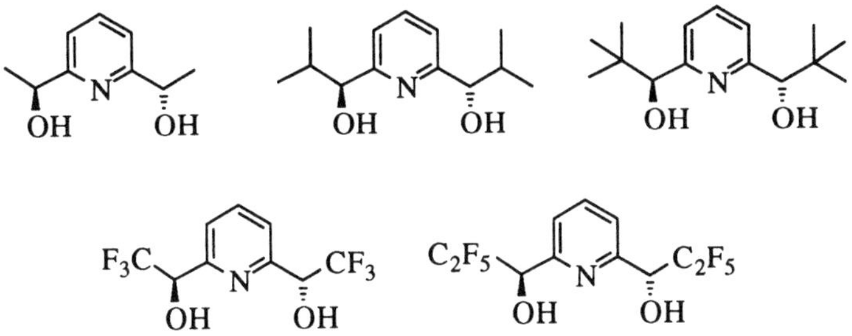

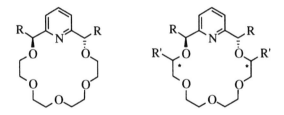

Figure 15. Synthesis of $C_2$-symmetric tridentate ligands

We have since synthesized pyridino-18-crown-6 from these enantiopure pyridyl diols. The possibility of varying the R and R' groups makes possible a systematic study of these macrocycles for enantiomer recognition involving host-guest chemistry (Figure 16) (Ramachandran, P. V., unpublished results).

Figure 16. Synthesis of enantiopure pyridino-18-crown-6

## Neighboring Group Effects in Asymmetric Reductions with Ipc₂BCl

While carrying out a systematic study of the effect of substituents in the phenyl ring of acetophenone on the reduction with Ipc₂BCl (Gong, B., unpublished results), we stumbled onto an unexpected phenomenon. We discovered a new neighboring group effect in asymmetric reductions involving Ipc₂BCl. We found that the presence of a protonic group or a lone pair of electrons in the proximity of the carbonyl group assists in achieving an intramolecular reduction of the ketone providing the product alcohols in high ee.

**Hydroxy Ketones.** The reaction of one molar equiv of $^d$Ipc₂BCl with *o*-hydroxyacetophenone in THF provides a reddish complex which reacts very slowly at rt and provides an "aromatic" intermediate. Due to the stability of this intermediate, the usual diethanolamine or acetaldehyde workups were unsuccessful and the phenolic alcohol was isolated by an oxidative workup in 90% ee in the *R*-isomer, compared to the *S*-isomer obtained from the reduction of *o*-methoxyacetophenone with the same isomer of the reagent. Figure 17 outlines the intramolecular reduction pathway (*35*).

The intramolecular nature of this reduction provides an opportunity to use Ipc₂BH, which is ordinarily a poor reducing agent, for the reduction of *o*-hydroxyacetophenones and -benzophenones in very high ee.

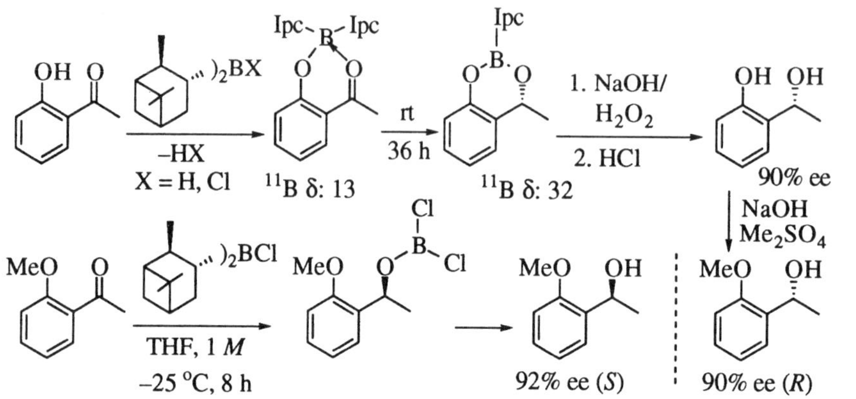

Figure 17. Neighboring group effect: reduction of o-hydroxyketones

This phenomenon was applied by a Ciba-Geigy group in the synthesis of CGS 26214, a potent LDL cholesterol lowering agent (36) (Figure 18).

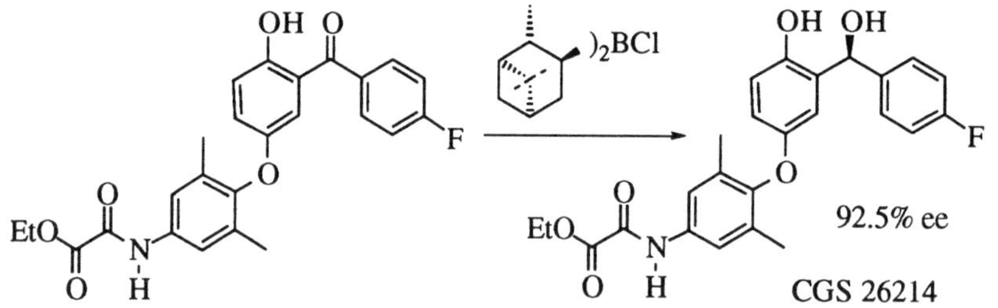

Figure 18. Neighboring group effect: synthesis of CGS 26214

The neighboring group effect has been extended to the reduction of aliphatic hydroxyketones as well. Thus, a series of α- and β-hydroxyketones were reduced with Ipc$_2$BCl or Ipc$_2$BH to 1,2- and 1,3-diols, respectively in excellent ee (Figure 19) (Lu, Z. H., unpublished results).

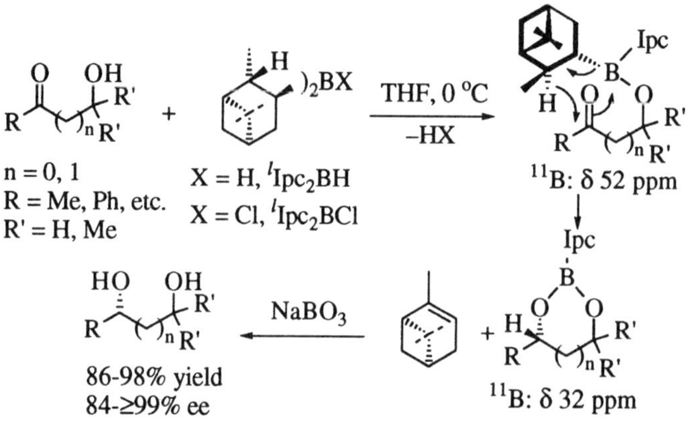

Figure 19. Neighboring group effect: reduction of 1,2- and 1,3-hydroxy ketones

**Keto Acids.** We extended this phenomenon to *o*-carboxylic aryl ketones and synthesized enantiomerically enriched 3-substituted-1(*3H*)-isobenzofuranones (3-substituted phthalides) (*37*). While 3-alkylphthalides are obtained in the highest ee (97-99%) by the intermolecular reduction of the *o*-carboxylic esters, enantiopure 3-arylphthalides are obtained by the intramolecular reduction of the free *o*-carboxylic acids (Figure 20). Ipc₂BH proved to be a better reagent for the intramolecular reduction since the equilibration of Ipc₂BCl with the HCl produced results in a decreased ee of the product, probably due to a competing intra- and intermolecular reductions.

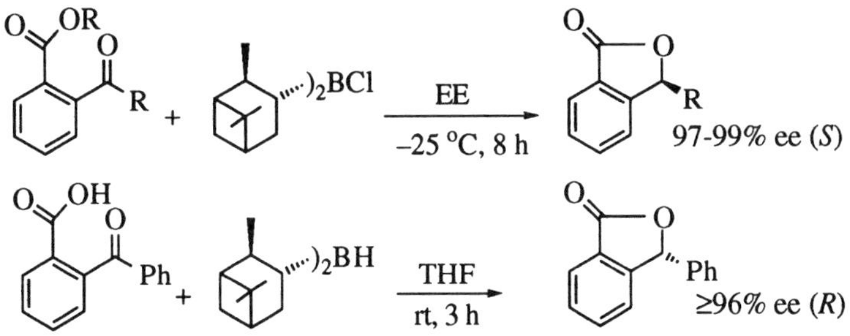

Figure 20. Neighboring group effect: reduction of *o*-carboxylic ketones

**Amino Ketones.** The extension of this phenomenon to the *o*-aminoacetophenones with Ipc₂BCl and Ipc₂BH revealed that while Ipc₂BH liberates one equiv of H₂ to form an intermediate with a B-N bond, followed by a slow reduction, the chloro-reagent reduces the ketone rapidly after an initial complexation (Figure 21). The configuration of the products from both these processes are the same (Malhotra, S. V., unpublished results).

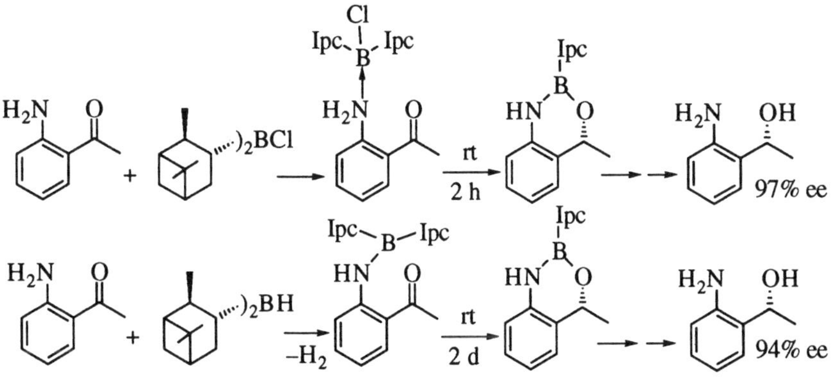

Figure 21. Neighboring group effect: reduction of *o*-aminoketones

In contrast to the 21% ee obtained for the reduction of 4-phenyl-3-butyn-2-one with Ipc₂BCl (*10*), Morita and coworkers reported 95% ee for the reduction of 4-(2-amino-3-bromo-5,6-difluorophenyl)-3-butyn-2-one (Figure 22) (*38*). The authors were puzzled by the results. We believe that this unusual effect may be due to a favorable positioning of the isopinocampheyl group for an intramolecular delivery of hydride. The results are unsatisfactory if the bromine atom in the phenyl ring is absent.

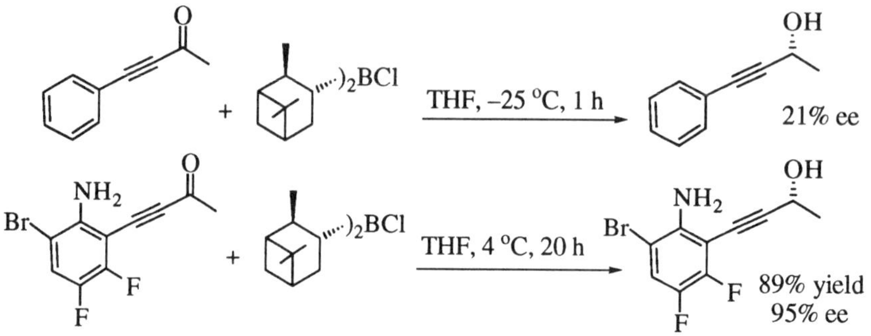

Figure 22. Neighboring group effect:
reduction of 4-(2-amino-3-bromo-5,6-difluorophenyl)-3-butyn-2-one

**2-Acylpyridines.** Although we had utilized two equiv of Ipc$_2$BCl for the reduction of 3-acetylpyridine to allow for the complexation of one mole with the reagent (*10*), our understanding of the neighboring group effect led to a study that revealed an enhanced rate of reduction of 2-acetylpyridine with only one mole of Ipc$_2$BCl. This phenomenon is general for all types of 2-acylpyridines that we have tested so far (Figure 23) (Malhotra, S. V.; Madappat, K. V., unpublished results).

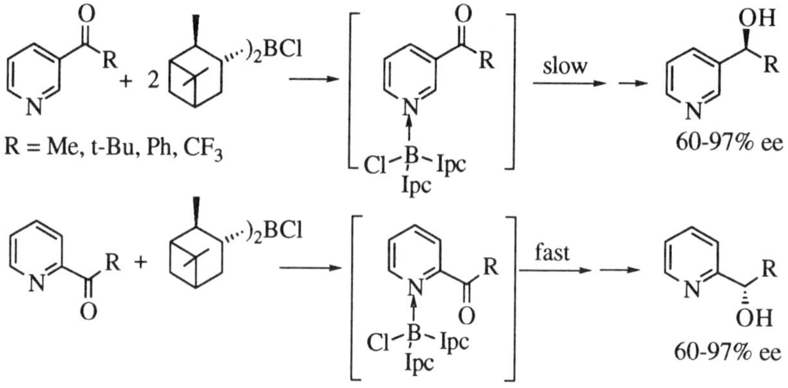

R = Me, t-Bu, Ph, CF$_3$

Figure 23. Neighboring group effect: reduction of 2-acetylpyridines

This neighboring group effect accounts for the considerably faster rate of reduction of 2-pivaloylpyridine in comparison with pivalophenone (Figure 24). Bolm and coworkers have not made any mention of the unusual fast rate in the reduction of a hindered ketone (*39*).

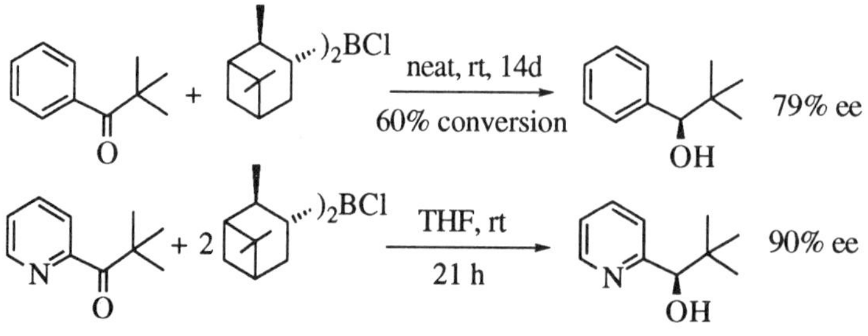

Figure 24. Neighboring group effect: reduction of 2-pivaloylpyridine

## Applications of Ipc₂BCl

Ipc₂BCl is frequently used in organic syntheses involving the reduction of aralkyl or α-hindered ketones (8,9). The application of this reagent in the first enantioselective synthesis of the currently widely used anti-depressant, N-methyl-γ-[4-trifluoromethyl]phenoxy]benzenepropanamine hydrochloride (fluoxetine hydrochloride, Eli Lilly: Prozac) and the analogs of this serotonin and norepinephrine uptake inhibitor, Tomoxetine and Nisoxetine, is outlined in Figure 25 (40).

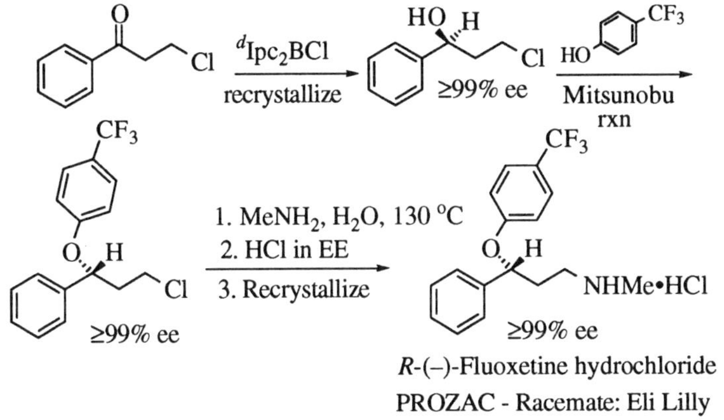

Figure 25. Application of Ipc₂BCl: synthesis of Prozac

This reagent has also been applied for the syntheses of both isomers of an analog of a potential antipsychotic agent BMS 181100 (Figure 26) (41) and all the four diastereomers of a bronchodilator, Eprozinol (Figure 27) (42).

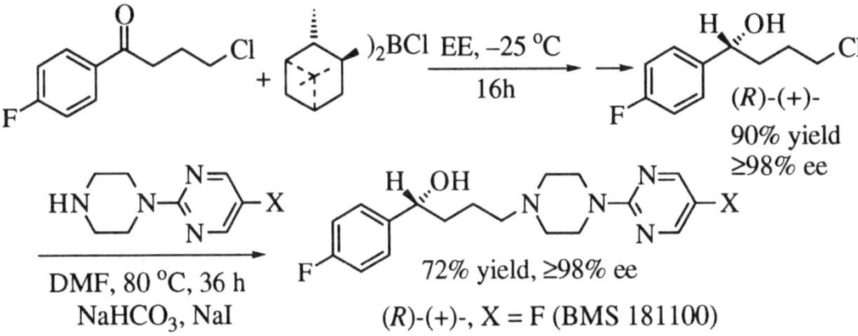

Figure 26. Application of Ipc₂BCl: synthesis of BMS-181100 analog

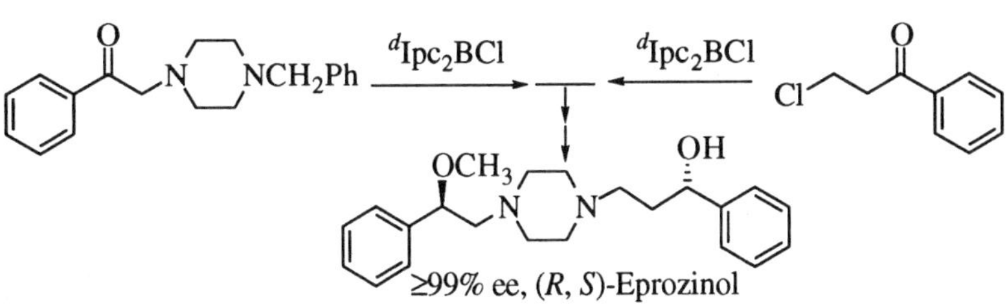

Figure 27. Application of Ipc₂BCl: synthesis of (R,S)-Eprozinol

## Conclusions

Ipc$_2$BCl is an excellent reducing agent for the asymmetric reduction of aralkyl, α-hindered and perfluoroalkyl ketones. It is used for the kinetic resolution of α-hindered ketones and invariably dictates the diastereomeric outcome of a double asymmetric reduction. A beneficial neighboring group effect allows the syntheses of diols, amino alcohols, hydroxy acids, and lactones in excellent ee via an intramolecular reduction with Ipc$_2$BCl or Ipc$_2$BH. It provides excellent diastereomeric and enantiomeric excess in the reduction of aromatic diketones, providing a route for the synthesis of C$_2$-symmetric diols. This reagent has been utilized in the synthesis of several pharmaceuticals and natural products.

Ipc$_2$BCl satisfies the requirements of a versatile chiral reducing agent, such as: (1) Both isomers are readily available in very high ee; enantiomeric enrichment of the commercial material is easily attained during hydroboration. (2) The preparation of the reagents and the reaction conditions are very simple and convenient. (3) The workup is easy. (4) The chiral auxiliary is readily recovered in an easily recyclable form without loss of any enantiomeric purity. (5) A tentative mechanism is known for the reaction which facilitates planning of modifications. (6) The configuration of the products can be predicted based on the mechansim; exceptions are rare. (7) Scaling up of the reactions is easy. (8) Most important of all, the enantiomeric excesses achieved in most reactions are usually high.

## Acknowledgement

The financial assistance from the United States Army Research Office is gratefully acknowledged.

## Literature Cited

1. Doering, W. von E.; Young, R. W. *J. Am. Chem. Soc.* **1950**, *72*, 631.
2. Jackman, L. M.; Mills, J. A.; Shannon, J. S. *J. Am. Chem. Soc.* **1950**, *72*, 4814.
3. Vavon, G.; Riviere, C.; Angelo, B. *Compt. rend.* **1946**, *222*, 959.
4. For an early extensive discussion of this topic, see: Morrison, J. D.; Mosher, H. S. *Asymmetric Organic Reactions*, Prentice Hall: Englewood Cliffs, N.J., 1971, pp 160-218.
5. Yamaguchi, S.; Mosher, H. S.; Pohland, A. *J. Am. Chem. Soc.* **1972**, *94*, 9254.
6. Noyori, R.; Tomino, I.; Tanimoto, Y.; Nishizawa, M. *J. Am. Chem. Soc.* **1984**, *106*, 6709.
7. Brown, H. C.; Ramachandran, P. V. *Acc. Chem. Res.* **1992**, *25*, 16.
8. Brown, H. C.; Ramachandran, P. V. In *Advances in Asymmetric Synthesis*; Hassner, A. Ed.; JAI Press: Greenwich, CT., 1995, Vol.1; pp 147-210.
9. Brown, H. C.; Ramachandran, P. V. *J. Organometal. Chem.* **1995**, *500*, 1.
10. Brown, H. C.; Chandrasekharan, J.; Ramachandran, P. V. *J. Am. Chem. Soc.* **1988**, *111*, 1539.
11. Mikhailov, B. M.; Bubnov, Yu. N.; Kiselev, V. G. *J. Gen. Chem. U. S. S. R. (Engl. Transl)* **1966**, *36*, 65.
12. Midland, M. M.; Greer, S.; Tramontano, A.; Zderic, S. A. *J. Am. Chem. Soc.* **1979**, *101*, 2352.
13. Midland, M. M. *Chem. Rev.* **1989**, 1553.
14. Midland, M. M.; Zderic, S. A. *J. Am. Chem. Soc.* **1982**, *104*, 525.
15. Brown, H. C.; Pai, G. G. *J. Org. Chem.* **1985**, *50*, 1384.
16. Midland, M. M.; McLoughlin, J. I.; Gabriel, J. *J. Org. Chem.* **1989**, *54*, 159.
17. Brown, H. C.; Ramachandran, P. V.; Chandrasekharan, J. *Heteroatom Chem.* **1995**, *6*, 117.

18. Ramachandran, P. V.; Teodorovic', A. V.; Rangaishenvi, M. V.; Brown, H. C. *J. Org. Chem.* **1992**, *57*, 2379.
19. Brown, H. C.; Chandrasekharan, J.; Ramachandran, P. V. *J. Org. Chem.* **1986**, *51*, 3394.
20. Bravo, P.; Resnati, G. *Tetrahedron: Asym.* **1990**, *1*, 661.
21. Ramachandran, P. V.; Teodorovic', A. V.; Brown, H. C. *Tetrahedron*, **1993**, *49*, 1725.
22. Ramachandran, P. V.; Gong, B.; Teodorovic', A. V.; Brown, H. C. *Tetrahedron: Asym.* **1994**, *5*, 1061.
23. Ramachandran, P. V.; Teodorovic', A. V.; Gong, B.; Brown, H. C. *Tetrahedron: Asym.* **1994**, *5*, 1075.
24. Ramachandran, P. V.; Gong, B.; Brown, H. C. *J. Org. Chem.* **1995**, *60*, 61.
25. King, A. O,; Corley, E. G.; Anderson, R. K.; Larsen, R. D.; Verhoeven, T. R.; Reider, P. J.; Xiang, Y. B.; Belley, M.; Leblanc, Y.; Labelle, M.; Prasit, P.; Zamboni, R. *J. Org. Chem.* **1993**, *58*, 3731.
26. Girard, C.; Kagan, H. B. *Tetrahedron: Asym.* **1995**, *6*, 1881.
27. Ramachandran, P. V.; Chen, G. M.; Brown, H. C. *J. Org. Chem.* **1996**, *61*, 88.
28. Ramachandran, P. V.; Chen, G. M.; Brown, H. C. *J. Org. Chem.* **1996**, *61*, 95.
29. Whitesell, J. K. *Chem. Rev.* **1989**, *89*, 1581.
30. Chong, J. M.; Clarke, I. S.; Koch, I.; Olbach, P. C.; Taylor, N. J. *Tetrahedron: Asym.* **1995**, *6*, 409.
31. Ramachandran, P. V.; Chen, G. M.; Lu, Z. H.; Brown, H. C. *Tetrahedron Lett.* **1996**, *37*, 3795.
32. Ishizaki, M.; Fujita, K.; Shimamoto, M.; Hoshino, O. *Tetrahedron: Asym.* **1994**, *5*, 411.
33. Jiang, Q.; Plew, D. V.; Murtuza, S.; Zhang, X. *Tetrahedron Lett.* **1996**, *37*, 797.
34. Déziel, R.; Malenfant, E.; Bélanger, G. *J. Org. Chem.* **1996**, *61*, 1875.
35. Ramachandran, P. V.; Gong, B.; Brown, H. C. *Tetrahedron Lett.* **1994**, *35*, 2141.
36. Shieh, W.; Cantrell, W. R.; Carlson, J. A. *Tetrahedron Lett.* **1995**, *36*, 3797.
37. Ramachandran, P. V.; Chen, G. M.; Brown, H. C. *Tetrahedron Lett.* **1996**, *37*, 2205.
38. Morita, S.; Otsubo, K.; Matsubara, J.; Ohtani, T.; Uchida, M. *Tetrahedron: Asym.* **1995**, *6*, 245.
39. Bolm, C.; Zehnder, M.; Bur, D. *Angew. Chem. Int. Ed. Engl.* **1990**, *29*, 205.
40. Srebnik, M.; Ramachandran, P. V.; Brown, H. C. *J. Org. Chem.* **1988**, *53*, 2916.
41. Ramachandran, P. V.; Gong, B.; Brown, H. C. *Tetrahedron: Asym.* **1993**, *4*, 2399.
42. Ramachandran, P. V.; Gong, B.; Brown, H. C. *Chirality* **1995**, *7*, 103.

# Chapter 6

# The Practical Enantioselective Reduction of Prochiral Ketones

Anthony O. King, David J. Mathre, David M. Tschaen, and Ichiro Shinkai

Process Research and Development, Merck Research Laboratories, Rahway, NJ 07065

The current literature describes many useful methods for the preparation of chiral *secondary* hydroxy compounds from prochiral ketones. Of these methods, we have examined and developed the Itsuno-Corey's OAB and Brown's B-chlorodiisopinocampheylborane ($Ipc_2BCl$) procedures for the large scale synthesis of chiral hydroxy intermediates for further elaboration to pharmaceutical products.

**Oxazaborolidine Catalyzed Enantioselective Reductions.**

Oxazaborolidines **1** derived from chiral ß-amino alcohols [equation 1] have proven to be an important class of reagents and catalysts for the enantioselective reduction of prochiral ketones [equation 2] (*1-3*).

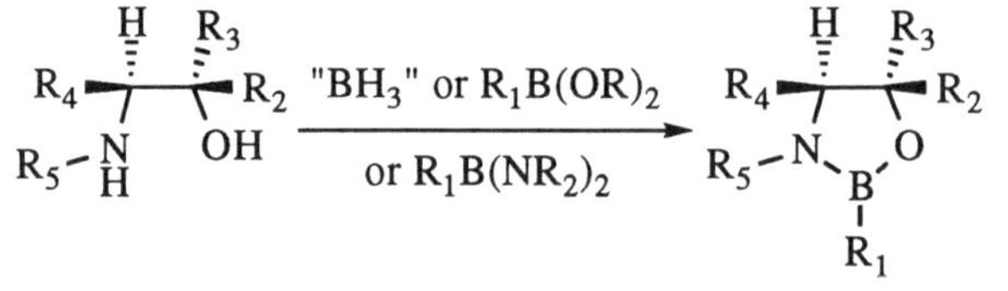

(equation 1)

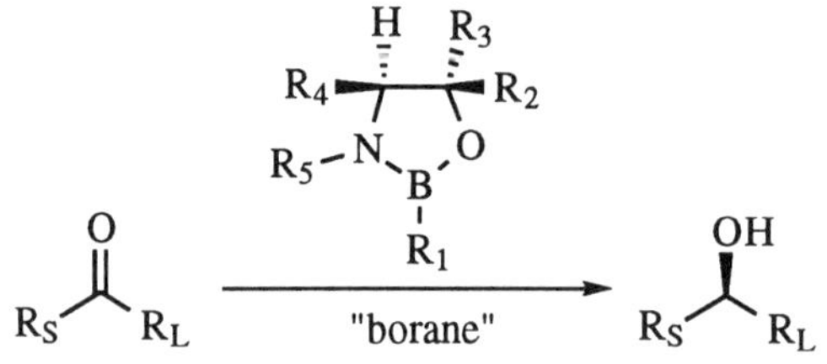

(equation 2)

Following the pioneering work of Itsuno (*4-10*) and Corey (*11-13*), well over 100

# 6. KING ET AL.   *Practical Enantioselective Reduction of Prochiral Ketones*

chiral oxazaborolidines [Figure 1] have been reported as reagents and catalysts for the enantioselective reduction of prochiral ketones (*1-9, 11-13*), imines (*14-18*), oximes (*7-8, 10, 15, 19-20*), α-ketophosphonates (*21-23*), cyclic meso-imides (*24*), and 2-pyranones to afford chiral biaryls (*25, 26*). We have been interested in the practical application of these catalysts for the large-scale preparation of the chiral pharmaceutical drug candidates MK-0417 (*27, 28*) and MK-0499 (*29-32*).

Figure 1. Chiral Oxazaborolidines.

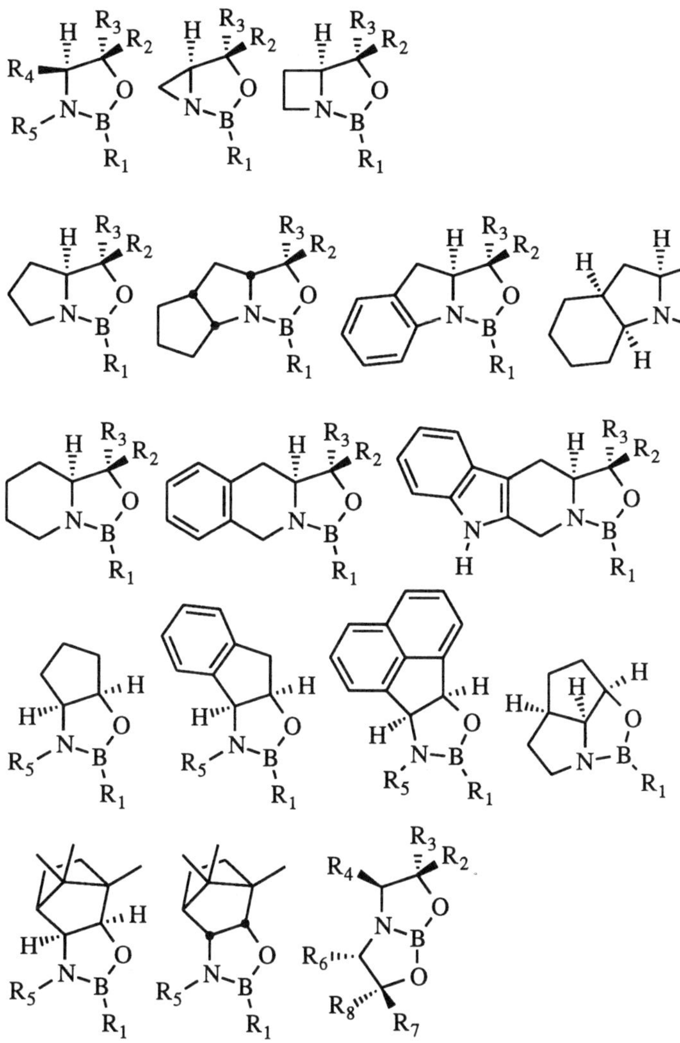

The key transformation required for the preparation of MK-0417 is the enantioselective reduction of the prochiral ketosulfone shown in equation 3 (*27*). After evaluating a number of enantioselective reducing agents, we found the *B*-methyloxazaborolidine derived from α,α-diphenylprolinol championed by Corey to be the most promising (12-13).

MK-0417

MK-0499

ketosulfone → hydroxysulfone (equation 3)

The first step to making this oxazaborolidine catalyst useful on a large, multi-kilogram scale was development of a practical synthesis of α,α-diphenylprolinol and its conversion to the corresponding B-methyloxazaborolidine. The synthesis of (S)-α,α-diphenylprolinol from natural (S)-proline was accomplished by Grignard addition to proline N-carboxyanhydride (proline-NCA) [Figure 2] (33). By going through proline-NCA, extra steps for protecting and deprotecting the nitrogen are avoided. This chemistry has been successfully employed to prepare 100 g quantities of various α,α-diarylprolinols in the laboratory, and 10+ kg quantities of enantiomerically pure (S)-α,α-diphenylprolinol in our pilot plant facilities. [For the synthesis of (R)-α,α-diphenylprolinol, chemistry developed by Beak (34-36) looks promising.]

Next, we developed a procedure using trimethylboroxine (TMB) to convert (S)-α,α-diphenylprolinol to the corresponding B-methyloxazaborolidine [Figure 2] (33). This method affords the pure oxazaborolidine, free of less enantioselective intermediates and by-products. This became especially important while optimizing the reduction of the ketosulfone using the oxazaborolidine catalyst. We discovered that the enantioselectivity is very dependent on water content – the less water present in the system, the higher the enantioselectivity (27). Water readily reacts with the strained bicyclic oxazaborolidine ring system. The TMB procedure to prepare the oxazaborolidine provides catalyst free of these water related impurities.

Although the free B-methyloxazaborolidine is stable (as a toluene or xylene solution) if scrupulously protected from moisture, we found that the crystalline oxazaborolidine-borane complex to be a significantly more stable form of the catalyst. [Figure 2] (37-39). Corey has reported the single crystal x-ray structure of the oxazaborolidine-borane complex (40).

With the availability of pure B-methyloxazaborolidine and the oxazaborolidine-borane complex, we have been able to investigate the stoichiometric catalytic cycle by low temperature $^1$H, $^{11}$B, and $^{13}$C NMR (41). Based on these studies, the catalytic cycle, shown in Figure 3, appears to possess features of the catalytic cycles proposed by Corey (13, 14) and Evans (42).

Figure 2. Practical Preparation of the Oxazaborolidine Catalyst from Proline

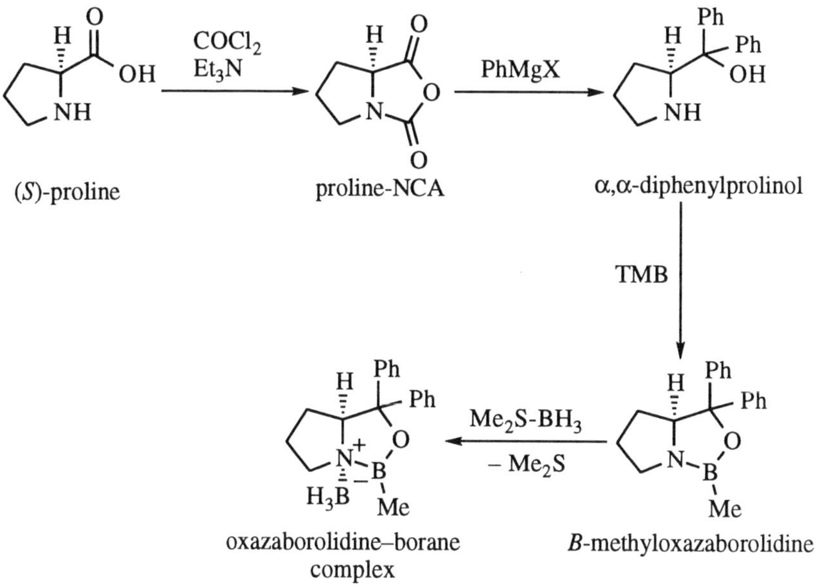

(S)-proline → proline-NCA (COCl$_2$, Et$_3$N) → α,α-diphenylprolinol (PhMgX) → B-methyloxazaborolidine (TMB) → oxazaborolidine–borane complex (Me$_2$S-BH$_3$, − Me$_2$S)

The first observable intermediate after the highly exothermic reaction between the oxazaborolidine-borane complex and ketone appears to have the structure first proposed by Corey (13), with the newly formed carbinol attached to the tetrahedral oxazaborolidine ring boron, and a coordinatively unsaturated BH$_2$ ($^{11}$B ~ -45 ppm) attached to the bridgehead nitrogen. This intermediate is only seen at very low (-60 to -100 °C) temperatures, and its lifetime appears to be dependent on the sterics of the starting ketone (1-tetralone t(1/2) ~ 10 min vs. t-butyl phenyl ketone t(1/2) ~ 3 h; both at -60 °C).

Normally, the first observed intermediate proceeds on to the second observed intermediate, which appears to have the structure first proposed by Evans (42). The BH$_2$ ($^{11}$B ~ 4 ppm) is now tetrahedral, attached to both the newly formed carbinol oxygen, and the bridgehead nitrogen. The monoalkoxyborane is not tightly bound to the oxazaborolidine, and based on peak broadening, appears to be easily exchanged with free oxazaborolidine. The monoalkoxyborane is not observed by NMR. It immediately disproportionates to the dialkoxyborane ($^{11}$B ~ 28 ppm) and borane. The borane formed then reacts with free oxazaborolidine ($^{11}$B ~ 34 ppm) to afford the oxazaborolidine-borane complex ($^{11}$B ~ 34 ppm and -15 ppm).

In the presence of excess ketone, or slower reacting ketones where the level of ketone can build up, the first observed intermediate, with the coordinatively unsaturated BH$_2$, can also act as a reducing agent. The enantioselectivity for this second reduction is not the same as the highly enantioselective first hydride transfer, and often can be much lower. As a result, the mode of addition can have a dramatic effect on the overall level of enantioselection.

With the simple, fast reacting ketone, 1-tetralone, slow addition of the substrate to the oxazaborolidine-borane complex (1 equiv) over a 6 h period at -20 °C, affords 1,2,3,4-tetrahydro-1-naphthol with an enantiomeric purity of 99.2%. Reversing the order of addition decreases the enantiomeric purity to 87.2%. The same effect is observed when the reaction is run using 5 mol% of the oxazaborolidine-borane complex and excess borane-methyl sulfide (99% ee vs. 90% ee).

Figure 3. Stoichiometric Oxazaborolidine Catalytic Cycle.

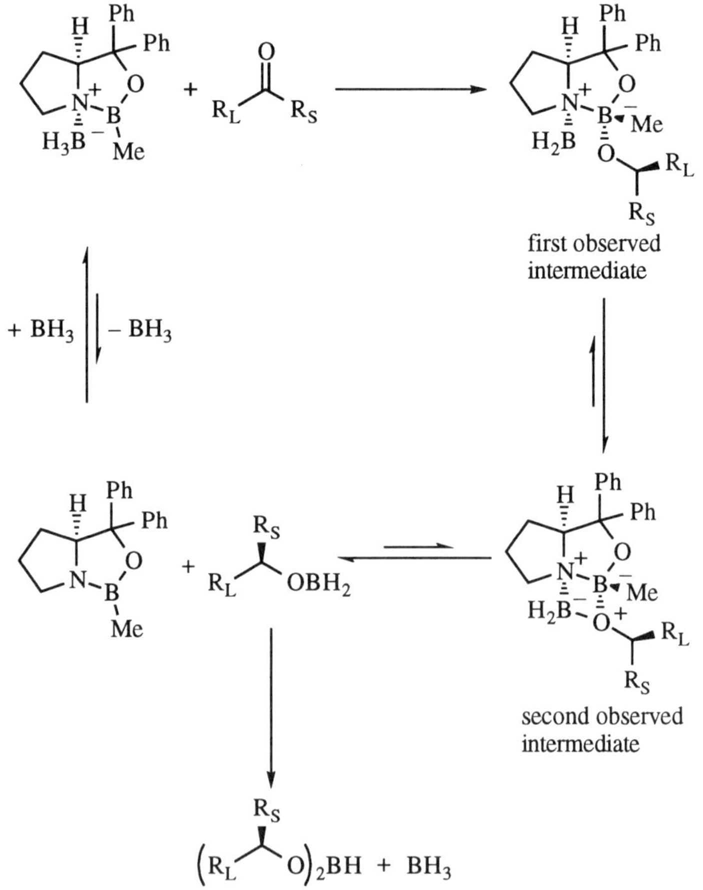

With the more hindered, slow reacting ketone, 2,2-diphenylcyclopentanone (*43*), the results were initially more erratic ranging from 50% to 95% ee. Based on a report by Stone (*44*), the reduction was run at higher temperatures (40 °C). In this case, slow addition of the substrate to the oxazaborolidine-borane complex (20 mol%) and excess borane affords 2,2-diphenylcyclopentanol with an enantiomeric purity of 92-95%. This reaction has been scaled-up as part of an *Organic Syntheses* preparation (*45*).

The enantioselective reduction of the ketone leading to MK-0499, shown in Figure 4, initially required stoichiometric amounts of the oxazaborolidine-borane complex to provide product of high (99+%) enantiomeric purity (*29-30*). With catalytic amounts of the oxazaborolidine and stoichiometric amounts of borane, the enantiomeric purity was only 90-92%. Addition of isopropanol to the reaction mixture was found to be a unique method to overcome this problem (*31*). Presumably, isopropanol (or an intermediate isopropoxyborane) is able to intercept the reactive intermediate responsible for the lower enantioselectivity in this system. Indeed, the enantioselectivity of the catalytic system is increased from 92% to 98%.

Figure 4. Enantioselective Reduction to Prepare MK-0499.

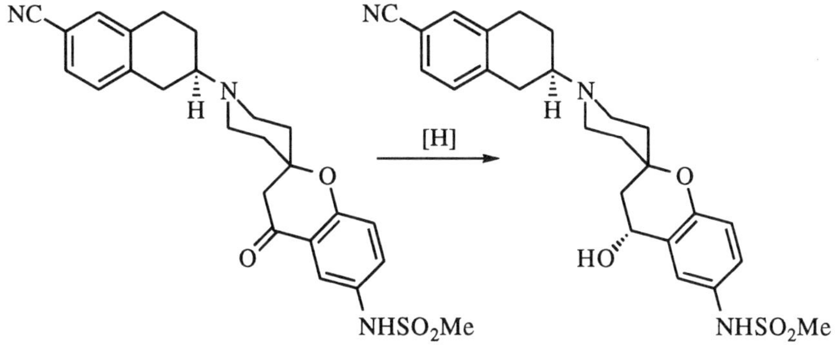

In a typical procedure (*31*), a solution of the ketone in dichloromethane is treated with isopropanol (1 equiv) and borane-methyl sulfide (2.5 equiv) for 0.5 h at -20 °C. At this point the amine-borane complex has formed, however, not all of the isopropanol has reacted to form isopropoxyborane (not observed by NMR) and diisopropoxyborane (observed by NMR). The oxazaborolidine-borane complex (0.1 equiv) is added, and the mixture slowly warmed to 15 °C over a 0.75 h period. The mixture is kept at this temperature until the reaction is complete (ca. 0.5 h) then quenched by the addition of methanol. The mixture is heated to 65 °C with distillation to completely break the amine-borane complex. The isolated yield is >90% with an enantiomeric purity of >98%. This reaction has been scaled up from the laboratory to the pilot plant for the preparation of 10+ kg of MK-0499.

## Chiral Ketone Reduction With Ipc$_2$BCl.

We were also interested in stereospecifically reducing iodoketo ester **2** to the corresponding hydroxy compound **3** (Figure 5). This intermediate was further elaborated to give MK-0287 which is a platelet-activating factor receptor antagonist and can be potentially used for treating asthma, inflammation, acute allergy, ischemia, and toxic shock (*46*). The (-) enantiomer of MK-0287 (the *S,S*-enantiomer) is more desirable than the (+) enantiomer since it is 20-fold more potent. Brown's *B*-chlorodiisopinocampheylborane (Ipc$_2$BCl) reagent was initially chosen because of its ease of preparation from commercially available starting materials and the overall ease of operation (*47*). A modification of Brown's procedure for the generation of the Ipc$_2$BCl was followed initially whereby diisopinocampheylborane (Ipc$_2$BH) was first prepared by the reaction of 98% ee (+)-α-pinene with BH$_3$-Me$_2$S in THF at 0 °C for 17-18 h and, without isolation, the Ipc$_2$BH intermediate was reacted with a THF solution of HCl at the same temperature to generate the Ipc$_2$BCl reducing agent along with a mole of H$_2$. Reduction of iodoketo ester **2** by the *in-situ* generated Ipc$_2$BCl at 0 °C for 24-48 h gave 89% yield of the chiral alcohol in 90% ee (*48*). The higher than optimal reaction temperature (better at -25 °C) was necessary due to the slow reduction rate for this substrate. Even at 0 °C the reduction required 24-48 h. Fortunately, after the saponification and lactonization steps, the ee of the lactone was readily upgraded to >99.5% during isolation of the crystalline lactone product **5**.

Figure 5. MK-0287 Synthetic Route.

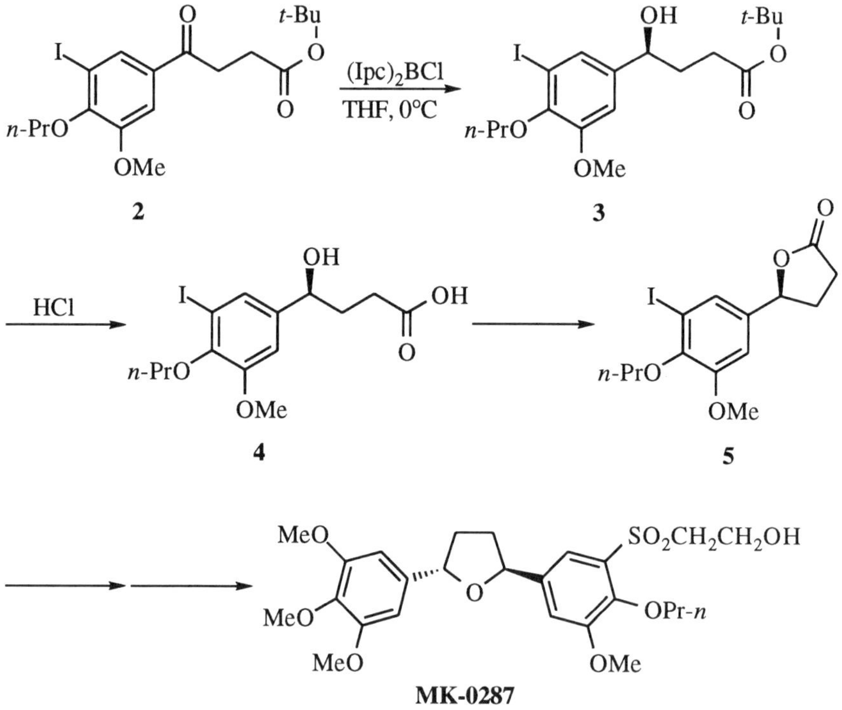

Although the procedure provides product of high ee and regenerating and recovering the α-pinene is possible, the cost of 98% ee (+)-α-pinene coupled with the absence of a commercial source for this grade of α-pinene are major prohibitive factors for using this procedure. Lower ee (-)- or (+)-α-pinene in the range of 72-85% is more readily available on a commercial scale and inexpensive. The generation of 100% ee Ipc$_2$BH from 91% ee α-pinene has been described in the literature and subsequent reaction of the 100% ee Ipc$_2$BH with HCl should provide 100% ee Ipc$_2$BCl but there is a major drawback with the procedure (*48*). The procedure requires aging the Ipc$_2$BH slurry in a slight excess of α-pinene without agitation at 0 °C for three days in order to upgrade the Ipc$_2$BH ee to 100% along with a filtration of the precipitated Ipc$_2$BH from the excess pinene. This time consuming procedure is not desirable for commercial scale operation.

To our surprise, using 91% ee (+)-α-pinene in the modified procedure above, even without a filtration to remove the excess pinene, gave alcohol **2** having essentially the same ee (90% vs. 89%) (*49*). In this case, the possibility of upgrading the Ipc$_2$BH enantiomeric purity during its formation to >91% cannot be discounted but subsequent studies with ClBH$_2$-Me$_2$S, which are discussed below, showed that the asymmetric amplification results observed are not simply due to upgrading of the Ipc$_2$BH enantiomeric purity during the hydroboration phase of the reaction.

Another chiral *secondary* hydroxy alcohol intermediate we needed for the synthesis of a $LTD_4$ antagonist drug candidate is the chlorohydroxy ester **8**. The intermediate chiral alcohol is ultimately converted to MK-0476 which is used for the treatment of asthma and other associated diseases [Figure 6] (*50*).

Figure 6. MK-0476 Synthetic Route.

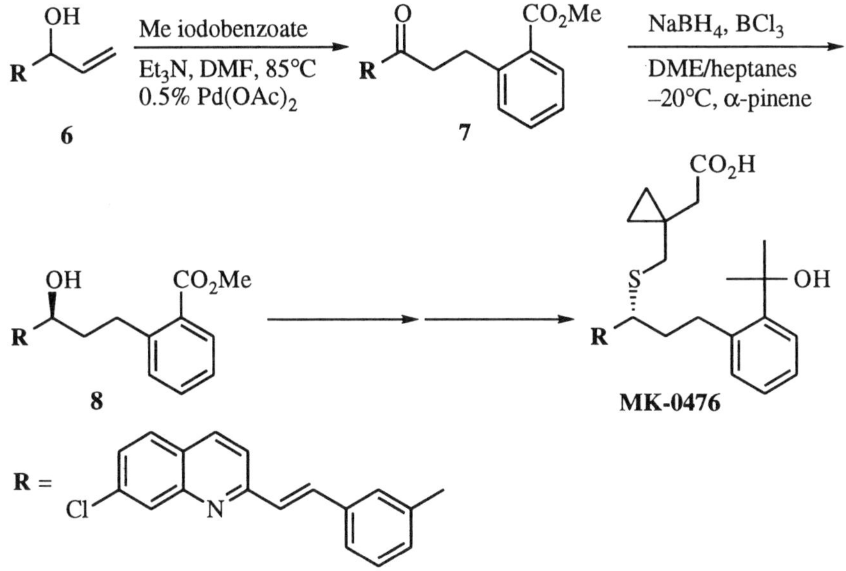

Although the 2-step procedure described above for the preparation of $Ipc_2BCl$ has been demonstrated in multi-kilogram scale reactions, the preparation, the HCl titer determination immediately prior to charging, and accurately charging the required amount of the unstable THF solution of HCl are time consuming. In order to bypass these negative factors associated with the use of HCl, the procedure involving $ClBH_2$-$Me_2S$ and α-pinene for the direct preparation of the $Ipc_2BCl$ reagent was examined (*51*). $ClBH_2$-$Me_2S$ is commercially available or can also be prepared readily by mixing $BH_3$-$Me_2S$ and $BCl_3$-$Me_2S$ in a 2:1 ratio (*52*). The $ClBH_2$-$Me_2S$ is actually a mixture approximately consisting of 13% each of $BH_3$-$Me_2S$ and $Cl_2BH$-$Me_2S$ and 74% of $ClBH_2$-$Me_2S$ as determined by $^{11}B$ NMR studies. Hydroboration of α-pinene by $ClBH_2$-$Me_2S$ in hexane at 0 °C to rt produces a mixture of isopinocampheylborane derivatives. Only when the mixture is heated to 40-50 °C for 30 min before further hydride/chloride redistribution and hydroboration of α-pinene to give exclusively $Ipc_2BCl$.

The faster reaction of chloroketo ester **7** with $Ipc_2BCl$ allows the use of a lower temperature (-25 °C) and resulted in a higher chiral induction than with iodoketo ester **2**. With 97% ee (+)-α-pinene, the reduction of chloroketo ester **10** gave 97.5% ee chlorohydroxy ester **8**. Carrying out the reaction at 0 °C dropped the hydroxy ester **8** ee from 97.5% to 93%. Further lowering the temperature to -35 °C did not give better chiral induction but slightly lowered the reaction rate. For this substrate, even with 70% ee (+)-α-pinene the chiral induction observed for the chlorohydroxy ester **8** was 94-95%.

The above method could have been the final process employed if not for the expense associated with borane and the need to handle Me$_2$S which is a stench compound. The least expensive hydrido boron compound available commercially is NaBH$_4$ and it would be ideal if Ipc$_2$BCl could be prepared directly from it. Although the formation of ClBH$_2$-Et$_2$O from LiBH$_4$ and BCl$_3$ in diethyl ether and its subsequent hydroboration of olefins to give R$_2$BCl•Et$_2$O has been described (53), our attempts to substitute the much more expensive LiBH$_4$ with NaBH$_4$ in the same solvent failed to provide Ipc$_2$BCl of acceptable quality. On the other hand, the reaction of NaBH$_4$ with BF$_3$ in THF in the presence of α-pinene for the preparation of Ipc$_2$BH has been reported by Brown (54), and by not using HCl to generate the Ipc$_2$BCl but instead using the requisite amount of BCl$_3$, we were able to make use of all of the hydride in NaBH$_4$ for the preparation of Ipc$_2$BCl. It was further found that BF$_3$ is not necessary for the reaction and Ipc$_2$BCl could be generated by using BCl$_3$ alone but satisfactory results were observed only in monoglyme (DME) and diglyme and not with THF (unpublished data from Zhao, M.; King, A. O.; Larsen, R. D.; Verhoeven, T. R.; Reider, P. J. of Merck Research Laboratories). Faster reaction of BCl$_3$ with THF may have caused the excessive decomposition of BCl$_3$ before it is converted to less Lewis acidic boron species. The solvent DME was ultimately chosen for the process mainly because of its lower boiling point which is important for solvent and also α-pinene recovery. Thus the final process for generating Ipc$_2$BCl entails the addition of a solution of BCl$_3$ in hexane to a slurry of NaBH$_4$ and 84-85% ee (+)-α-pinene in DME at <0 °C followed by heating the slurry to 40 °C for 1 h. A THF solution of the chloroketo ester **7** is then added at <-25 °C and the mixture is aged at -25 °C until reaction completion (8-10 h) (55), and 96% ee of chlorohydroxy ester **8** is produced. The product is isolated from a mixture of hexane/THF/DME in 92% yield and >99.5% optical purity.

The chiral induction in the reduction of chloroketo ester **7** with Ipc$_2$BCl prepared from (+)-α-pinene of varying stereochemical enrichment and NaBH$_4$ were examined [Table I]. Higher asymmetric amplification was observed with lower ee (+)-α-pinene. At no time did the chiral induction drop below the enantiomeric purity of the (+)-α-pinene [Graph I]. The asymmetric amplification also progressively increased with lower ee α-pinene until nearly racemic α-pinene was used.

Table I. Chiral Reduction of Chloroketo Ester 7 with Ipc$_2$BCl Prepared from (+)-α-Pinene of Varying Enantiomeric Purity

| % ee of (+)-α-pinene | % ee of 8 | % Asymmetric Amplification |
|---|---|---|
| 97 | 97.5 | 0.5 |
| 91 | 95.9 | 5.4 |
| 85 | 96.2 | 13.2 |
| 81 | 95.6 | 18.0 |
| 70 | 89.3 | 27.6 |
| 60 | 85.6 | 42.7 |
| 50 | 79.8 | 59.6 |
| 40 | 71.8 | 79.5 |
| 30 | 59.7 | 99.0 |
| 20 | 43.7 | 118.5 |
| 10 | 23.1 | 131.0 |
| 0 | 0.0 | 0.0 |

The asymmetric amplification results observed with Ipc$_2$BCl can be partly understood using the following analysis. In the hydroboration reaction involving 'ClBH$_2$' and <100% α-pinene, a mixture of diastereomeric Ipc$_2$BCl which is composed of (+)(-)-Ipc$_2$BCl, (+)(+)-Ipc$_2$BCl and (-)(-)-Ipc$_2$BCl is generated. When 1 equiv of Ipc$_2$BCl generated from racemic α-pinene was used, 47% conversion of **7** to

**8** was observed in 3 h. The reaction became quite sluggish at this point, only reaching 52% conversion after 6 h. The meso (+)(-)-Ipc$_2$BCl appeared to be much less reactive towards the reduction of the chloroketo ester **7** when compared to the other two Ipc$_2$BCl diasteromers. Assuming that the hydroboration reaction with α-pinene is random, the asymmetric amplification that one can obtain with α-pinene samples of varying enantiomeric purity can be calculated and the results are presented in Table II.

**Graph I. Chlorohydroxy Ester 8 Asymmetric Amplification % vs. Pinene ee %.**

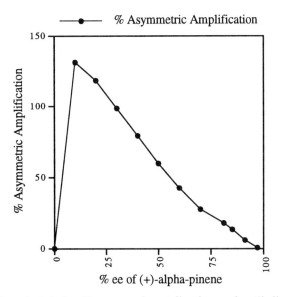

Comparing the calculated ee % vs. experimentally observed ee % did not show a good correlation with the asymmetric amplification theory (Table II). Other as yet unknown factors must also be involved. Non-random hydroboration resulting in the slight preference for the formation of (+)(-)-Ipc$_2$BCl is a possibility.

A similar phenomenon has also been subsequently described in the literature in which $R,R$-isomer **9** was found to react more rapidly with PhCHO than the corresponding $S,R$-isomer **10**. The optical purity of the boronic ester **11** was therefore enhanced from 93% ee to >99% ee [equation 4] (55).

**Isolation.** Problems encountered during the isolation of chiral hydroxy products after the Ipc$_2$BCl reduction have been addressed in several publications (33, 47). In order to recover most of the α-pinene used in a process, the regeneration of α-pinene from unreacted Ipc$_2$BCl is necessary. Although (+)(-)-Ipc$_2$BCl reacts very sluggishly with ketones, it reacts rapidly with aldehydes (for example acetaldehyde and benzaldehyde) at rt to 40 °C to regenerate two moles of α-pinene per mole of (+)(-)-Ipc$_2$BCl with the concomitant formation of the alcohol product from the aldehyde. Likewise, IpcBCl(OCHR$^1$R$^2$) also reacts with an aldehyde to give one mole of α-pinene. Subsequent product isolation largely depends on the physical properties of the products. Liquid products that are both insoluble in water and have similar boiling points as α-pinene or the solvents, or non distillable, oily alcohols due to high boiling points or thermal instability, posed the most difficult scenario for the isolation of clean products on a large scale. One way to overcome this difficulty is to transform these products into a water soluble form. For iodohydroxy ester **3**, which is a thermally unstable, oily compound, saponification to the iodohydroxy acid **4** under acidic conditions followed by basification enabled the extraction of the carboxylate salt into the aqueous phase. The α-pinene remains in the organic layer and is separated and recovered by fractional distillation of the organic solution.

**Table II. Comparison of Calculated Chiral Induction vs. Observed Chiral Induction**

| Pinene ee% | Chiral Induction % | |
|---|---|---|
| | Calculated | Observed |
| 97 | 99.9 | 97.5 |
| 91 | 99.6 | 95.9 |
| 85 | 98.7 | 96.2 |
| 81 | 97.8 | 95.6 |
| 70 | 94.0 | 89.3 |
| 60 | 88.2 | 85.6 |
| 50 | 80.0 | 79.8 |
| 40 | 69.0 | 71.8 |
| 30 | 55.0 | 59.7 |
| 20 | 38.5 | 43.7 |
| 10 | 19.8 | 23.1 |
| 0 | 0.0 | 0.0 |

Some products that are crystalline can be easily crystallized from appropriate solvents and, in most cases, the α-pinene, the appropriately chosen aldehyde, and its corresponding alcohol will be completely soluble in the solvent and removed with the filtrate. The chlorohydroxy ester **8** was purified in this fashion using benzaldehyde for regenerating the α-pinene. Although acetaldehyde may be a better choice in other cases, for chlorohydroxy ester **8** the use of benzaldehyde provided better isolated yield of the product. Recovery of α-pinene from the filtrate was again carried out via distillation.

$$\begin{array}{c} \text{OCH}_2\text{Ph} \\ \text{B–Ipc} \\ \textbf{9}\ 96.5\% \end{array} + \begin{array}{c} \text{OCH}_2\text{Ph} \\ \text{B–Ipc} \\ \textbf{10}\ 3.5\% \end{array} \xrightarrow{\text{PhCHO}} \begin{array}{c} \text{OCH}_2\text{Ph} \\ \text{B–OCH}_2\text{Ph} \\ \textbf{11} >99\%\ \text{ee} \end{array} \text{Product} + \begin{array}{c} \text{OCH}_2\text{Ph} \\ \text{B–Ipc} \\ 71\%\ \text{ee} \end{array} \text{Unreacted} \quad \text{(equation 4)}$$

**Comparison of Brown's & Corey's Procedure.** Even though the reduction of chloroketo ester **9** with Ipc$_2$BCl was chosen initially, Itsuno/Corey's oxazaborolidine method was also considered during the early stages of development. Comparison of the two methods using chloroketo ester **7** as the substrate showed that both procedures provide product with high chiral induction. With Corey's method, an impurity **12**, which was derived from the reduction of the ethylene bridge, was formed to an extent of 1-3% when 0.6 equiv of the oxazaborolidine-borane complex was used for the reduction (stoichiometric reaction).

**12**

Performing the reaction with 0.1 equiv of the oxazaborolidine-borane and 0.5 equiv of $BH_3$-$Me_2S$ according to literature procedures (*11-13*) gave up to 15% of the overreduced impurity 12. The overreduction was traced to the presence of residual Pd in the substrate from the previous Pd-catalyzed Heck coupling step [Figure 6]. Carefully reducing the residual Pd to <10 ppm by recrystallization of the chloroketo ester 7 from DMF before submitting to Corey's catalytic chiral reduction procedure reduced the amount of overreduction product 12 to <1%. For comparison, no reduction of the olefin was observed using $Ipc_2BCl$ even with chloroketo ester 7 contaminated with as much as 900 ppm of Pd. With the availability of both enantiomer of α-pinene, the asymmetric amplification observed, the ease of preparation of $Ipc_2BCl$ from inexpensive reagents, the readily recoverable α-pinene, and the tolerance for residual Pd (or other transition metals) in the substrate makes Brown's method more enticing and practical for the large scale reduction of chloroketo ester 6 and other aryl alkyl ketones.

**Acknowledgement.** This work represents the efforts of several of our colleagues in Process Research Department whose names are mentioned in the references.

**References**

(1)  Wallbaum S.; Martens, J. *Tetrahedron Asymmetry* **1992**, *3*, 1475.
(2)  Singh, V. K. *Synthesis* **1992**, 605.
(3)  Deloux, L.; Srebnik M. *Chem. Rev.* **1993**, *93*, 763.
(4)  Itsuno, S.; Hirao, A.; Nakahama, S.; Yamazaki, N. *J. Chem. Soc., Perkin Trans. 1* **1983**, 1673.
(5)  Itsuno, S.; Ito, K.; Hirao, A.; Nakahama, S. *J. Chem. Soc., Chem. Commun.* **1983**, 469.
(6)  Itsuno, S.; Ito, K.; Hirao, A.; Nakahama, S. *J. Org. Chem.* **1984**, *49*, 555.
(7)  Itsuno, S.; Nakano, M.; Miyazaki, K.; Masuda, H.; Ito, K.; Hirao, A.; Nakahama, S. *J. Chem. Soc., Perkin Trans. 1* **1985**, 2039.
(8)  Itsuno, S.; Nakano, M.; Ito, K.; Hirao, A.; Owa, M.; Kanda, N.; Nakahama, S. *J. Chem. Soc., Perkin Trans. 1* **1985**, 2615.
(9)  Itsuno, S.; Ito, K.; Maruyama, T.; Kanda, N.; Hirao, A.; Nakahama, S. *Bull. Chem. Soc. Jpn.* **1986**, *59*, 3329.
(10) Itsuno, S.; Sakurai, Y.; Ito, K.; Hirao, A.; Nakahama, S. *Bull. Chem. Soc. Jpn.* **1987**, *60*, 395.
(11) Corey, E. J.; Bakshi, R. K.; Shibata, S. *J. Am. Chem. Soc.* **1987**, *109*, 5551.
(12) Corey, E. J.; Bakshi, R. K.; Shibata, S.; Chen, C. P.; Singh, V. K. *J. Am. Chem. Soc.* **1987**, *109*, 7925.
(13) Corey, E. J.; Helal, C. J. *Tetrahedron Lett.* **1995**, *36*, 9153; and references contained therein.
(14) Cho, B. T.; Chun, Y. S. *Tetrahedron Asymmetry* **1992**, *3*, 1583.

(15) Cho, B. T.; Ryu, M. H.; Chun, Y. S.; Dauelsberg, C.; Wallbaum, S.; Martens, J. *Bull. Korean Chem. Soc.* **1994**, *15*, 53.
(16) Ling, I.; Podanyi, B.; Hamori, T.; Solyom, S. *J. Chem. Soc., Perkin Trans. 1* **1995**, 1423.
(17) Shimizu, M.; Kamei, M.; Fujisawa, T. *Tetrahedron Lett.* **1995**, *36* 8607.
(18) Sakai, T.; Yan, F.; Kashino, S.; Uneyama, K. *Tetrahedron*, **1996**, *52*, 233.
(19) Cho, B. T.; Ryu, M. H. *Bull. Korean Chem. Soc.* **1994**, *15*, 191.
(20) Tillyer, R. D.; Boudreau, C.; Tschaen, D.; Dolling, U.-H.; Reider, P. J. *Tetrahedron Lett.* **1995**, *36*, 4337.
(21) Gajda, T. *Tetrahedron: Asymmetry* **1994**, *5*, 1965.
(22) Meier, C.; Laux, W. H. G. *Tetrahedron: Asymmetry* **1995**, *6*, 1089.
(23) Meier, C.; Laux, W. H. G.; Bats, J. W. *Liebigs. Ann.* **1995**, 1963.
(24) Romagnoli, R.; Roos, E. C.; Hiemstra, H.; Moolenaar, M. J.; Speckamp, W. N.; Kaptein, B.; Schoemaker, H. E. *Tetrahedron Lett.* **1994**, *35*, 1087.
(25) Bringmann, G.; Hartung, T. *Angew. Chem., Int. Ed. Engl.*, **1992**, *31*, 761.
(26) Bringmann, G.; Vitt, D. *J. Org. Chem.* **1995**, *60*, 7674.
(27) Jones, T. K.; Mohan, J. J.; Xavier, L. C.; Blacklock, T. J.; Mathre, D. J.; Sohar, P.; Jones, E. T. T.; Reamer, R. A.; Roberts, F. E.; Grabowski, E. J. J. *J. Org. Chem.* 1991, *56*, 763.
(28) Shinkai, I. *J. Heterocycl. Chem.* **1992**, *29*, 627.
(29) Claremon, D. A.; Baldwin, J. J.; Elliott, J. P.; Ponticello, G. S.; Selnick, H. G.; Lynch, J. J; Sanguinetti, M. C. Perspect. Med. Chem. 383-404. Edited by: Testa, B. Verlag Helvetica Chim. Acta: Basel, Switz. (Eng) **1993**.
(30) Cai, D.; Tschaen, D.; Shi, Y.-J.; Verhoeven, T. R.; Reamer, R. A.; Douglas, A. W. *Tetrahedron Lett.* **1993**, *34*, 3243.
(31) Shi, Y. J.; Cai, D.; Dolling, U.-H.; Douglas, A. W.; Tschaen, D. M.; Verhoeven, T. R. *Tetrahedron Lett.* **1994**, *35*, 6409.
(32) Tschaen, D. M.; Abramson, L.; Cai, D.; Desmond, R.; Dolling, U.-H.; Frey, L.; Karady, S.; Shi, Y.-J.; Verhoeven, T. R. *J. Org. Chem.* **1995**, *60*, 4324.
(33) Mathre, D. J.; Jones, T. K.; Xavier, L. C.; Blacklock, T. J.; Reamer, R. A.; Mohan, J. J.; Jones, E. T. T.; Hoogsteen, K.; Baum, M. W.; Grabowski, E. J. J. *J. Org. Chem.* 1991, *56*, 751.
(34) Kerrick, S. T.; Beak, P. *J. Am. Chem. Soc.* **1991**, *113*, 9708.
(35) Gallagher, D. J.; Beak, P. *J. Org. Chem*, **1995**, *60*, 7092.
(36) Gallagher, D. J.; Wu, S.; Nikolic, N. A.; Beak, P. *J. Org. Chem*, **1995**, *60*, 8148.
(37) Blacklock, T. J.; Jones, T. K.; Mathre, D. J.; Xavier, L. C. U. S. Patent 5,039,802 (**1991**).
(38) Blacklock, T. J.; Jones, T. K.; Mathre, D. J.; Xavier, L. C. U. S. Patent 5,189,177 (**1993**).
(39) Mathre, D. J.; Thompson, A. S.; Douglas, A. W.; Hoogsteen, K.; Carroll, J. D.; Corley, E. G.; Grabowski, E. J. J. *J. Org. Chem.* **1993**, *58*, 2880.
(40) Corey, E. J.; Azimioara, M.; Sarshar, S. *Tetrahedron Lett.* **1992**, *33*, 3429.
(41) Douglas, A. W.; Tschaen, D. M.; Reamer, R. A.; Shi, Y.-J. *Tetrahedron Asymmetry* **1996**, in press.
(42) Evans, D. A. *Science* **1988**, *240*, 420.
(43) Denmark, S. E.; Schnute, M. E.; Marcin, L. R.; Thorarensen, A. *J. Org. Chem.* **1995**, *60*, 3205.
(44) Stone, G. B. *Tetrahedron: Asymmetry* **1994**, *5*, 465.
(45) Denmark, S. E. et. al. *Organic Syntheses*, in press.
(46) Hanahan, D. J. *Ann. Rev. Biochem.* **1986**, *55*, 483. Braquet, P.; Touqui, L.; Shen, T. Y.; Varagftig, B. B. *Pharmacol. Rev.* **1987**, *39*, 98. Braquet, P.; Godfroid, J. J. *Trends. Pharmacol. Sci.* **1986**, *7*, 397.
(47) Brown, H. C.; Chandrasekharan, J.; Ramachandran, P. V. *J. Org. Chem.*

**1985**, *50*, 5446. Brown, H. C.; Park, W. S.; Cho, B. T.; Ramachandran, P. V. *J. Org. Chem.* **1987**, *52*, 5406. Brown, H. C.; Ramachandran, P. V. *Acc. Chem. Res.* **1992**, *25*, 16.

(48) Simpson, P.; Tschaen, D.; Verhoeven, T. R. *Syn. Comm.* **1991**, *21*, 1705. Thompson, A. S.; Tschaen, D. M.; Simpson, P.; McSwine, D. J.; Reamer, R. A.; Verhoeven, T. R., Shinkai, I. *J. Org. Chem.* **1992**, *57*, 7044.

(49) Brown, H. C.; Joshi, J. J. *J. Org. Chem.* **1988**, *53*, 4059. Brown, H. C.; Dhokte, U. P. *J. Org. Chem.* **1994**, *59*, 2365.

(50) Gauthier, J. Y.; Jones, T.; Champion, E.; Charette, L.; Dehaven, R.; Ford-Hutchinson, A. W.; Hoogsteen, K.; Lord, A.; Masson, P.; Piechuta, H.; Pong, S. S.; Springer, J. P.; Therien, M.; Zamboni, R.; Young, R. N. *J. Med. Chem.* **1990**, *33*, 2841. Zamboni, R.; Belley, M.; Champion, E.; Charette, L.; Dehaven, R.; Frenette, R.; Gauthier, J. Y.; Jones, T. R.; Leger, S.; Masson, P.; McFarlane, C. S.; Metters, K.; Pong, S. S.; Piechuta, J.; Rockach, J.; Therien, M.; Williams, H. W. R.; Young, R. N. *J. Med. Chem.* **1992**, *35*, 3832.

(51) Brown, H. C.; Ravindran, N. *J. Am. Chem. Soc.* **1976**, *98*, 1798. Brown, H. C.; Ravindran, N. *J. Org. Chem.* **1977**, *42*, 2533. Brown, H. C.; Ravindran, N. *Synthesis* **1977**, 695. Brown, H. C.; Ravindran, N.; Kulkarni, S.U. *J. Org. Chem.* **1979**, *44*, 2417. Bir, G.; Kaufmann, D. *Tetrahedron. Lett.* **1987**, *28*, 777. King, A. O.; Corley, E. G.; Anderson, R. K.; Larsen, R. D.; Verhoeven, T. R.; Reider, P. J.; Xiang, Y. B.; Belley, M.; Leblanc, Y.; Labelle, M.; Prasit, P.; Zamboni, R. J. *J. Org. Chem.* **1993**, *58*, 3731.

(52) Brown, H. C.; Ravindran, N. *Inorg Chem.* **1977**, *16*, 2938.

(53) Brown, H. C.; Tierney, P. A. *J. Am. Chem. Soc.* **1958**, *80*, 1552. Brown, H. C.; Ravindran, N. *J. Am. Chem. Soc.* **1972**, *94*, 2112.

(54) Brown, H. C.; Ayyangar, N. R.; Zweifel, G. *J. Am. Chem. Soc.* **1964**, *86*, 393.

(55) Joshi, N. N.; Pyun, C.; Mahindroo, V. K.; Singaram, B.; Brown, H. C. *J. Org. Chem.* **1992**, *57*, 504.

## Chapter 7

# Diphenyloxazaborolidine for Enantioselective Reduction of Ketones

George J. Quallich, James F. Blake, and Teresa M. Woodall

Process Research and Development, Central Research Division,
Pfizer, Inc., Groton, CT 06340

**Abstract.** A systematic investigation of oxazaborolidine structure and enantioselectivity relationships resulted in the identification of diphenyloxazaborolidine as an efficient catalyst for enantioselective reduction of prochiral ketones. The primary requirement for a good oxazaborolidine catalyst to obtain high enantiomeric excess in prochiral ketone reductions was to completely block one face of the oxazaborolidine. Alkyl substitution of nitrogen in erythro oxazaborolidines provided lower enantiomeric excess due to retarded rate of borane coordination and competition from the *endo* transition state. Relative enantioselectivities were correctly predicted by *ab initio* calculations, and the calculated structures, energetics, and the existence of key intermediates were demonstrated to be sensitive to the level of theory applied. *Ab initio* calculations and utility of diphenyloxazaborolidine are described in detail.

**Introduction.** Our interest in enantioselective ketone reduction resulted from the need to prepare a variety of drug candidates, wherein a specific alcohol enantiomer had been selected for development. A number of good methods existed for asymmetric reduction of prochiral ketones with stoichiometric reagents that provided the product alcohol in high enantiomeric excess (ee) (1). However, limitations to the use of stoichiometric reagents for ketone reductions are availability, cost, ease of product purification, and chiral auxilliary recovery on large scale (2). Corey's oxazaborolidines, **2**, appeared particularly appropriate because of their catalytic nature and the high ee afforded of predictable absolute stereochemistry (3). In addition, a catalytic reduction was envisioned to be more environmentally benign and less expensive than use of a stoichiometric reagent. Continued use of oxazaborolidines derived from unnatural D-aminoacids prompted a systematic investigation of their structure and enantioselectivity relationships which resulted in the discovery of diphenyloxazaborolidine.

**Applications of Oxazaborolidine 2.** One of the earliest examples requiring a specific alcohol enantiomer was oxazole **6**, equation 1. At that time, oxazaborolidine catalyzed reduction was limited to compounds containing carbon, hydrogen, oxygen, halogen, and sulfur (4). No published examples of high enantiomeric excess were reported when competing nitrogen coordination sites existed in the substrate (5). In addition, researchers at Merck found that small quantities of diphenylprolinol, methylboronic acid, triphenylboroxine, and benzophenone decreased enantioselectivity in the ketone reduction (6). We were concerned that nitrogen, contained in the oxazole **4**, would coordinate borane resulting in non-oxazaborolidine catalyzed reduction, leading to lower enantioselectivity. Initial attempts to reduce oxazole **4** under standard conditions, 5 mole% catalyst **5** and 0.7 equivalents of borane, reproducibly led to an 85:15 mixture of recovered **4** and the alcohol **6**. We hypothesized that the oxazole nitrogen was coordinating borane competitively with the catalyst and, once coordinated, the amine-borane was inert under the reaction conditions. To test this possibility, a tetrahydrofuran solution of oxazole **4** and catalyst **5** was titrated with borane. Complete reaction was obtained with 1.7 eq of borane, and the desired (S) alcohol **6** was obtained in 94% ee with the CBS catalyst **5**. Additional examples of prochiral ketones containing nitrogen capable of coordinating borane were reduced catalytically with oxazaborolidines to afford the corresponding alcohols with high ee's (7). In these examples, when nitrogen coordinated borane competitively with the oxazaborolidine, high ee resulted because the catalyzed path was faster than reduction by the amine-borane, and free borane concentrations are minimized by this coordination.

The success of oxazaborolidine reduction of ketones containing a nitrogen heteroatom was short-lived; nearly every compound entering development required the unnatural D-proline derived oxazaborolidine **5** to secure the desired absolute stereochemistry as exemplified in equation 1. In addition, CBS catalysts were not limited to alcohol enantiomer targets, as the initial oxazaborolidine reduction products were displaced with nitrogen (8,9) carbon (10), and sulfur nucleophiles (11), as depicted in Scheme 1. Because of the overall value of this technology and the concerns of catalyst precursor availability, an investigation into the oxazaborolidine structure and enantioselectivity relationships was undertaken.

**Scheme 1**

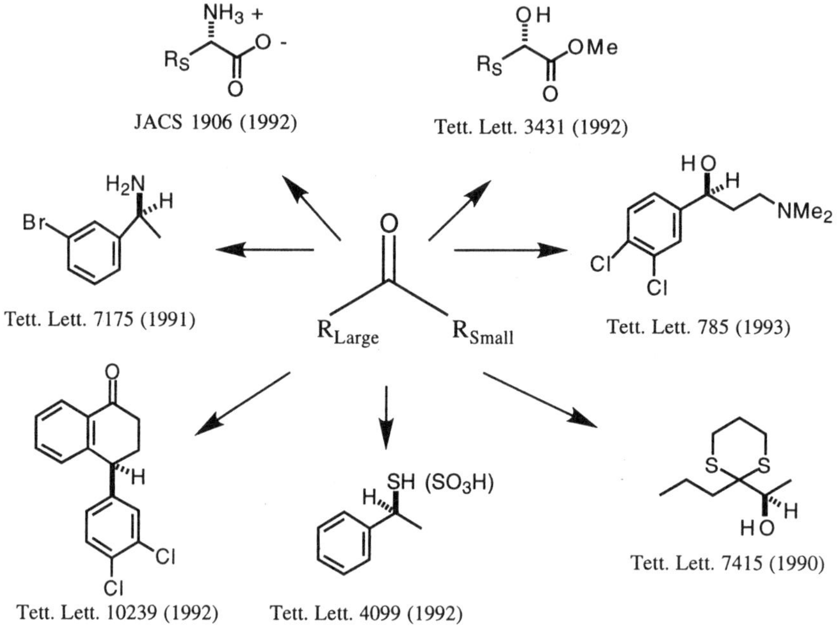

JACS 1906 (1992)

Tett. Lett. 3431 (1992)

Tett. Lett. 7175 (1991)

Tett. Lett. 785 (1993)

Tett. Lett. 10239 (1992)

Tett. Lett. 4099 (1992)

Tett. Lett. 7415 (1990)

**Oxazaborolidine Prior Art.** Itsuno and coworkers pioneering investigations of 1,2-aminoalcohols in combination with borane demonstrated that high levels of enantioselection could be achieved in ketone reduction (12). Geminal diphenyl substitution proved optimum with their studies culminating with diphenyl valinol as the catalyst precursor for **1**. Corey and coworkers unambiguously determined the structure of Itsuno's oxazaborolidine **1**, and proposed a mechanism to account for the observed enantioselectivity (3a). They proceeded to develop oxazaborolidines derived from diphenylprolinol, CBS catalysts **2 / 5**, and detailed their numerous applications including preparation of α-aminoacids, α-hydroxy acids, deuterated primary alcohols, and a variety of natural products (13). Jones *et al.* studied the effect of electron density on boron using substituted phenyl derivatives, catalyst **3**, and concluded that there was no significant electronic effect (6). Thus, geminal diphenyl substitution evolved from Itsuno's survey of amino alcohol derivatives secured from amino acids. Investigations into alternative geminal substitutions of **2**, *i.e.*, isopropyl, α-naphthyl, β-naphthyl, *t*-butyl, or substituted phenyl resulted in either slight improvement or a decrease in ee (13c,14). Numerous catalysts had been developed, but none were superior to the CBS catalyst, **2** (15). In an attempt to discover alternative catalysts not dependent on the use of D-aminoacids and capable of providing enantiomeric excess comparable to the CBS catalyst, a systematic study to provide an understanding of the relationship between oxazaborolidines structure and enantioselectivity was initiated.

**Synthetic Evaluation of Oxazaborolidine Structure and Enantioselectivity Relationships.** Our approach was to evaluate various substitutions on the oxazaborolidine ring, and determine the resulting enantiomeric excess obtained with the prototypical ketone, α-tetralone. The reaction conditions for screening were THF at 25 °C using 5 mole% catalyst and borane methyl sufide complex as hydride reagent. Five amino alcohols were converted into their corresponding oxazaborolidines **7-11**, and reduction of α-tetralone performed, equation 2. Three intriguing points from the preliminary study, Table I, emerged: 1) **7** and **11** provide modest ee, 2) in contrast to oxazaborolidines **2** and **3**, ee fell precipitously when comparing B-phenyl substituted **8** to B-methyl substituted **7**, and 3) stoichiometric use of **7** provided a higher ee than 5 mole%, suggesting possible leakage through a non-catalyzed pathway. The hypothesis from the preliminary experiments was that *the primary requirement of a good oxazaborolidine catalyst to obtain high enantiomeric excess in prochiral ketone reductions was to completely block one face of the oxazaborolidine*. This was based on the proposed mechanism of the reduction which entailed a two point binding on the oxazaborolidine of borane and the carbonyl oxygen atom (3). Blocking one face of the oxazaborolidine in conjunction with two point binding was envisaged to afford a highly enantioselective reduction. (1*S*, 2*R*)-(+)-Erythro 2-amino-1,2-diphenylethanol was chosen to challenge the hypothesis. The erythro 2-amino-1,2-diphenylethanol enantiomers were investigated four decades earlier as potential analgesic agents (16), and were inexpensive (17). All boron substitutions in **12-15** generated high ee, as shown in Table I (18).

A broad survey of ketones reduced with 5 mole% of the oxazaborolidine **13** and its enantiomer **24** is depicted in Table II. High enantiomeric excess was secured when the groups flanking the ketone were of significantly different size (*e.g.*, compare entry A and E). A variety of heteroatom-containing ketones were compatible with the reduction conditions, entries C, D, G, N, P. 1,2-Aminoalcohols could be prepared with good enantiomeric excess employing 1.7 equivalents of borane as required with any substrate containing a basic nitrogen, entry M (7). By contrast, the low ee obtained with dimethylaminoacetophenone, entry I, was a result of two groups of comparable size flanking the ketone. Catalyst **24** provided high ee in the reduction of 1,3-aminoalcohols, entry P, and offers an alternative to the use of Chirald with Mannich bases (19). The erythro diphenyloxazaborolidines were compatible with oxygen, sulfur, halogen, nitriles, amines, and other nitrogen-containing compounds. Of particular note is the higher ee obtained with catalyst **24** relative to the CBS catalyst in the reduction of acetylpyridines (entries N, O). Other than the acetylpyridine case, the diphenyloxazaborolidines and the CBS catalysts were quite comparable for these examples.

As mentioned earlier, these catalysts operate via a two point binding of borane and the carbonyl oxygen atom on the oxazaborolidine. The absolute stereochemistry of the alcohol is predicted by the six-membered ring *exo* transition state depicted on the next page. This model illustrates the reduction of pinacolone by **24**. The diphenyl substituents must be orthogonal to the oxazaborolidine ring, otherwise their ortho protons would contact the other aromatic ring. Due to the orthogonal arrangement, the phenyl groups synergistically augment each other's steric requirements, effectively shielding one face of the oxazaborolidine. Predictable absolute stereochemistry of the product alcohols, substantiated by examples in Table II, results from the *exo* transition state illustrated on page 118.

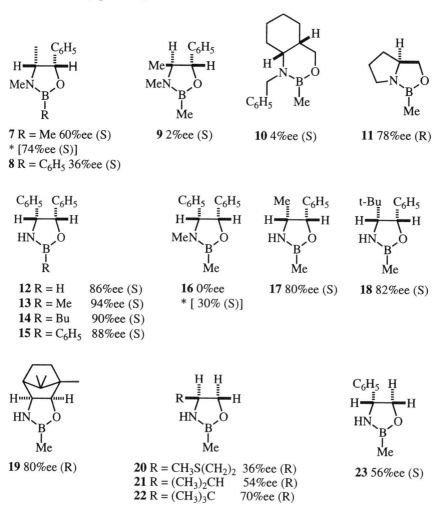

**Table I.** Enantiomeric Excess Obtained with Oxazaborolidine Catalyzed Reduction of α-Tetralone (equation 2)

**7** R = Me  60%ee (S)
* [74%ee (S)]
**8** R = $C_6H_5$  36%ee (S)

**9** 2%ee (S)

**10** 4%ee (S)

**11** 78%ee (R)

**12** R = H     86%ee (S)
**13** R = Me    94%ee (S)
**14** R = Bu    90%ee (S)
**15** R = $C_6H_5$  88%ee (S)

**16** 0%ee
* [ 30% (S)]

**17** 80%ee (S)

**18** 82%ee (S)

**19** 80%ee (R)

**20** R = $CH_3S(CH_2)_2$  36%ee (R)
**21** R = $(CH_3)_2CH$    54%ee (R)
**22** R = $(CH_3)_3C$     70%ee (R)

**23** 56%ee (S)

* Stoichiometric oxazaborolidine use.
Source: Reprinted with permission from ref. 21. Copyright 1994.

**Table II.** Enantiomeric Excess Obtained in the Catalytic Reduction of Prochiral Ketones with Oxazaborolidines **13** and **24**

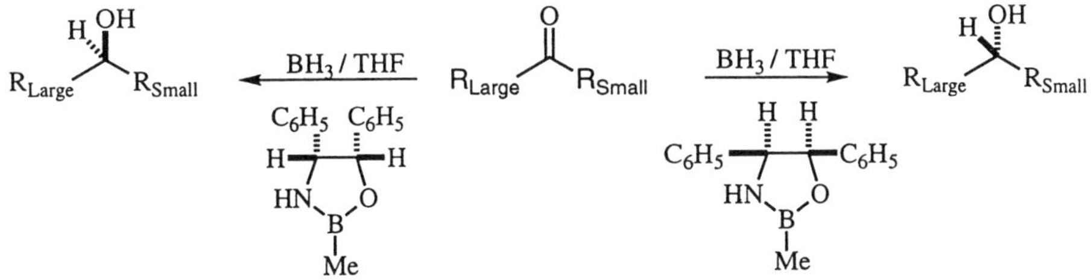

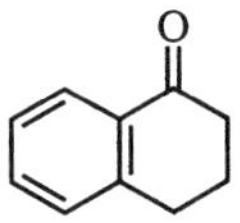

A, 94 (R) **24**

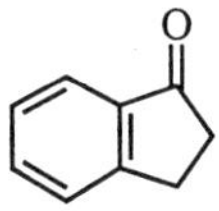

B, 90 (R) **24**

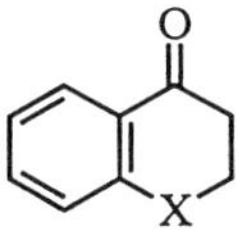

C, X=O, 97 (R) **24**
D, X=S, 90 (S) **13**

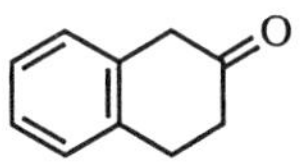

E, 30 (R) **24**

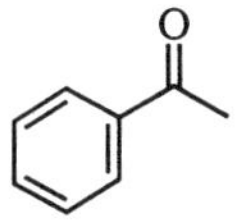

F, 92 (R) **24**

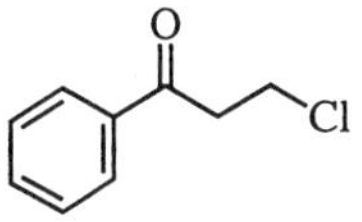

G, 84 (S) **13**

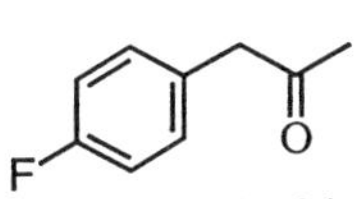

H, 50 (R) **24**

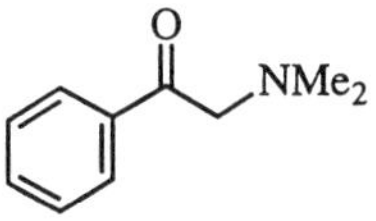

I, 22 (R) **13**

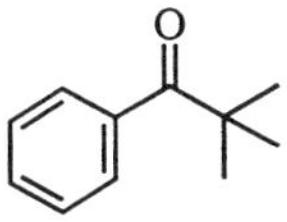

J, 92 (S) **24**

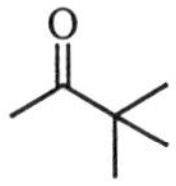

K, 88 (S) **13**

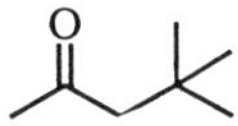

L, 76 (R) **24**

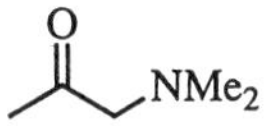

M, 87 (S) **13**

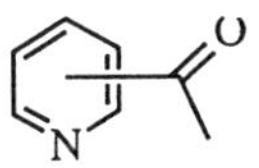

N, 3-acetyl, 80 (R) **24**
O, 4-acetyl, 76 (R) **24**

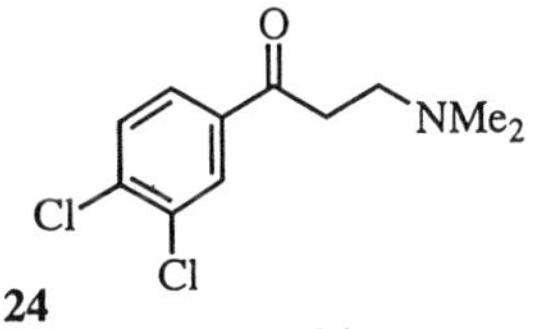

P, 94 (R) **24**

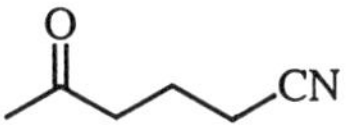

Q, 66 (R) **24**

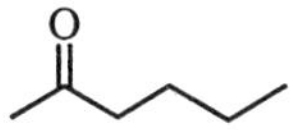

R, 47 (S) **13**

Table code: entry letter, %ee, (absolute configuration), catalyst enantiomer used in **bold**.
Source: Reprinted with permission from ref. 18. Copyright 1993.

[Structural diagrams showing Exo and Endo conformations of oxazaborolidine-borane-ketone complexes with labeled bond distances 1.559, 1.672 (Exo) and 1.546, 1.667 (Endo)]

Alkyl substitution on the nitrogen of erythro substituted oxazaborolidines proved to be detrimental to the enantiomeric excess, Table I. A methyl group on nitrogen in the diphenyl catalyst, **16**, afforded no selectivity (*i.e.*, 0% ee). In this case, 20% of the α-tetralone remained on workup employing standard reduction conditions. If a stoichiometric quantity of **16** was used, a 30% ee of tetralol was obtained. On the basis of this trend, catalyst **17** derived from norephedrine should and did afford a higher ee than the ephedrine oxazaborolidine **7**. Employing a stoichiometric quantity of oxazaborolidine **7** or **16** mitigated non-catalyzed reduction in the ephedrine based system, but not in the erythro diphenyl case. Similarly, the camphor N-H oxazaborolidine **19** provided higher ee when compared to what has been reported for the N-alkyl catalyst **25** in the reduction of acetophenone, equation 3 (20). Thus, nonbonded interactions that develop between the alkyl nitrogen substituent and the proximal carbon substituent (C4) resulted in increased non-catalyzed reduction of the ketone by free borane due to retarded complexation of borane by the oxazaborolidine and/or unselective reduction by the catalyst. Therefore, the erythro substituted oxazaborolidines tolerate a variety of substituents on boron, but not on nitrogen.

[Equation 3: acetophenone reduced with BH$_3$/Me$_2$S/THF, catalysts **19** R = H 88%ee, **25** R = CH$_2$-(N-methylpyrrole) 73%ee, giving chiral 1-phenylethanol] (3)

A series of catalysts derived from amino acids were evaluated to determine whether the face of the oxazaborolidine could be effectively shielded by a single substituent on the carbon atom proximal (C4) to the nitrogen atom that coordinates borane (3a). The trend observed for catalysts **20-23** (Table I) reflects the A values with the *t*-butyl substituent yielding a 70% ee. Finally, the erythro substituted *t*-

butyl/phenyl oxazaborolidine **18** provided good ee, but not better than the diphenyl system **13**. Thus, the primary requirement of a good catalyst was to completely block one face of the oxazaborolidine which was best achieved with the erythro diphenyl oxazaborolidines **12-15** (21).

*Ab Initio* **Calculations.** In tandem with the synthetic approach, *ab initio* calculations were performed to determine the level of theory necessary for reproducing the experimentally determined enantioselectivities and to gain insight into possible reaction mechanisms, including the existence of various proposed intermediates (22). *Ab initio* calculations reported herein were carried out with the Gaussian 92 series of programs (23) with the 3-21G basis set, and full details of the computational methodology are described elsewhere (21). As discussed above, the catalyst was prepared via coordination of borane to an oxazaborolidine. The computed structures for borane coordination to catalysts **13** and **17** displayed significant pyramidalization of the oxazaborolidine nitrogen. Not surprisingly, there was also a large preference for *trans* coordination of borane over *cis*, (with respect to the erythro substituents) consistent with the hypothesis of blocking one face of the oxazaborolidine (11.3 and 7.5 kcal/mol, respectively). The next step in the catalytic mechanism required complexation of the ketone (pinacolone) to the oxazaborolidine boron, *syn* to coordinated borane. In all cases, the ketone complexed in an *anti* fashion (with respect to the *t*-butyl group of pinacolone) to the boron of the oxazaborolidine ring, consistent with prior *ab initio* investigations (24). There was a clear preference for the *endo* mode of complexation for the borane adducts **13** and **17**, though the intrinsic driving force for *endo* complexation was not overwhelming, typically less than 4 kcal/mol (The inversion of *exo* and *endo* complexes was not considered; given the similarities in complexation energies and geometries to other Lewis acid complexes, barriers of *ca.* 10 kcal/mol, or more were anticipated for the process). Characteristics of the tighter *endo* complexes were shorter oxazaborolidine boron to carbonyl oxygen (B---O=C) distances, and smaller B---O=C angles. Complexation of the ketone appeared to have a relatively minor effect on most of the geometrical parameters of the oxazaborolidine-borane adducts. The most striking geometrical change among all the complexes was the pyramidilization of the oxazaborolidine boron (N3-B2-O1-CH3 torsion) for both **13** and **17**, which was distorted from a planar arrangement in the borane adduct to *ca.* 130° in the ketone complexes.

Our initial hypothesis was that the "large" and "small" groups attached to the ketone determined the mode of complexation to the borane adduct, due to nonbonded interactions with the substituent on the oxazaborolidine boron. If steric repulsion between the boron substituent and the "large" group of the ketone were a major determinant in the mode of complexation, one would anticipate the *endo* complexes and transition states to be higher in energy than the *exo* forms. The opposite trend was observed, with the *endo* complexes and transition states predicted to be lower in energy. This was consistent with the relative lack of enantioselective sensitivity displayed by various boron substituents in catalysts **12 - 15**. While the *endo* transition state was shown to be lower in energy than the *exo*, the overall barrier associated with the *exo* pathway is lower, leading to the observed enantioselectivity. Steric repulsion was important in the preference for *anti* complexation of ketones, with respect to the "large" group, to the Lewis acid, which ultimately determined the nature of the product.

Alkyl substitution on the oxazaborolidine nitrogen tended to effect the *exo* transition state more than the corresponding *endo* transition state, the net result was a decrease in the predicted enantioselectivity. This effect was more pronounced for catalysts with larger substituents on C4, which was demonstrated by comparing **16** and **7**. A unique striking conformational change in the *exo* transition state of **16** in comparison to **13** resulted from the unfavorable nonbonded interactions between the N-methyl substituent and the proximal (C4) phenyl group. The envelope of the five-membered oxazaborolidine ring **16** was "flipped" relative to previous conformations, forcing the phenyl substituent into a higher energy conformation in order to accommodate the geometric requirements of the *exo* transition state. Thus, significant loss of enantioselectivity was accounted for by the conformational change that occurred in the *exo* transition state. The unfavorable interactions increased from norephedrine **17** / ephedrine **7** to the erythro diphenyl N-H / N-CH$_3$ (catalysts **13** and **16**) derived oxazaborolidines, such that the catalyzed path was overridden by the non-catalyzed borane reduction and competition from the *endo* transition state in the latter. By contrast, the trisubstituted amine in the CBS catalysts **2** readily formed the borane complex due to pyramidalization of nitrogen imposed by the [3.3.0] ring system. Hence, the geminal diphenyl substitutions in **1** and **2** functioned by placing the phenyl group *cis* to the isopropyl or pyrrolidine ring, which blocked one face of the catalyst. This effect was achieved by the erythro catalysts **12-15** wherein the two phenyl groups, orthogonal to the oxazaborolidine ring, synergistically augment their steric requirements. The second phenyl group in **1** and **2**, *trans* to the isopropyl group or pyrrolidine ring, was either not required, or functioned to provide additional nonbonded interactions with the ketone to enhance the enantioselectivity. The results of the *ab initio* calculations correctly predicted the relative enantioselectivities in the erythro substituted series of oxazaborolidines. Additionally, less than absolute enantioselectivity in the oxazaborolidine catalyzed reduction was attributed to at least two variables: non-catalyzed borane reduction (due to slower rate of borane and ketone coordination to the catalyst as exemplified for **7** versus **17** and **16** versus **13**), and competition from the *endo* transition pathway.

*In Situ* **Oxazaborolidines.** Emphasis now shifted to practical considerations in performing enantioselective reductions with the erythro diphenyl oxazaborolidines **12 - 15**. Oxazaborolidines containing the B-H, B-alkyl, and B-aryl moieties had been prepared over two decades ago for gas chromatography and gas chromatography-mass spectrometry analysis (25). Conversion of 1,2-aminoalcohols to form the corresponding oxazaborolidines was shown to require heating. For instance, ephedrine and pseudoephedrine were heated to 120 °C, otherwise only the more acidic hydroxyl group reacted with borane (26). The Itsuno catalyst **1** (3a) was prepared at 35 °C while the Corey oxazaborolidine **2** (B-H) was formed at elevated pressure and temperature apparently due to ring strain in the [3.3.0] ring system (27). Due to the air and moisture sensitivity of the B-H oxazaborolidines **1** and **2**, methods to generate more chemically stable B-alkyl, and B-aryl catalysts have been reported. This trend toward isolating catalysts of greater stability has evolved to the amine-borane complex of **2** (B-Me) which was reported to be a stable crystalline solid (28). While this tack was considered for the erythro diphenyl oxazaborolidines, large scale use suggested a more efficient solution. In other words, oxaborolidine synthesis and isolation as the amine-borane complex was not as direct as Itsuno's original method. Thus, investigation of borane both to form the oxaborolidine ring *in situ*, and in sufficient excess to reduce prochiral ketones was investigated.

    Combination of commercially available (1*S*, 2*R*)-2-amino-1,2-diphenylethanol with excess borane-methyl sulfide complex in THF at ambient temperature generated the oxazaborolidine **12** in 8-10 hrs, equation 4. One equivalent of hydrogen evolved in the first hour, and over the next few hours an additional equivalent formed. Addition of prochiral ketones to this solution of oxazaborolidine **12** containing excess borane yielded the alcohol products in ≥90% yield and good ee as depicted in Table III (29). Recognition of the steric demand of a methyl versus a butyl group in the case of 2-hexanone is noteworthy, 75% ee. Generally, THF as solvent gave alcohols with the highest ee's, although in some cases toluene was comparable. The *in situ* B-H

**Table III.** Enantiomeric Excess Produced in the Catalytic Reduction of Prochiral Ketones with *In Situ* Generated Oxazaborolidine Catalysts **12** and **26** (equation 4)

| Ketone | catalyst | solvent | %ee (configuration) |
|---|---|---|---|
| t-BuCOCH$_3$ | 12 | THF | 82 (S) |
| t-BuCOCH$_3$ | 12 | THF | *92 (S) |
| t-BuCOCH$_3$ | 26 | THF | 78 (S) |
| t-BuCOCH$_3$ | 26 | THF | *88 (S) |
| t-BuCOCH$_3$ | 12 | toluene | 44 (S) |
| t-BuCOCH$_3$ | 26 | toluene | 38 (S) |
| t-BuCOCH$_3$ | 12 | CH$_2$Cl$_2$ | 46 (S) |
| t-BuCOCH$_3$ | 26 | CH$_2$Cl$_2$ | 20 (S) |
| t-BuCOC$_6$H$_5$ | 12 | THF | 96 (R) |
| t-BuCOC$_6$H$_5$ | 26 | THF | 80 (R) |
| 2-hexanone | 12 | THF | 72 (S) |
| 2-hexanone | 12 | THF | *75 (S) |
| 2-hexanone | 26 | THF | 68 (S) |
| 2-hexanone | 26 | THF | *69 (S) |
| α-tetralone | 12 | THF | 94 (S) |
| α-tetralone | 26 | THF | 82 (S) |
| α-tetralone | 12 | toluene | 42 (S) |
| α-tetralone | 26 | toluene | 32 (S) |
| α-tetralone | 12 | CH$_2$Cl$_2$ | 60 (S) |
| α-tetralone | 26 | CH$_2$Cl$_2$ | 54 (S) |
| acetophenone | 12 | THF | 94 (S) |
| acetophenone | 26 | THF | 84 (S) |
| acetophenone | 12 | toluene | 52 (S) |
| acetophenone | 12 | CH$_2$Cl$_2$ | 52 (S) |
| 4-MeO-acetophenone | 12 | THF | 94 (S) |
| 4-MeO-acetophenone | 12 | toluene | 48 (S) |
| 4-MeO-acetophenone | 12 | CH$_2$Cl$_2$ | 66 (S) |
| 4-CF$_3$-acetophenone | 12 | THF | 92 (S) |
| 4-CF$_3$-acetophenone | 12 | toluene | 46 (S) |
| 4-CF$_3$-acetophenone | 12 | CH$_2$Cl$_2$ | 20 (S) |
| 3-acetylpyridine | 12 | THF | 96 (S) |
| 3-acetylpyridine | 26 | THF | 90 (S) |
| 3-acetylpyridine | 12 | toluene | 94 (S) |
| 3-acetylpyridine | 26 | toluene | 90 (S) |
| 3-acetylpyridine | 12 | CH$_2$Cl$_2$ | 92 (S) |
| 3-acetylpyridine | 26 | CH$_2$Cl$_2$ | 84 (S) |
| C$_6$H$_5$COCH$_2$Cl | 12 | THF | 92 (R) |
| C$_6$H$_5$COCH$_2$Cl | 26 | THF | 88 (R) |
| C$_6$H$_5$COCH$_2$Cl | 12 | toluene | 80 (R) |
| C$_6$H$_5$COCH$_2$Cl | 26 | toluene | 56 (R) |
| C$_6$H$_5$COCH$_2$Cl | 12 | CH$_2$Cl$_2$ | 80 (R) |
| C$_6$H$_5$COCH$_2$Cl | 26 | CH$_2$Cl$_2$ | 70 (R) |

5 mole % *in situ* generated catalyst employed unless * 10 mole %.
Source: Reprinted with permission from ref. 29. Copyright 1993.

derivative results (ee's) were similar to the B-alkyl (**13, 14**) and B-aryl (**15**) derivatives in the erythro diphenyl system, and were higher than with isolated B-H oxazaborolidine **12**. Recovery of the aminodiphenylethanol was achieved by extraction of the aminoalcohol into aqueous acid and precipitation with ammonium hydroxide. Using Sure Seal bottles, **12** and its enantiomer were formed in THF or toluene and were stable for months at ambient temperature. When the catalysts were prepared in toluene, high ee could be obtained by subsequently adding THF as cosolvent.

To determine whether the π-overlap stabilization distance in **12** accelerated the oxazaborolidine formation, norephedrine was investigated because it was commercially available and the derived B-Me catalyst **17** produced (*S*)-1,2,3,4-tetrahydro-1-naphthol in 80% ee. The norephedrine-based oxazaborolidine **26**, formed *in situ* by the conditions described above, evolved two equivalents of hydrogen during preparation similar to **12**, equation 4. Thus no significant difference in the rate of formation of **12** versus **26** was observed. Catalyst **26** was not as enantioselective as **12** with the ketones evaluated (Table III), although good enantioselectivity was obtained in most cases. Both **12** and **26** provided higher ee for aliphatic ketones when 10 mole% of the catalyst was employed. Aromatic ketones which are more Lewis basic did not show this trend.

$$\text{(4)}$$

**12** R = C$_6$H$_5$
**26** R = CH$_3$

**Synthesis of C$_2$-Symmetric Diols.** A general method to prepare a number of C$_2$-symmetric diols in an enantioselective manner was initiated to support other synthetic studies. Due to the limitations of the present methods (30), oxazaborolidine reduction of diketones was evaluated as a rapid access to these diols. Diketones were first reduced with borane to provide authentic samples of the meso and racemic diols, Table IV. The meso isomer was generated in significant quantities in all uncatalyzed borane reductions. *In situ* formed (29) and isolated catalysts (21) were then evaluated for the enantioselective preparation of diols, equation 5.

$$\text{(5)}$$

**12** X = H
**13** X = Me

All oxazaborolidine catalyzed reductions generated less of the meso isomer compared to the borane reductions (31). Benzil reduction demonstrated high enantiomeric excess although the meso isomer was still substantially produced, Table IV entry 1 (32). When two, three, or four methylene groups were present between the benzoyl groups, high enantiomeric excess and less meso isomer was formed (entries

**Table IV.** Enantiomeric Excess Obtained in the Reduction of Diketones with Oxazaborolidines **12** and **13** (equation 5)

| Entry | Diketone R; n | Catalyst X / mole % | Meso / R,R+S,S | %ee (R,R or S,S) |
|---|---|---|---|---|
| 1 | R = $C_6H_5$; n=O (benzil) | none | 83 / 17 | |
| | | H / 5 | 59 / 41 | 86 (R,R) |
| | | H / 10 | 48 / 52 | 85 (R,R) |
| | | Me / 10 | 61 / 39 | 70 (R,R) |
| 2 | R = Me; n = (indanyl) | none | 85 / 15 | |
| | | Me / 10 | 36 / 64 | 92 (S,S) |
| 3 | R = Me; n = 2 | none | 43 / 57 | |
| | | H / 10 | 31 / 69 | 17 (S,S) |
| 4 | R = $C_6H_5$; n = 2 | none | 61 / 39 | |
| | | H / 10 | 21 / 79 | 94 (S,S) |
| | | H / 100 | 16 / 84 | 99 (S,S) |
| | | H / 100 | 16 / 84 | 99 (S,S)* |
| 5 | R = t-Bu; n = 2 | none | 67 / 33 | |
| | | H / 10 | 20 / 80 | 97 (S,S) |
| 6 | R = $C_6H_5$; n = 3 | none | 57 / 43 | |
| | | H / 10 | 14 / 86 | 99 (S,S) |
| | | H / 10 | 9 / 91 | 99 (S,S)* |
| | | H / 100 | 2 / 98 | 99 (S,S) |
| | | Me / 10 | 16 / 84 | 96 (S,S) |
| 7 | R = $C_6H_5$; n = 4 | none | 50 / 50 | |
| | | H / 10 | 12 / 88 | 94 (S,S) |
| | | Me / 10 | 11 / 89 | 99 (S,S)* |

* Two equivalents of borane methyl sulfide complex were employed per diketone and all other examples used 1.4 equivalents.
Source: Reprinted with permission from ref. 31. Copyright 1995.

4, 5, 6, 7). Low enantiomeric excess with 2,5-hexanedione was the result of similar steric demands of the groups flanking the carbonyl group. In this case, the methylene groups dominate over the methyl group as demonstrated previously in the reduction of 2-hexanone (18). The *t*-butyl group in the 1,4-diketone system, entry 5, yielded high enantiomeric excess. In these catalyzed reductions, the meso isomer was derived from hydride addition opposite to the catalyzed path for one of the two diketone carbonyl groups. The result was diketone reduction in high ee at the expense of meso isomer formation.

Additives (33), stoichiometric oxazaborolidine, and stoichiometric use of borane (28a) (relative to the carbonyl group) have provided higher ee in reductions with the CBS catalyst **2**. Less meso isomer was generated when one equivalent of borane (in place of the typical 0.7 equivalents) per carbonyl group was used, entry 6. Thus, reduction of the diketone by uncatalyzed borane was not responsible for all the meso isomer. Only 2% of the meso isomer was formed in the reduction of dibenzoylpropane with stoichiometric oxazaborolidine. Crystallization of this product yielded the 1,5-diol with <1% meso isomer in 99% ee. Oxazaborolidine catalyzed reduction of appropriately substituted diketones generated $C_2$-symmetric diols with high ee.

**Enantioselective alkynylation.** In addition, to oxazaborolidines **12-15** being effective catalysts for hydride addition to ketones, they have also been employed for alkynylation of aldehydes, equation 6 (34).

$$R_1 \!\!\!\equiv\!\!\! SnBu_3 \xrightarrow{Me_2BBr} \underset{R_2CHO}{\overset{\begin{smallmatrix}H\;\;H\\C_6H_5\!-\!\!\!-\!\!\!-\!\!\!-C_6H_5\\HN\;\;\;O\\\diagdown B \diagup\\|\\C_6H_5\end{smallmatrix}}{\xrightarrow{\hspace{2cm}}}} R_1\!\!\!\equiv\!\!\!\overset{OH}{\underset{R_2}{\diagup\!\!\!\diagdown}} \quad (6)$$

85-97%ee

**Conclusion.** An understanding of the relationship between oxazaborolidine structure and the enantioselectivity obtained in prochiral ketone reductions has been presented. The primary requirement for a good catalyst is a blocked face. Two point binding of borane and the ketone carbonyl group on the catalyst then converge to give high ee unless borane coordination forces nonbonded interactions between the nitrogen and proximal carbon substituents. These nonbonded interactions slow the rate of borane coordination, resulting in non-catalyzed borane reduction and competition from the *endo* pathway effectively competing or surpassing the rate of the *exo* catalyzed path. Thus the molecular recognition of borane and the ketone carbonyl group by an appropriately substituted oxazaborolidine is the key element in the enantioselective reduction. For the erythro series of oxazaborolidines, the relative enantioselectivities are correctly predicted by the *ab initio* calculations. Additionally, new insights into the factors influencing the observed enantioselectivites have been reported.

Diphenyloxazaborolidines (catalysts **12-15**, Table I) possess the attributes important for an efficient catalyst, and high enantiomeric excess is achieved in reduction of prochiral ketones including those containing nitrogen which coordinates borane. *In situ* formation of the oxazaborolidines **12** and **26** using borane is a practical method for the reduction of ketones, and recovery of the 1,2-aminoalcohol catalyst precursors is straightforward. These catalysts are also useful in the synthesis of $C_2$-symmetric diols which have been employed for a variety of enantioselective transformations. Enantioselective alkynylation of aldehydes proceeds in high ee employing substoichiometric quantities of the diphenyloxazaborolidines. Since erythro 2-amino-1,2-diphenylethanol enantiomers are inexpensively prepared from benzoin, and thus not "chiral pool" dependent, the diphenyloxazaborolidines are an important contribution to enantioselective synthesis.

### Acknowledgments

This chapter is dedicated in memory of a friend and colleague, Charles William (Bill) Murtiashaw III (1951-1995).

### Literature Cited

1. (a) Brown, H. C.; Chandrasekharan, J.; Ramachandran, P. V. *J. Am. Chem. Soc.* **1988**, *110*, 1539-46. (b) Noyori, R.; Tomino, I.; Tanimoto, Y.; Nishizaw, M. *J. Am. Chem. Soc.* **1984**, *106*, 6709-16. (c) Noyori, R.; Tomino, I.; Yamada, M.; Nishizaw, M. *J. Am. Chem. Soc.* **1984**, *106*, 6717-25.
2. For a report on a large scale perspective in the synthesis of optically active compounds see Crosby, J. *Tetrahedron* **1991**, 4789-4846.
3. (a) Corey, E. J.; Bakshi, R. K.; Shibata, S. *J. Am. Chem. Soc.* **1987**, *109*, 5551-3. (b) Corey, E. J.; Bakshi, R. K.; Shibata, S.; Chen, C.; Singh, V. K. *J. Am. Chem. Soc.* **1987**, *109*, 7925-6. (c) US patent # 4,943,635 (1990). (d) Corey, E. J.; Bakshi, R. K.; Shibata, S. *J. Org. Chem.* **1988**, *53*, 2861-3.
4. Researchers at Abbott employed an acyl dithiane unit to sterically augment the enantioselectivity of the reduction. DeNinno, M. P.; Perner, R. J.; Lijewski, L. *Tetrahedron Lett.* **1990**, *31*, 7415-18.
5. The following publication reported the reduction of substrates which contain a basic nitrogen atom, but did not detail the enantiomeric excess or stoichiometry. Labelle, M.; et. al. *Bioorgan. Med. Chem. Lett.* **1992**, *2*, 1141-6.
6. (a) Mathre, D.; et. al. *J. Org. Chem.* **1991**, *56*, 751-62. (b) Jones, T. J; et. al. *J. Org. Chem.* **1991**, *56*, 763-9.
7. Quallich, G. J.; Woodall, T. M. *Tetrahedron Lett.* **1993**, *34*, 785-8.
8. Chen, P-C.; Prasad, K.; Repic, O. *Tetrahedron Lett.* **1991**, *32*, 7175-8.
9. Corey, E. J.; Link, J. O. *J. Am. Chem. Soc.* **1992**, *114*, 1906-8.
10. Quallich, G. J.; Woodall, T. M. *Tetrahedron* **1992**, *48*, 10239-48.
11. Corey, E. J.; Cimprich, K. A. *Tetrahedron Lett.* **1992**, *33*, 4099-102.
12. (a) Hirao, A.; Itsuno, S.; Nakahama, S.; Yamazaki, N. *Chem. Soc. Chem. Commun.* **1981**, 315-7. (b) Itsuno, S.; Ito, K.; Hirao, A.; *J. Chem. Soc. Chem. Commun.* **1983**, 469-70. (c) Itsuno, S.; Ito, K. *J. Org. Chem.* **1984**, *49*, 555-7. (d) Itsuno, S.; Sakuri, Y.; Hirao, A.; Nakahama, S. *Bull. Chem. Soc. Jpn.* **1987**, *60*, 395-6.
13. (a) Corey, E. J.; Link, J. O. *J. Am. Chem. Soc.* **1992**, *114*, 1906-8. (b) Corey, E. J.; Link, J. O. *Tetrahedron Lett.* **1992**, *33*, 3431-4. (c) Corey, E. J.; Link, J. O. *Tetrahedron Lett.* **1989**, *30*, 6275-8. (d) Corey, E. J.; Rao, K. S. *Tetrahedron Lett.* **1991**, *32*, 4623-6. (e) Corey, E. J.; Yi, K. Y.; Matsuda, S. P. T. *Tetrahedron Lett.* **1992**, *33*, 2319-22.
14. Corey, E. J.; Link, J. O. *J. Org. Chem.* **1991**, *56*, 442-4.
15. (a) Martens, J. *Tetrahedron: Asymmetry* **1992**, *3*, 1475-1504. (b) Deloux, L.; Srebnik, M. *Chem. Rev.* **1993**, *93*, 763-84. (c) Kim, Y. H.; Park, D. H.; Byun, I. S. *J. Org. Chem.* **1993**, *58*, 4511-12. (d) Didier, E.; Loubinoux, B.; Tombo, G. M. R.; Rihs, G. *Tetrahedron* **1991**, *47*, 4941-58.

16. Weijlard, J.; Pfister, K.; Swanezy, E. F.; Robinson, C. A.; Tishler M. *J. Am. Chem. Soc.* **1951**, *73*, 1216-8.
17. Enantiomers of 2-amino-1,2-diphenylethanol were purchased from Aldrich and TCI America. Yamakawa has generously provided a complimentary sample.
18. Quallich, G. J.; Woodall, T. M. *Tetrahedron Lett.* **1993**, *34*, 4145-48.
19. Deeter, J.; Frazier, J.; Stanten, G.; Staszak, M.; Weigel, L. *Tetrahedron Lett.* **1990**, *31*, 7101-4.
20. Tanaka, K.; Matsui, J.; Suzuki, H. *J. Chem. Soc. Chem. Commun.* **1991**, 1311-12.
21. Quallich, G. J.; Blake, J. F.; Woodall, T. M. *J. Am. Chem. Soc.* **1994**, *116*, 8516-25.
22. Previous *ab initio* investigations focused on model systems or employed semi-empirical treatments. (a) Nevalainen, V. *Tetrahedron: Asymmetry* **1991**, *2*, 63-74. (b) Nevalainen, V. *Tetrahedron: Asymmetry* **1991**, *2*, 429-35. (c) Nevalainen, V. *Tetrahedron: Asymmetry* **1991**, *2*, 827-42. (d) Nevalainen, V. *Tetrahedron: Asymmetry* **1991**, *2*, 1133-55. (e) Nevalainen, V. *Tetrahedron: Asymmetry* **1992**, *3*, 921-32. (f) Nevalainen, V. *Tetrahedron: Asymmetry* **1992**, *3*, 933-45. (g) Nevalainen, V. *Tetrahedron: Asymmetry* **1992**, *3*, 1441-53. (h) Nevalainen, V. *Tetrahedron: Asymmetry* **1992**, *3*, 1563-72. (i) Nevalainen, V. *Tetrahedron: Asymmetry* **1993**, *4*, 1597-1602.
23. Frisch, M. J. ; Trucks, G. W.; Head-Gordon, M.; Gill, P. M. W.; Wong, M. W.; Foresman, J. B.; Johnson, B. G.; Schlegel, H. B.; Robb, M.; Replogle, E. S.; Gomperts, R.; Andres, J. L.; Raghavachari, K.; Binkley, J. S.; Gonzalez, C.; Martin, R. L.; Fox, D. J.; DeFrees, D. J.; Baker, J.; Stewart, J. J. P.; Pople, J. A. *Gaussian 92 Version C*, Pittsburgh, PA, Gaussian, Inc. 1992.
24. LePage, T. J.; Wiberg, K. B. *J. Am. Chem. Soc.* **1988**, *110*, 6642-50.
25. (a) Brooks, C. J. W.; Middleditch, B. S.; Anthony, G. M. *Organic Mass Spectrometry* **1969**, *2*, 1023-32. (b) Anthony, G. M.; Brooks, C. J. W.; Maclean, I.; Sangster, I. J. *Chromo. Sci.* **1969**, *7*, 623-31.
26. Brown, J. M.; Leppard, S. W.; Lloyd-Jones, G. C. *Tetrahedron Asymmetry:* **1992**, 261-66.
27. Ganem, B. *Chemtracts-Organic Chemistry* **1988**, 40-42.
28. (a) Mathre, D. J.; Thompson, A. S.; Douglas, A. W.; Hoogsteen, K.; Carroll, J. D.; Corley, E. G.; Grabowski, E. J. J. *J. Org. Chem.* **1993**, *58*, 2880-8. (b) US patent # 5,189,177 (1993). (c) US patent # 5,264,574 (1993).
29. Quallich, G. J.; Woodall, T. M. *Syn Lett.* **1993**, *12*, 929-30.
30. (a) Kim, M. J.; Lee, I. S. *Syn Lett.* **1993**, 767-8. (b) Nagai, H.; Morimoto, T.; Achiwa, K. *Syn Lett.* **1994**, 289-90. (c) Burk, M. J. *J. Am. Chem. Soc.* **1993**, *115*, 10125-38.
31. Quallich, G. J.; Keavey, K. N.; Woodall, T. M. *Tetrahedron Lett.* **1995**, *36*, 4729-32.
32. (*R,R*)-1,2-diphenyl-1,2-ethane diol has been prepared by asymmetric dihydroxylation. Wang, Z-M.; Sharpless, K. B. *J. Org. Chem.* **1994**, *59*, 8302-3.
33. (a) Cai, D.; Tschaen, D. W.; Shi, Y.-J.; Verhoeven T. R.; Reamer, R. A.; Douglas, A. W. *Tetrahedron Lett.* **1993**, 3243-6. (b) Shi, Y.-J.; Cai, D.; Dolling, U.-H.; Douglas, A. W.; Tschaen, D. W.; Verhoeven, T. R. *Tetrahedron Lett.* **1994**, 6409-12.
34. Corey, E. J.; Cimprich, K. A. *J. Am. Chem. Soc.* **1994**, *116*, 3151-2.

## Chapter 8

# Diastereo- and Enantioselective Hydride Reductions of Ketone Phosphinyl Imines to Phosphinyl Amines
## Synthesis of Protected Amines and Amino Acids

**Robert O. Hutchins, Qi-Cong Zhu, Jeffrey Adams[1], Samala J. Rao, Enis Oskay, Ahmed F. Abdel-Magid, and MaryGail K. Hutchins**

Department of Chemistry, Drexel University, Philadelphia, PA 19104

Three protocols involving asymmetric reductions of ketone phosphinyl imines (**1**) to enantiomeric or diastereomeric phosphinyl amines (**2**), (which represent protected versions of primary amines), are presented, including the use of substrates containing diastereotopic faces to stereoselectively generate diastereomers, employment of chiral reagents for enantioselective reductions, and the utility of chiral auxiliary attachments for diastereoselective reductions followed by release of enantiomeric amines.

$$R''R'C=N-P(O)R_2 \qquad R''R'CH-NH-P(O)R_2$$
$$\mathbf{1} \qquad\qquad\qquad \mathbf{2}$$

Recent investigations (*1-4*) have demonstrated that reductions of *N*-diphenylphosphinyl imines (**1**) with hydride reagents to the corresponding *N*-diphenylphosphinyl amines (**2**) provide attractive and effective entrances to protected versions of primary amines. Thus, derivatives **1** are easily prepared, often stable, but highly reactive intermediates which are readily reduced by most hydride reagents (including very bulky examples), from which the free primary amines can be released by mild acid hydrolysis.
Of particular importance in this area are reductive protocols which afford stereoisomeric amine derivatives using processes that afford high stereoselection for particular diastereomers or enantiomers. This report presents three diversified applications directed toward describing the utility of intermediates **1** as reductive precursors to enantiomeric and diastereomeric amine stereoisomers. These are:
  • Reduction of cyclic derivatives bearing diastereotopic faces with bulky trialkylborohydrides to provide highly diastereoselective production of cyclic amine diastereomers.
  • Enantioselective reductions of α-iminoester derivatives with chiral hydride reagents to afford protected α-amino acids in high enantiomeric excesses.
  • Use of a chiral auxiliary group residing in the phosphinyl group (derived from a camphoryl *exo*–diol) to induce asymmetry *via* reduction with achiral reagents and subsequent removal of the auxiliary group to afford chiral amines.
Each of these approaches will be addressed separately below.

[1]Current address: BIOMOL Research Laboratories, Inc., 5100 Campus Drive, Plymouth Meeting, PA 19462

## Results and Discussion

**Diastereoselective Reductions of Cyclic N-Diphenylphosphinyl Imines.** Although considerable effort has been devoted to uncovering synthetically useful and stereo controllable reductions of cyclic ketones to diastereomeric alcohols (see *4* for cited references), relatively few investigations have focused on developing processes for the corresponding stereoselective reductions of cyclic imines to amines (*5, 6*). Particularly sparse are studies directed toward securing primary amines *via* reductive protocols with hydride reagents (*3–5*). In this regard, cyclic N-diphenylphosphinyl imines (**4**) were considered attractive intermediates for explorations aimed at developing successful stereoselective conversions to imine diastereomers. The requisite imines (**4**) were prepared from corresponding oximes (**3**) as outlined below.

<chemical scheme: oxime 3 (R-substituted cyclohexanone oxime) → [N-O-PPh2 intermediate] → N-P(O)Ph2 imine 4, with Ph2PCl, Et3N, -40°C>

Considerable experimentation revealed that while "small" unhindered reagents (e.g. LiAlH$_4$, NaBH$_4$, NaBH$_3$CN, amine boranes) provided very poor stereoselection, the very bulky reagent lithium tri–*sec*–butylborohydride afforded excellent diastereoselectivity for production of axial amines (with cyclohexyl systems), resulting from equatorial approach. This parallels results obtained with analogous ketones (*7*) and imines (*5*) and prompted a thorough study of the scope and generality of such highly stereoselective reductions using a variety of substituted ring systems. In all cases, $^{31}$P NMR was utilized to determine diastereomer ratios. For comparison (and to insure that both diastereomers gave separate $^{31}$P signals), the corresponding reductions were also conducted using NaBH$_4$. The results and data are presented in Tables I-IV for variously substituted cyclohexyl, cyclopentyl and bicyclo ring derivatives.

**Reductions of 2, 3, 4 and/or 5-Substituted Cyclohexyl Derivatives.** The results demonstrate that reductions of 3, 4 and/or 5-substituted cyclohexyl derivatives **5** with tri-*sec*-butylborohydride (Table I) in all cases provide excellent stereoselection (>97%) for the corresponding axial phosphinyl amine derivatives **6**. Likewise, 2-substituted derivatives **7** gave equally superior (>97%) diastereoselection for the axial derivatives **8** (Table II).

As seen in Tables I and II, reductions with borohydride were, in general, relatively non-stereoselective and led to mixtures of diastereomers. Further discussions of the trends observed with this reagent are beyond the scope of this chapter but may be found in reference *4*.

A notable exception involves the 2-*t*-butyl derivative **7e** (Table II) in which borohydride afforded the same high *cis* diastereoselectivity (>97/3) as obtained with tri–*sec*–butylborohydride. This probably arises because of unusual conformational situations created by the proximity of the 2-*t*-butyl group to the phosphinyl imine. Thus, with the phosphinyl analog **7e**, both chair conformations experience severe destabilizing steric interactions (in the axial *t*-butyl conformer) or severe allylic strain repulsions (*5, 8*) (in the equatorial *t*–butyl conformer) The likely result is that the 2–

$t$-butyl phosphinyl imine **7e** resides predominantly in non-chair conformations such as **9** ("boat") and/or **10** ("twist") both of which are concave and present incoming reagents with bowl-like conformations.

Table I. Reduction of 3 and 4–Substituted Cyclohexylidene $N$–Diphenylphosphinyl Imines

| Entry | 5 | R | LiBH($s$-Bu)$_3$ | NaBH$_4$ |
|---|---|---|---|---|
| | | | ratio of equatorial/axial attack of 5 | |
| | | | (% Yield)[a] | (% Yield)[a] |
| 1 | a | 4–$t$–Bu | >97/3 (64) | 26/74 (65) |
| 2 | b | 4–$i$–Pr | >97/3 (58) | 32/68 (60) |
| 3 | c | 4–Et | >97/3 (65) | 34/66 (64) |
| 4 | d | 4–Me | >97/3 (72) | 41/59 (72) |
| 5 | e | 4–Ph | 95/5 (59) | 52/48 (55) |
| 6 | f | $t$–3,4–diMe | >97/3 (69) | 37/63 (73) |
| 7 | g | $c$–3,5–diMe | >97/3 (70) | 40/60 (57) |
| 8 | h | 3,3,5–triMe | >97/3 (71) | 94/6 (66) |
| 9 | i | 3–Me | >97/3 (63) | 42/58 (79) |

[a] yields are for isolated, purified products calculated from oxime.

Thus, attack by both large and small hydride reagents suffers severe steric interactions in approach toward the inside the "bowl" (leading to the *trans* diastereomer) while attack from outside the "bowl" (which provides the *cis* isomer) is relatively un-encumbered and leads almost exclusively to the *cis* product as observed (>97% equatorial attack). The "twist" conformation **10** is probably preferred because it minimizes eclipsing interactions between the $t$–butyl group and the vicinal hydrogen (both in the "gunwale" position).

Table II. Reduction of 2-Substituted Cyclohexylidene N-Diphenylphosphinyl Imines

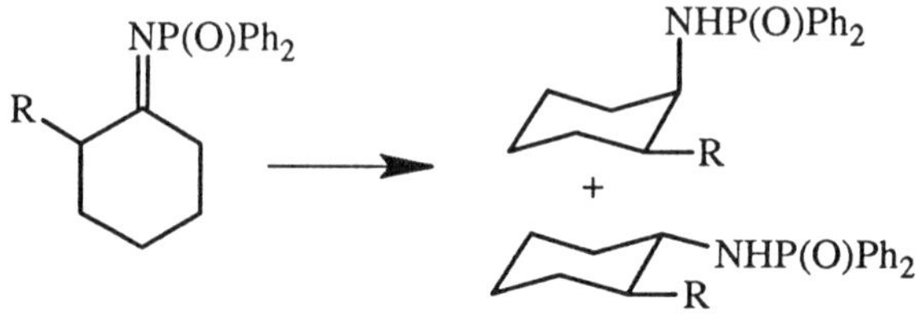

| Entry | 7 | R | LiBH(s–Bu)₃ | NaBH₄ |
|---|---|---|---|---|
| | | | ratio of equatorial/axial attack of 7 | |
| | | | (% Yield)[a] | (% Yield)[a] |
| 1 | a | 2–Me | >97/3 (64) | 50/50 (63) |
| 2 | b | 2–Et | >97/3 (67) | 58/42 (63) |
| 3 | c | trans–2–i–Pr–5–Me | >97/3 (60) | 60/40 (60) |
| 4 | d | 2–Ph | >97/3 (63) | 40/60 (57) |
| 5 | e | 2–t–Bu | >97/3 (57) | >97/3 (79) |
| 6 | f | 2–MeO | >97/3 (71) | 75/25 (71) |

[a] yields are for isolated, purified products calculated from oxime.

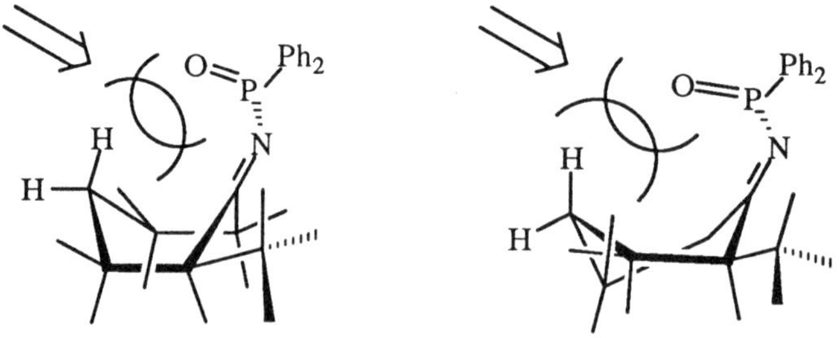

**Reduction of Substituted Cyclopentyl and Other Cyclic Derivatives.** The highly stereoselective reduction results obtained with cyclohexyl phosphinyl imines prompted an extension to cyclopentyl and bicyclic analogs. Table III presents results obtained for substituted cyclopentyl derivatives. As evident, reduction of 2-substituted derivatives with tri-sec-butylborohydride (entries 1 and 2) afforded highly stereoselective discrimination to produce the cis diastereoisomers in analogy to similar results with 2–alkyl cyclopentanones (7). Here again, allylic strain may force a normally pseudo equatorial group to adopt a pseudo axial orientation (**11a**), thus favoring approach of the bulky reagent from the face opposite the alkyl group. Reductions of 3-substituted derivatives with tri-s-butylborohydride (and NaBH$_4$) gave poor stereoselectivities (entries 3 and 4). This was not unexpected since preferred half–chair conformation of cyclopentyl systems provides essentially no bias for attack at either face (C$_2$ symmetry axis), unlike the situation in cyclohexyl systems. The 3-alkyl substituents probably reside in pseudo equatorial positions and thus offer no steric resistance to an incoming reagent (See **11b**).

Table III. Reduction of Cyclopentylidene N–Diphenylphosphinyl Imines

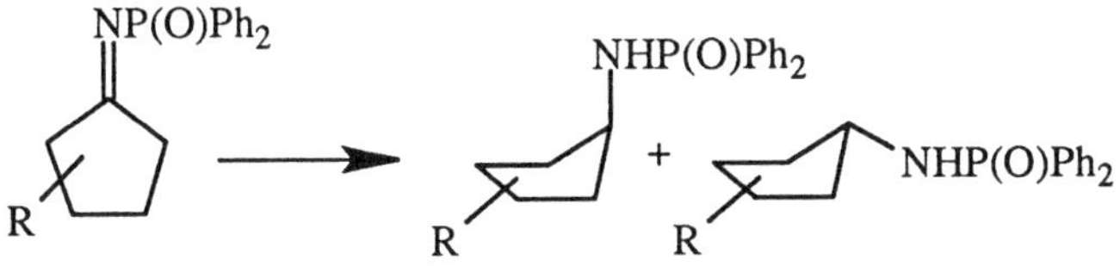

| Entry | 9 | R | LiBH(s-Bu)$_3$ | NaBH$_4$ |
|---|---|---|---|---|
|   |   |   | % cis attack of 9[a] | |
|   |   |   | (% Yield)[b] | (% Yield)[b] |
| 1 | a | 2–Me | <3 | 73 |
|   |   |   | (64) | (65) |
| 2 | b | 2–Et | <3 | 66 |
|   |   |   | (60) | (62) |
| 3 | c | 3–Me | 37 | 36 |
|   |   |   | (62) | (66) |
| 4 | d | 3–Et | 39 | 40 |
|   |   |   | (66) | (75) |
| 5 | e | cis–3,4–diMe | <3 | 10 |
|   |   |   | (66) | (62) |

[a] *cis* attack is approached from the same side of the ring as the substituent(s).
[b] yields are for isolated, purified products calculated from oxime.

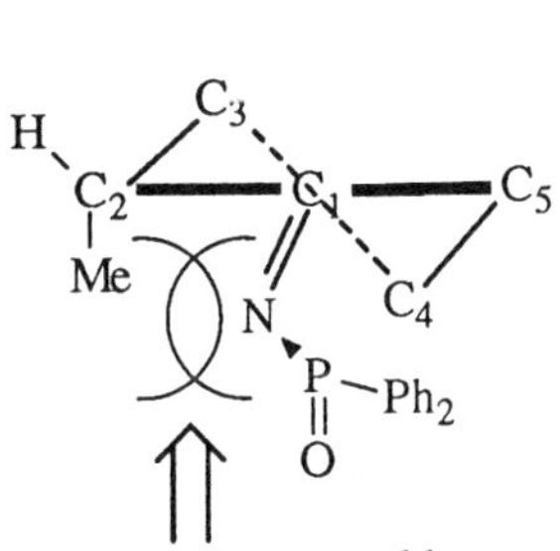

**11a** 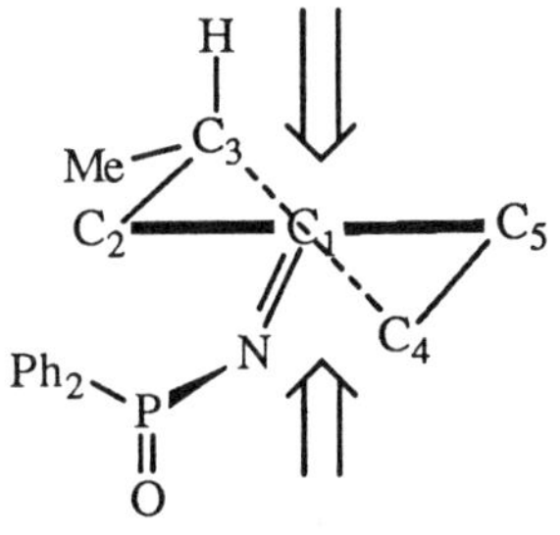  **11b**

The N–diphenylphosphinyl imines derived from norbornanone and camphor were also subjected to the reductive protocol. The results are listed in Table IV and closely parallel observations observed with the corresponding ketones. Thus, norbornyl imine **12** experiences predominant *exo* attack with both reagents, believed to result from steric hindrance from the *endo* hydrogens. The camphor imine **13** is attacked mainly from the *endo* face by both large and small reagents. Steric hindrance to *exo* attack provided by the *syn* $C_7$ methyl group is more severe than the encumbrance to *endo* attack provided by the *endo* hydrogens (Table IV).

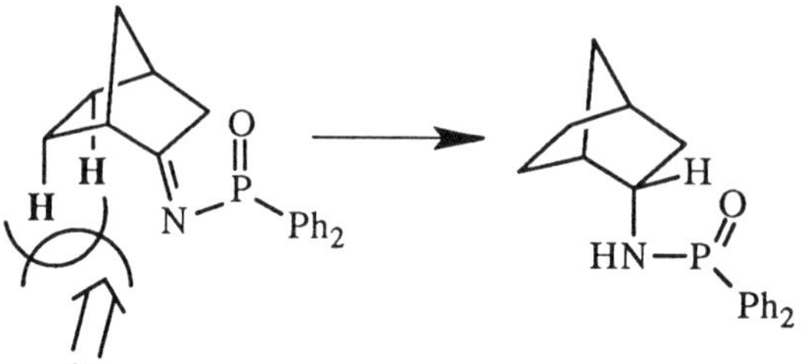

Table IV. Reduction of Norbornyl and Camphoryl *N*–Diphenylphosphinyl Imines

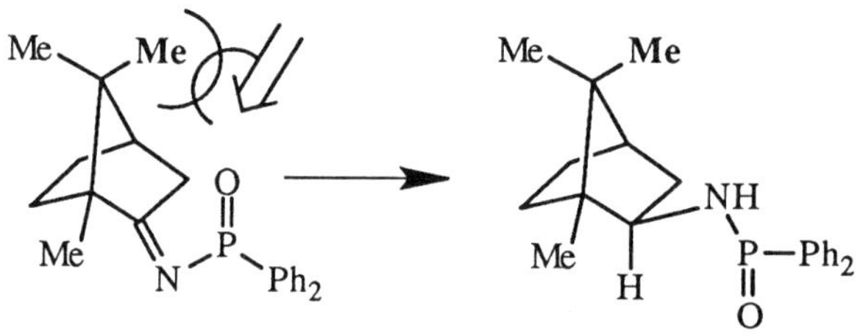

| Entry | R | LiBH(s–Bu)$_3$ % of the predominate isomer (% Yield)[a] | NaBH$_4$ (% Yield)[a] |
|---|---|---|---|
| 12 | Norbornyl | >97 *endo* (68) | 80 *endo* (77) |
| 13 | Camphoryl | >97 *exo* (63) | 90 *exo* (70) |

[a] yields are for isolated, purified products calculated from oxime.

In summary, tri-*s*-butylborohydride reductions of a wide variety of cyclohexyl phosphinyl imines including 2,3, and 4 and/or 5-substituted derivatives (**5** and **7**) provide excellent stereoselective entries to the corresponding axial phosphinyl amines resulting from equatorial attack. Likewise, similar reductions of 2-alkyl and 3,4-dialkyl cyclopentyl derivatives (**9a,b,e**) are also highly stereoselective with the directions of approach controlled by steric impedance to attack by pseudo axial alkyl groups. Steric effects also control the high diastereoselectivities observed in reductions of norbornyl and camphor phosphinyl imines **12** and **13**, with the former giving nearly exclusive *exo* attack and the latter almost entirely *endo* approach.

**Enantioselective Reductions of α-Iminoester Derivatives with Chiral Hydride Reagents to Afford Protected Phenyl Glycines.** Asymmetric reductions of carbonyl derivatives with chiral, non-racemic reagents to chiral alcohols have attracted attention for over 45 years and have resulted in the introduction of a number of successful reagent systems (9) for a variety of structural types. However, reports of applications of chiral reagents for the reductions of imines or imine derivatives to chiral amines are far more sparse and general processes remain elusive (2, 9).
Recent investigations have indicated that $N$-diphenylphosphinyl imines provide attractive intermediates for enantioselective reductions to protected primary amine derivatives using chiral regents (2, and cited references). This prompted a study of the utility of such intermediates for the enantioselective preparation of α-amino acids. As a preliminary incursion into this area, ring substituted aryl α-diphenylphosphinyl imino esters **14** (derived from corresponding α-keto phenyl acetic esters) were chosen as entrances to chiral, non–racemic substituted phenylglycine esters **15**. This choice of substrates also allowed a determination of the effects of aromatic substituents on the asymmetric inductions available.

A survey of various available chiral reducing reagents (4, 9) revealed that the most successful (identified) reagent for this category of phosphinyl imines (**14**) was **16**, generated from chiral, non-racemic bis-2-naphthol, lithium aluminum hydride and ethanol (BINAL–H, Noyori's Reagent, 10).

(R)–(+)–BINAL–H
**16**

Determination of the enantiomeric excesses of the product amine esters **15** was accomplished via $^{31}P$ NMR analysis of the methylphosphonothionic diamide derivatives as reported by Feringa (11) (following cleavage with 10% p–toluenesulfonic acid/ methanol to the free amines) while the product absolute configurations of **15** were determined by an optical rotation technique originally developed by Rabin and coworkers (12). This latter process involved conversion of the freed amines to the corresponding p–toluenesulfonyl–2,4–dinitrosulfenyl amides and recording the rotation sign at the sodium D wavelength. Determinations of both the absolute configurations and % enantiomeric excesses were confirmed using examples of known configurations and enantiomeric excesses.

Reduction of a variety of examples of phosphinyl imines **14** with reagent **16** gave good to excellent enantiomeric excesses of the corresponding α–amino phosphinyl amine esters **15** as presented in Table V. These data demonstrate that for unsubstituted or meta or para substituted ring derivatives, the configurations of the product amine esters **15** predictably afforded R enantiomers from S-bis-2-naphthol. This is consistent with a model [**17**, related to that put forth for corresponding carbonyl reductions (10)] involving a six–membered ring transition state in which the bulkier aromatic ring assumes a pseudo–equatorial orientation and complexation of the pseudo–axial carbonyl with a lithium ion as shown.

**17**

**Table V. Reductions of Substituted Aryl α-Diphenylphosphinyl Imino Esters to Substituted Phenylglycine Esters with BINAL–H**

| X | % Yield 14 | Configuration | % e.e. |
|---|---|---|---|
| H | 64 | S | 77 |
| p–Br | 80 | S | 87 |
| p–Cl | 85 | S | 96 |
| p–CH$_3$[a] | 69 | S | 93 |
| p–CH$_3$O | 69 | S | 91 |
| m–CH$_3$[a] | 70 | R[b] | 90 |
| m–CH$_3$O | 64 | S | 93 |
| o–CH$_3$ | 86 | R | 52 |
| o–CH$_3$O | 63 | R | 92 |
| o–Cl | 68 | R | 90 |

[a] ethyl ester
[b] (S)–BINAL–H

On the other hand, reduction of the *ortho* ring substituted derivatives of **14**, while still providing moderate to excellent enantiomeric excesses of amino esters **15** (Table V), afforded the opposite configurational isomers (i.e., the R enantiomer of bis–2–naphthol gave the S enantiomers of **15**). In order to ensure that the method of configurational assignment was not flawed and had led to erroneous results, the *o*-chloro amino derivative of **15** (obtained by reduction of the corresponding derivative of **14** and cleavage with 10% *p*–toluenesulfonic acid/methanol) was de-chlorinated with $NaBH_4/PdCl_2$ (*13*) to the unsubstituted derivative of known configuration. Thus, the reduction of *ortho* substituted derivatives affords the opposite chiral sense with equally high enantiomeric excesses of the corresponding phosphinyl amines, (except for the *o*-methyl derivative).

The reason for this reversal may be related to a combination of increased steric interactions encountered by the *ortho*-substituted rings located in the pseudo–equatorial position which is destabilizing (i.e. **18**) and/or increased complexation by electronegative *ortho* ring substituents with the lithium ion (i.e. **19**). Noteworthy in this regard is the observation that the two examples with complexable *ortho* substituents (Cl and $OCH_3$) provide superior % enantiomeric excesses compared to the result with the *ortho*-substituted, non-complexing methyl derivative of **15** (52% e.e. for methyl versus 90–92% e.e. for Cl and $OCH_3$).

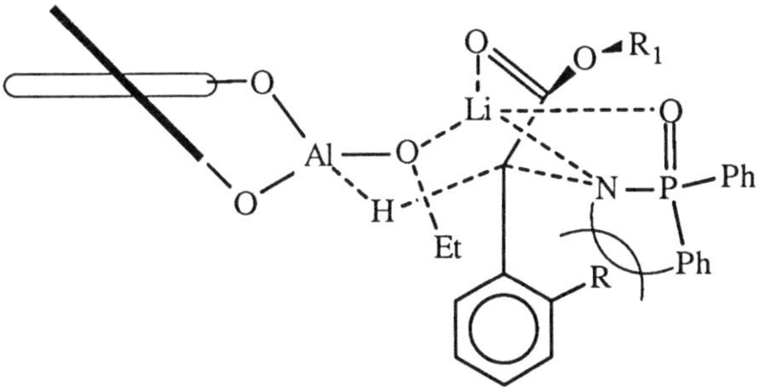

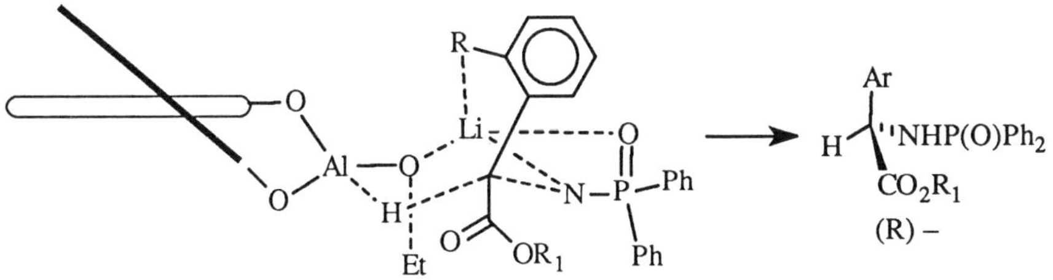

| R | % e.e. |
|---|---|
| $OCH_3$ | 92 |
| Cl | 90 |
| $CH_3$ | 52 |

In summary, the reduction of ring substituted aryl α–diphenylphosphinyl imino esters **14** with chiral BINAL–H (**16**) provides, in most cases, efficient and highly enantioselective entrances to the corresponding substituted phenylglycines (**15**). The enantiomeric senses obtained with *meta* and *para* substituted derivatives is accurately accounted for by assuming a cyclic, six–membered ring transition state (**17**) in which the bulkier aryl ring is favored in a pseudo equatorial orientation. To account for the reversal in the enantiomeric sense observed with *ortho* substituted derivatives of **14**, further considerations are required as depicted in **18** and **19**.

**Use of a Chiral Phosphorus Auxiliary Group for Generation of Chiral Phosphinyl Imines and Subsequent Reduction to Protected Amine Enantiomers.** Another conceptual approach to chiral, non-racemic protected amines involves the employment of a chiral auxiliary group incorporated in the phosphorus segment of phosphinyl imines. In this regard the utility of the chiral camphoryl phosphinyl imines derived from the corresponding chlorophosphine **20** was explored. Thus, diastereoselective reduction of imines **21** (which, by $^{31}$P NMR appear to be single diastereomers, presumed to correspond to that depicted in **21**, although the intermediates were too labile for isolation and further characterization), with NaBH$_4$/THF to the product phosphinyl imines **22** followed by cleavage with acidic methanol afforded the corresponding amine salts **23** along with *exo*-camphordiol **24** for recycling to **20**. The % enantiomeric excesses were determined by comparison of the diastereomer ratios ($^{31}$P NMR) and were confirmed, except for **23g** and **h**, by comparisons with literature rotation values of the free amines (*14*) or the amine salts (*15*). The results are presented in Table VI and indicate that while good enantiomeric excesses (76–87%) were obtained with dialkyl examples (**21a–d**), aryl derivatives (**21e–h**) gave much poorer asymmetric selections (26–29% e.e.). Perhaps coincidentally, we had observed a similar structure relation in the diasteroselective reduction of other phosphinyl imines with chiral reagents (*2*).

In summary, reductions of dialkyl phosphinyl imines bearing a chiral auxiliary group derived from camphordiol **24** provide entrance to chiral, non-racemic amines with good (76-87 %) enantioselectivities but relatively poor results (26-29% ee's) with aromatic derivatives. Current investigations include the search for other auxiliary attachments and the utility of other reducing agents.

## Table VI. Asymmetric Reduction of Camphoryl Phosphinyl Imines. Preparation of Chiral Primary Amines.

| 3 | R | R' | % Yield 22[a] | % e.e. 23 (configuration) |
|---|---|---|---|---|
| a | $CH_3$ | $C_2H_5$ | 72 | 87 (R) |
| b | $CH_3$ | $CH(CH_3)_2$ | 61 | 82 (R) |
| c | $CH_3$ | $(CH_2)_5CH_3$ | 61 | 76 (R) |
| d | $CH_3$ | $C(CH_3)_3$ | 65 | 76 (R) |
| e | $CH_3$ | $C_6H_5$ | 75 | 28 (R) |
| f | $C_2H_5$ | $C_6H_5$ | 80 | 29 (R) |
| g | $CH(CH_3)_2$ | $C_6H_5$ | 64 | 29 |
| h | $C(CH_3)_3$ | $C_6H_5$ | 70 | 26 |

[a] Yields are for isolated, purified products calculated from the oximes

## References

1. Krzyzanowska, B.; Stec, W. J. *Synthesis* **1982**, 270.
2. Hutchins, R. O.; Abdel-Magid, A. F.; Stercho, Y. P.; Wambsgans, A. *J. Org. Chem.* **1987**, *52*, 702 and cited references.
3. Hutchins, R. O.; Rutledge, M. C. *Tetrahedron Lett.* **1987**, *28*, 5619.
4. Hutchins, R. O.; Adams, J.; Rutledge, M. C. *J. Org. Chem.* **1995**, *60*, 7396.
5. Hutchins, R. O.; Su, W-Y.; Sivakumar, R.; Cistone, F.; Stercho, Y. P. *J. Org. Chem.* **1983**, *48*, 3412.
6. Review: Hutchins, R. O.; Hutchins, M. K. In *Comprehensive Organic Synthesis*; Trost, B., Ed., Pergamon, New York: 1991; Vol. 8, Ch. 1.2; other reductive protocols are discussed in chapters 1.3-1.8.
7. Brown, H. C.; Krisnamurthy, S. *J. Amer. Chem. Soc.* **1972**, *94*, 7159. Brown, H. C.; Krisnamurthy, S.; Kim, S. C. *J. Chem. Soc. Chem. Commun.* **1973**, 391.
8. Johnson, F. *Chem. Rev.* **1968**, *68*, 375.
9. Review: Nishizawa, M.; Noyori, R. In *Comprehensive Organic Synthesis*; Trost, B., Ed., Pergamon, New York: 1991; Vol. 8, Ch. 1.7.
10. Noyori, R.; Tomino, I.; Tanimoto, M.; Nishizawa, M. *J. Am. Chem. Soc.* **1984**, *106*, 1609. Noyori, R.; Tomino, I.; Tanimoto, M.; Nishizawa, M. *J. Am. Chem. Soc.* **1984**, *106*, 6717.
11. Feringa, B. L.; Strijtveen, B.; Kellogg, R. M. *J. Org. Chem.* **1986**, *51*, 5484.
12. Raban, M.; Moulin, C. P.; Lauderback, S. K.; Swilley, B. *Tetrahedron Lett.*. **1984**, *25*, 3419; Raban, M.; Lauderback, S. K. *J. Org. Chem.* **1980**, *45*, 2636.
13. Bosin, T. R.; Raymond, M. G.; Buckpitt, A. R. *Tetrahedron Lett.* **1973**, 4699; Satoh, T.; Mitsuo, N.; Nishiki, M.; Nanba, K.; Suzuki, S. *Chem. Lett.* **1981**, 1029.
14. Landor, S. R.; Chan, Y. M.; Sonola, O. O.; Tatchell, A. R. *J. Chem. Soc. Perkin Trans. I*, **1984**, 493.
15. Moss, N.; Gauthier, J.; Ferland, J.-M. *Synlett.* **1995**, 142.

## Chapter 9

# Remote Acyclic Diastereocontrol in Hydride Reductions of 1,n-Hydroxy Ketones

**Bruce E. Maryanoff, Han-Cheng Zhang, Michael J. Costanzo, Bruce D. Harris, and Cynthia A. Maryanoff**

R. W. Johnson Pharmaceutical Research Institute, Spring House, PA 19477

We have discovered asymmetric reductions in hydroxy ketone systems (see **I**) with high diastereoselectivity between remote (e.g., 1,5 and 1,6) carbon stereocenters. The favorable results are presumably achieved through the formation of bicyclic metal chelates that involve the intra-chain nitrogen atom. For example, reduction of **I** (R = Ph, m = n = 1) with $R$-Alpine-Hydride or $Zn(BH_4)_2$ in $CH_2Cl_2$ gave high *anti* stereo-selectivity (**II:III** = *anti*:*syn* = 10:1 or 13:1, respectively). Reduction of **I** (R = Ph, m = 1, n = 2) with $R$-Alpine-Hydride in $CH_2Cl_2$ afforded a preponderance of *anti* diastereomer (**II:III** = *anti*:*syn* = 12:1). Evidence is presented to support a chelation-controlled mechanism involving a bicyclic metal complex. Our studies with a range of substrates and reducing conditions are discussed.

There has been a great deal of interest in the stereochemical control of reactions involving conformationally flexible, acyclic molecules. Indeed, major advances have occurred in the generation of new stereogenic centers with high asymmetric induction for molecules in which the new centers are proximal to an existing stereogenic center, such as 1,2 and 1,3 diastereocontrol (*2*, *3*). However, examples of effective stereo-control (>90% diastereoselectivity) between remote centers (>1,3) in acyclic systems are still quite the exception (4). Certainly, the discovery and investigation of new types of reactions with high stereocontrol in this area would be a valuable contribution to organic chemistry. We encountered such an event in a medicinal chemistry project dealing with nebivolol (**1**) (*1*), a cardiovascular drug taken into clinical development for the treatment of hypertension by the Janssen Research Foundation (*5*).

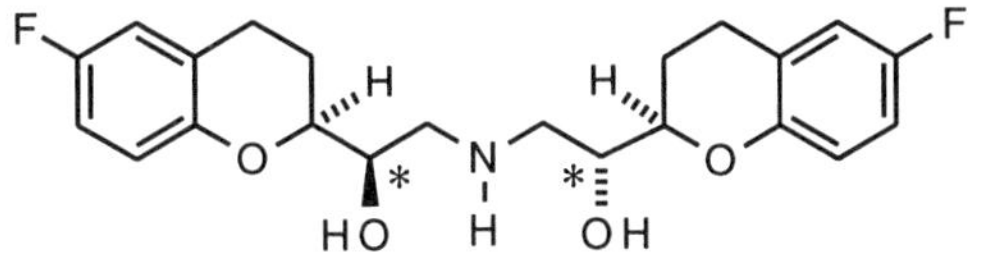

Nebivolol (**1**)

Racemic mixture of (*S,R,R,R*) and (*R,S,S,S*) enantiomers

(*S,R,R,R*) is a potent, selective β$_1$-adrenergic blocker

Although there are 16 possible isomers for a molecule with four stereocenters ($2^4$ isomers), the full set of isomers represented by nebivolol (**1**) contains only 10 (4 *dl* and 2 *meso* forms) because of the inherent symmetry. The investigational drug substance is a racemic mixture of the (*S,R,R,R*) and (*R,S,S,S*) forms, with the (*S,R,R,R*) enantiomer being a potent, selective β$_1$-adrenergic blocker (*5*). In the study of nebivolol analogues, we wanted to decrease the number of stereoisomers by eliminating the stereogenic centers in the heterocycles. Thus, we set out to synthesize bis-benzofuran analogue **2** (Scheme 1), which is comprised of just one *dl* pair and one *meso* form! A key step in the reaction sequence leading to *dl*-**2** (Scheme 1) is the selective reduction of racemic δ-hydroxy ketone **4** to *dl*-diol **5**. We prepared **4** from 2-bromoacetylbenzofuran and 2-(1-hydroxy-2-benzylaminoethyl)benzofuran (**6**) by using (*i*-Pr)$_2$NEt in THF (2 equiv, 5 → 23 °C; 16 h), with salicylaldehyde and chloromethylacetone serving as the original starting materials (Scheme 1).

**Scheme 1. Synthesis of 2 and 4.**

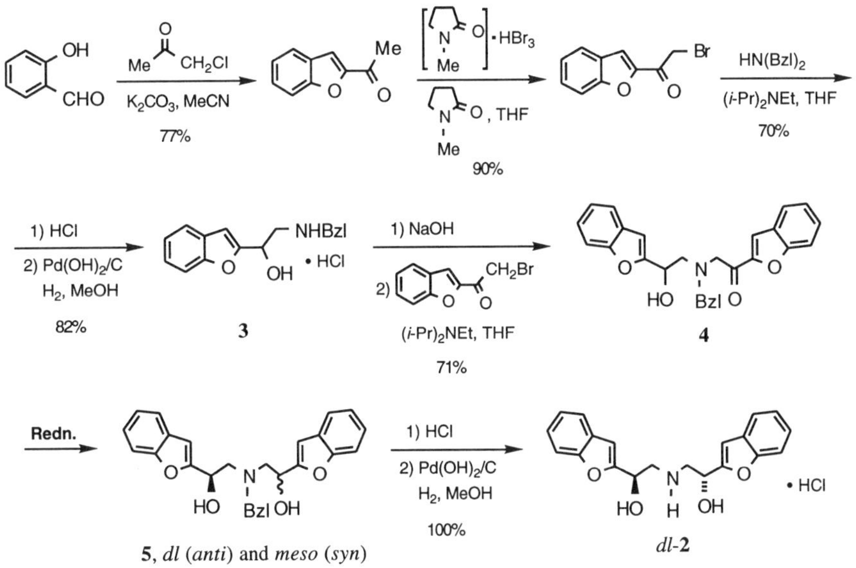

## Reductions of 1,n-Hydroxy Ketones

We hoped to identify suitable conditions for effectively controlling the stereochemistry of the reduction of **4** to obtain the desired *dl* form of **5**. Unfortunately, there was no real precedent in the literature for success with such an acyclic 1,5 reduction. Indeed, examples of high asymmetric induction by a singular, remote stereogenic center bearing a hydroxyl group in nucleophilic addition to hydroxy ketones are rare (7). In one case, Maier et al. (8) reported that the reduction of δ-hydroxy ketone **6** with LiAlH$_4$ in Et$_2$O provides the corresponding 1,5 diols with a *anti:syn* ratio of just 55:45. However, at the outset we thought the diastereoselectivity for the reduction of **4** with LiAlH$_4$ might be better because of the amine nitrogen atom in the chain, which could cooperate with the hydroxyl to direct hydride delivery.

When **4** was reduced with LiAlH$_4$, the level of asymmetric induction in **5** was negligible (*dl:meso* = 1:1), and the bulkier LiAlH(O-*t*-Bu)$_3$ showed no selectivity, as well (Scheme 1; Table I, entries 7 and 8). The *dl* and *meso* products were isolated as a mixture and the diastereomeric *anti:syn* ratios were determined by proton NMR in CDCl$_3$ (at 300 or 400 MHz). The two diastereotopic benzyl protons of the *anti* isomer appear as a pair of doublets at δ 3.75 and 3.92 (J = 13.7 Hz), while the two enantiotopic benzyl protons in the *syn* isomer appear as a singlet at δ 3.79.

In an attempt to achieve an excess of *anti* over *syn* diols, we studied 14 additional reducing agents (Table I). Reactions were conducted at -78 °C, then slowly warmed to 23 °C over 18 h, unless otherwise noted. Most of the results in Table I show virtually no selectivity. However, interestingly, with *R*-Alpine-Hydride (**7**; Table I, entry 16) (9) we obtained a useful level of diastereoselectivity (7:1) and isolated analytically pure *dl*-**5** in 43% yield (mp 103-104 °C; C, H, N analysis). This *dl* product had no enantiomeric enrichment, as assessed by 400-MHz proton NMR (CDCl$_3$) with (*S*)-(+)-2,2,2-trifluoromethyl-1-(9-anthryl)ethanol (*10*). Modest 1,5 asymmetric induction was observed with Zn(BH$_4$)$_2$ (entry 15) and lithium thexyl-(*R*)-limonylborohydride (**9**; entry 14), and none of the reducing agents produced more *syn* than *anti* diols. NB-Enantride (**8**), which is structurally similar to *S*-Alpine-Hydride and reported to be more effective for the asymmetric reduction of ketones (*11*), was ineffective in our case (entry 13).

The results with **4** suggest the intermediacy of a metal chelate with external hydride delivery, as discussed by Baker et al. for the 1,3 asymmetric reduction of a δ-hydroxy ketone with LiEt$_3$BH (12). Thus, we reasoned that the stereoselectivity might be enhanced by changing conditions to favor a rigid chelate species, for example, by using a less coordinating solvent and by lowering the reaction temperature.

Table I. Reduction of 4 to dl- (anti) and meso- (syn) 5

| entry | reagent | solvent | dl : meso |
|---|---|---|---|
| 1[a] | Pd(OH)$_2$, H$_2$ | MeOH | 1:1[b] |
| 2[c] | NaBH$_4$ | MeOH | 1:1[d] |
| 3[c] | NaBH$_4$/CeCl$_3$ | MeOH | 1:1[d] |
| 4 | NaBH(OAc)$_3$ | THF | 1:1[d] |
| 5 | Red-Al | CH$_2$Cl$_2$ | 1:1[e] |
| 6 | K-Selectride | THF | 1:1[e] |
| 7 | LiAlH$_4$ | Et$_2$O | 1:1[d] |
| 8 | LiAlH(O-t-Bu)$_3$ | THF | 1:1[e] |
| 9 | DIBAL | CH$_2$Cl$_2$ | 1:1[d] |
| 10 | BH$_3$•THF | THF | 1:1[e] |
| 11 | BH$_3$•pyridine | THF | 1:1[e] |
| 12 | BH$_3$•Me$_2$S | THF | 1:1[e] |
| 13 | 8 | THF | 1:1[e] |
| 14 | 9 | THF | 2:1[d] |
| 15 | Zn(BH$_4$)$_2$ | Et$_2$O | 2.5:1[d] |
| 16[f] | 7 | THF | 7:1[d] |

(a) At 23 °C for 2 h. (b) The N-debenzylated products, 2, were obtained; the ratio was determined by $^{13}$C NMR. (c) At 5 °C for 2 h. (d) The ratio was determined by proton NMR. (e) The ratio was estimated by TLC. (f) At -98 °C with slow warming to 23 °C over 18 h. (Adapted from ref. 1.)

In follow-up studies we used 10 (13), a simpler, more readily available δ-hydroxy ketone, as the substrate (Eq 1; Table II). A typical reduction procedure follows. To a solution of 10 (0.22 mmol) in CH$_2$Cl$_2$ (5 mL) at -78 °C was added dropwise R-Alpine-Hydride (0.46 mmol, 0.5 M in THF) via syringe over 35 min. The mixture was stirred at -78 °C for 26 h and quenched with water. The mixture was extracted (CH$_2$Cl$_2$) and the organic layer was washed (water, then brine) and dried (Na$_2$SO$_4$). The volatiles were removed in vacuo and the residue was separated by preparative TLC (EtOAc-hexanes, 1:2.5) to give 11, as a mixture of diastereomers. The products were isolated as a mixture and the *anti:syn* ratios were determined by proton NMR in CDCl$_3$ (at 300 or 400 MHz). The diastereotopic benzyl protons of the *anti* isomer appear as a pair of doublets at δ 3.66 and 3.94 (J = 13.5 Hz), whereas the two protons in the *syn* isomer appear as a singlet at δ 3.77.

$$\underset{10}{\text{Ph}\diagdown\underset{\text{HO}}{\text{CH}}\text{-}\underset{\text{Bzl}}{\text{N}}\text{-}\underset{\text{O}}{\text{CH}_2}\text{-Ph}} \xrightarrow{[\text{H}^-]} \underset{dl\ (anti)\ 11}{\text{Ph}\diagdown\underset{\text{HO}}{\text{CH}}\text{-}\underset{\text{Bzl}}{\text{N}}\text{-}\underset{\text{OH}}{\text{CH}}\text{-Ph}} + \underset{meso\ (syn)\ 11}{\text{Ph}\diagdown\underset{\text{HO}}{\text{CH}}\text{-}\underset{\text{Bzl}}{\text{N}}\text{-}\underset{\text{OH}}{\text{CH}}\text{-Ph}} \quad (1)$$

**Table II. Reduction of 10 to *dl*- (*anti*) and *meso*- (*syn*) 11** [a]

| entry | reagent | temp (°C) | time (h) | solvent | dl : meso | % yield |
|---|---|---|---|---|---|---|
| 1 | R-Alpine-Hydride | -98 → 0 | 16 | THF | 6:1 | 57 |
| 2 | S-Alpine-Hydride | -98 → 0 | 16 | THF | 6:1 | 62 |
| 3 | R-Alpine-Hydride | -78 | 5.5 | THF | 6:1 | 17 |
| 4 | R-Alpine-Hydride | -40 | 5 | THF | 4:1 | 37 |
| 5 | R-Alpine-Hydride | 0 | 2 | THF | 3:1 | 62 |
| 6 | R-Alpine-Hydride | -40 | 5 | $CH_2Cl_2$ | 6:1 | 40 |
| 7 | R-Alpine-Hydride | -78 | 26 | $CH_2Cl_2$ | 10:1 | 40 |
| 8 | $LiEt_3BH$ | -78 | 26 | $CH_2Cl_2$ | 8:1 | |
| 9 | $LiBH_4$ | -78 | 26 | $CH_2Cl_2$ | 5:1 | |
| 10 | $Zn(BH_4)_2$ | 0 | 2 | THF | 2:1 | |
| 11 | $Zn(BH_4)_2$ | -40 | 5 | THF | 4:1 | |
| 12 | $Zn(BH_4)_2$ | -78 | 26 | $CH_2Cl_2$ | 13:1 | 47 |

(a) Ratios were determined by $^1$H NMR; yields are for mixtures isolated from preparative TLC. (Adapted from ref. 1.)

The following observations are noteworthy. (1) When **10** was reduced with *R*-Alpine-Hydride under the conditions employed for **4**, similar diastereoselectivity was achieved (Table II; entry 1). (2) The 1,5 asymmetric induction is independent of reagent chirality, as the *R* and *S* forms of **7** furnished the same ratio of *dl*:*meso* **11** (entries 1 and 2). (3) The *anti*:*syn* selectivity increases as temperature decreases (entries 3-5). (4) The structure of the hydride anion plays an important role (entries 7-9; also see Table I). (5) Replacement of THF by noncoordinating $CH_2Cl_2$ significantly enhances 1,5 diastereoselectivity (cf. entries 3 and 7; 11 and 12). The best selectivity of *anti* vs. *syn* diols was obtained for the reduction of **10** with $Zn(BH_4)_2$ in $CH_2Cl_2$ at -78 °C (*anti*:*syn* = 13:1, entry 12).

$$\underset{\text{OH}\ \ \text{Bzl}}{\text{Me}\diagdown\text{CH-NH}} + \underset{\text{O}}{\text{Cl}\diagdown\text{C}\text{-Me}} \xrightarrow[\substack{\text{THF} \\ 60\%}]{(i\text{-Pr})_2\text{NEt}} \underset{\mathbf{12}}{\text{Me}\diagdown\underset{\text{HO}}{\text{CH}}\text{-}\underset{\text{Bzl}}{\text{N}}\text{-}\underset{\text{O}}{\text{C}}\text{-Me}}$$

$$\xrightarrow[\substack{CH_2Cl_2 \\ -78 \rightarrow 0\ °C \\ 84\%}]{R\text{-}7} \underset{\mathbf{13}\ anti\ (\pm)}{\text{Me}\diagdown\underset{\text{HO}}{\text{CH}}\text{-}\underset{\text{Bzl}}{\text{N}}\text{-}\underset{\text{OH}}{\text{CH}}\text{-Me}} + \underset{\mathbf{13}\ syn\ (meso)}{\text{Me}\diagdown\underset{\text{HO}}{\text{CH}}\text{-}\underset{\text{Bzl}}{\text{N}}\text{-}\underset{\text{OH}}{\text{CH}}\text{-Me}} \quad (2)$$

Dimethyl 1,5-hydroxy ketone **12** was prepared and reduced with *R*-Alpine-Hydride at -78 °C in $CH_2Cl_2$ to give **13** with an *anti:syn* ratio of 7.8:1 (Eq 2), similar to the 10:1 ratio obtained with the corresponding diphenyl system (Table II, entry 7).

The issue of enantioselectivity due to the use of a chiral reducing agent, Alpine-Hydride, was investigated with **10**. Reduction of **10** with *R*-Alpine-Hydride at -98 °C produced *dl*-**11** that had no measurable optical rotation and no enantiomeric excess (e.e.) [400-MHz proton NMR with (*S*)-(+)-2,2,2-trifluoromethyl-1-(9-anthryl)ethanol (*10*)]. This result agrees with that obtained for the reduction of **4**, mentioned earlier. As a control, we reduced $PhC(O)CH_2N(Bzl)_2$ with *R*-Alpine-Hydride at -98 °C to get the corresponding alcohol with merely 10% e.e. (proton NMR assay).

$$\underset{\substack{\textbf{14:}\ X=CH_2\\ \textbf{16:}\ X=S}}{\underset{HO\quad\quad O}{Ph\diagdown\diagup X\diagdown\diagup Ph}} \xrightarrow[\text{-78°C}]{\text{R-7}\atop\text{THF}} \underset{\substack{anti\ (\pm)\\ \textbf{15:}\ X=CH_2\\ \textbf{17:}\ X=S}}{\underset{HO\quad\quad OH}{Ph\diagdown\diagup X\diagdown\diagup Ph}} + \underset{syn\ (meso)}{\underset{HO\quad\quad OH}{Ph\diagdown\diagup X\diagdown\diagup Ph}} \qquad (3)$$

To determine the influence of the amine nitrogen (actually N-Bzl), we sought to replace the N-benzyl moiety with carbon and sulfur groups (Eq 3). Thus, δ-hydroxy ketone **14** (*14*) was reduced in THF at -78 °C with *R*-Alpine-Hydride or $Zn(BH_4)_2$ to give 1,5-diols **15** with a *anti:syn* ratio of 1:1 (estimated by 75-MHz $^{13}C$ NMR in $CDCl_3$ from the peak intensities of the $C_3$ methylene carbons at δ 22.2 and 22.3). Additionally, reduction of sulfur analogue **16** (*15*) to **17** (*16*) with *R*-Alpine-Hydride in THF at -78 °C gave poor diastereoselectivity [*anti:syn* = 1.4:1; estimated by $^{13}C$ NMR from the peak intensities for δ 41.6 (*anti* $CH_2$), 42.1 (*syn* $CH_2$); δ 72.4 (*anti* CH), 73.1 (*syn* CH)], indicating that divalent sulfur is not effective as a coordinating entity in this situation. The nearly 1:1 *anti:syn* ratio for these reductions certainly underscores the importance of the amine nitrogen (N-substituted) in **4** and **10** for achieving high 1,5 asymmetric induction.

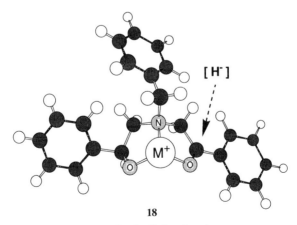

**18**

5,5-Bicyclic Chelate Complex

To rationalize our results, we propose a 5,5-bicyclic chelate structure, such as depicted in *model* **18**, in which the lithium or zinc ion is complexed with the hydroxyl, amine, and ketone groups in **4** or **10**. In the conformationally rigid array, endo attack by the hydride species is sterically unfavored, so attack is preferred from the less hindered exo side, leading to the *anti* 1,5-diols. Thus, when the reaction is performed in THF, this good donor solvent competes with the heteroatom groups in **4** or **10** for complexation of the metal center and thereby counteracts chelate formation. However, the relatively non-coordinating $CH_2Cl_2$ enhances chelate participation. The absence of stereoselectivity with NB-enantride (**8**) may be associated with the presence of a coordinating ether group in the reagent itself, which could inhibit formation of the necessary chelate complex. The results with amine replacement are consistent with this scenario: NBzl → $CH_2$ (**14**) gave no selectivity and N → S (**16**) gave little selectivity.

We have an interesting example of high 1,5 asymmetric induction in an acyclic system, involving hydride reduction of a prochiral ketone ($sp^2$ center) five atoms removed from an existing stereocenter bearing a hydroxyl group. The hydroxyl and intrachain nitrogen appear to participate in a metal chelate that serves to direct hydride addition ("5,5-bicyclic chelation control"). Given these intriguing results, we set out to explore for high 1,6 and 1,7 acyclic diastereoselection in related reductions of hydroxy ketones (i.e., 5,6-, 6,6-, and 5,7-bicyclic chelation control).

1,6-Hydroxy ketone **19** was readily prepared and reduced in $CH_2Cl_2$ at -78 °C to 1,6-diols **20** (Eq 4). We obtained *anti*:*syn* ratios for **20** of 2:1 with $LiBH_4$ (77% yield), 7.5:1 with $Zn(BH_4)_2$ (83%), and 12:1 with *R*-Alpine-Hydride (83%). Such high diastereoselectivity, 12:1, for a 1,6 reduction in an acyclic substrate is quite remarkable! It is interesting to note that reductions of **19** proceeded much more rapidly and in higher yields than reductions of **10**.

$$(4)$$

The diastereomeric ratios for **20** were nicely quantitated by 400-MHz proton NMR ($CDCl_3$). There were two different pairs of doublets for the aliphatic benzylic protons ("AB quartets") centered at δ 3.75 (*anti*) and δ 3.73 (*syn*), with no overlap of the signals because of a large chemical shift difference between the pair of doublets for the *anti* isomer, Δδ(*anti*) = ca. 0.50 ppm, and a small difference between the pair of doublets for the *syn* isomer, Δδ(*syn*) = ca. 0.12 ppm. Unambiguous assignment of the benzyl resonances required the synthesis of an authentic sample of *syn*- or *anti*-**20**, which we pursued by two routes (Scheme 2). Since the route involving the (*S*) or (*R*) chlorohydrin furnished either the *syn* or *anti* product in rather low yield, we preferred synthesizing the *anti* (*R,R*) isomer of **20** by means of (*R*)-styrene oxide.

**Scheme 2. Synthesis of Authentic Samples of 20.**

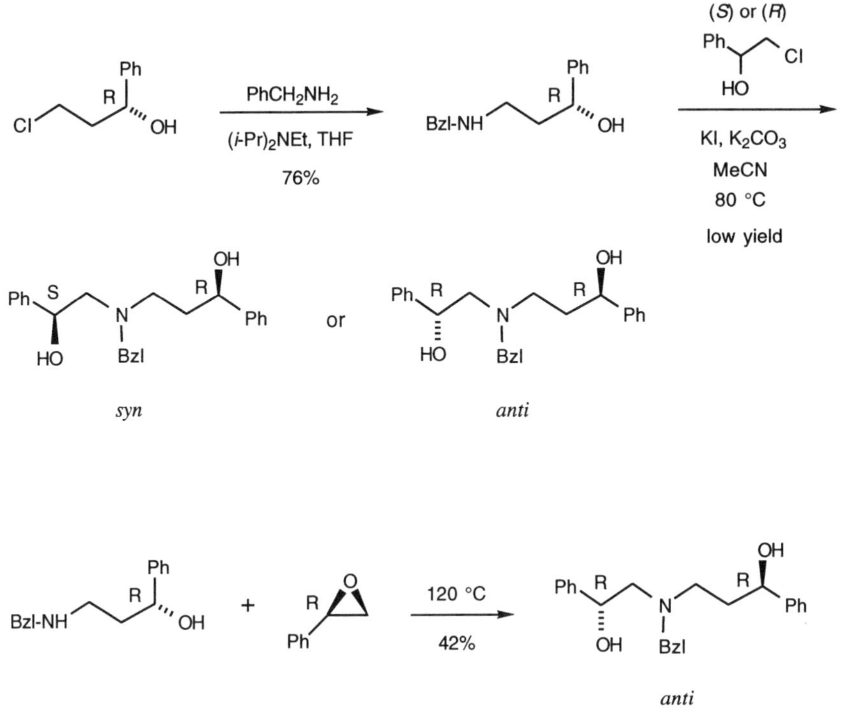


We propose a 5,6-bicyclic chelate structure, such as in model **21**, to account for the outstanding 1,6 acyclic diastereocontrol. The lithium or zinc ion would be complexed with the hydroxyl, amine, and ketone groups in **19** to establish a conformationally rigid array, and the hydride species would attack the carbonyl in the 6-membered ring from an axial direction to allow an equatorial phenyl ring to develop in the transition state. This leads to the major *anti* 1,6-diol product.

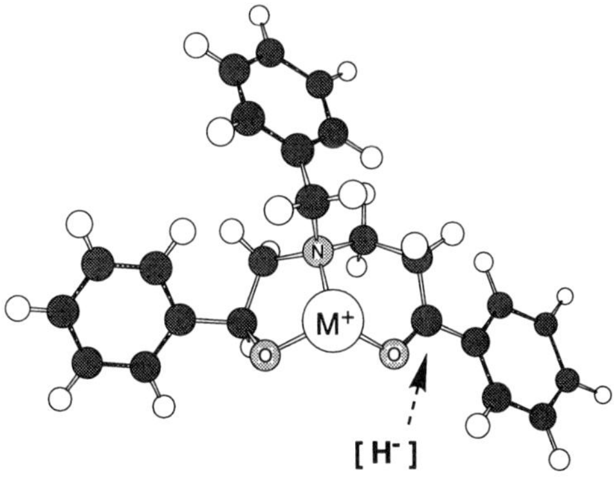


**21**

5,6-Bicyclic Chelate Complex

1,6-Hydroxy ketone **22** (*R* enantiomer only) was prepared and reduced with *R*-Alpine-Hydride or $Zn(BH_4)_2$ in $CH_2Cl_2$ at -78 °C to give 1,6-diols **20** in 75% or 72% yield, respectively (Eq 5). The *anti:syn* ratios of 5:1 and 3:1, respectively, for this "reversed" 1,6 acyclic system are significant, albeit not remarkable. Nevertheless, where 1,6 acyclic stereocontrol is concerned, a 5:1 ratio is quite meaningful. From this result, it seems that the degree of 1,6 diastereoselectivity is dependent on the distance between the hydroxyl and amine groups, with the β-hydroxy amine arrangement being preferred over the γ-hydroxy amine.

To test for the importance of a free hydroxyl, we prepared methoxy amino ketone **23**, by a Boc protection sequence, and subjected it to *R*-Alpine-Hydride at -78 °C in $CH_2Cl_2$ (Eq 6). Methoxy amino alcohols **24** were isolated in 80% yield as a 4:1 mixture of *anti* and *syn* forms, and the isomer assignment was established by the synthesis of an authentic sample of *anti*-**24** (Eq 7). Clearly, methylation of the hydroxyl group in **19** caused some loss in stereoselectivity (*anti:syn* = 12:1 for the parent reaction, **19** → **20**); however, this substitution was not completely detrimental. If one considers a 5,6-bicyclic chelate model, such as **21**, as being key to the 1,6 diastereoselectivity, then the methoxy group must be reasonably effective as a ligand for lithium in this reaction (but not as effective as a hydroxyl).

Given the favorable results with the 1,6 system, we decided to push our luck and examine related 1,7-hydroxy ketone cases. Hydroxy ketone **25** (*R* enantiomer), corresponding to a 6,6-bicyclic chelate model, was prepared and reduced with *R*-Alpine-Hydride in CH$_2$Cl$_2$ at -78 °C to give 1,7-diols **26** in 80% yield (Eq 8). The *anti:syn* (*dl:meso*) ratio was a mere 1.2:1. The *dl* and *meso* products were assigned by proton NMR: the two diastereotopic benzyl protons of the *anti* (*dl*) isomer appear as a pair of doublets at δ 3.32 and 3.67 (J = 13.7 Hz), while the two enantiotopic benzyl protons in the *syn* (*meso*) isomer appear as a singlet at δ 3.53.

The more promising 1,7-hydroxy ketone case is that of **27** (Eq 9), due to its β-hydroxy amino system. Reduction of **27** in CH$_2$Cl$_2$ at -78 °C with *R*-Alpine-Hydride or Zn(BH$_4$)$_2$ afforded 1,7-diols **28** in 84% or 63% yield, with *anti:syn* ratios of 3:1 or 1:2, respectively. The outcome with Zn(BH$_4$)$_2$ represents the first case in which a *syn* isomer was favored, although the degree of 1,7-stereocontrol in either direction is not impressive. The isomer assignment for **28** is based on proton NMR, in analogy with **20**. Mixture **28** displays two different pairs of doublets for the aliphatic benzylic protons ("AB quartets") centered at δ 3.72 (*anti*) and δ 3.70 (*syn*), with no overlap of the signals because of a large chemical shift difference between the pair of doublets for the *anti* isomer, Δδ(*anti*) = ca. 0.45 ppm, and a small difference between the pair of doublets for the *syn* isomer, Δδ(*syn*) = ca. 0.30 ppm. Considering the less than exciting diastereoselectivities for the reductions of **25** and, especially, **27**, we did not pursue the remaining 1,7 example that contains a δ-hydroxy amine.

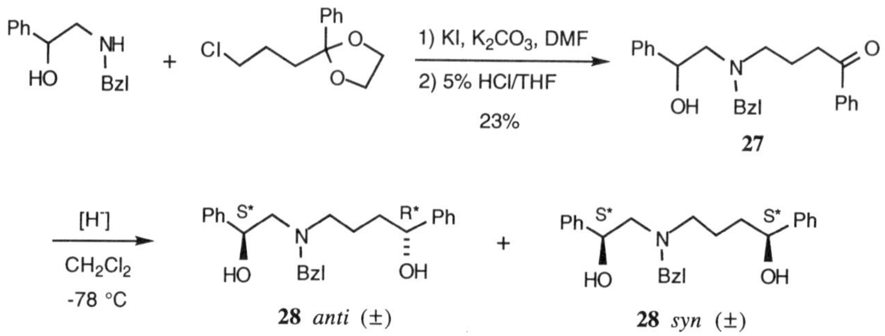

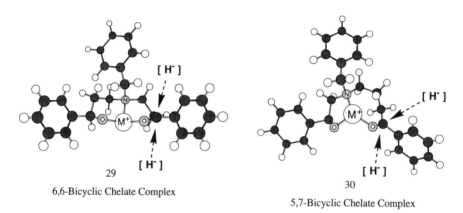

(9)

The lack of stereocontrol for the 1,7-hydroxy ketones, **25** and **27**, might be viewed with a consideration of 6,6- and 5,7-bicyclic chelate models, **29** and **30**. Presumably, attack of the hydride species has little facial preference from steric factors.

6,6-Bicyclic Chelate Complex

5,7-Bicyclic Chelate Complex

### Discussion on Remote Acyclic Diastereocontrol

As mentioned earlier, Maier et al. (*8*) reported that the reduction of δ-hydroxy ketone **6** with LiAlH$_4$ in Et$_2$O provides 1,5-diols with a *anti:syn* ratio of just 1.2:1. The main body of their work involved diastereocontrol in the double reduction of 1,2-, 1,3-, 1,4- and 1,5-diketones **31-34** with LiAlH$_4$ in Et$_2$O at 23 °C. The reactions yielded mixtures of the corresponding *dl* and *meso* diols, and the level of diastereoselectivity decreased gradually with increasing distance. The *meso:dl* ratios for double reduction of **31-34** were 4.9:1, 1:3.8, 3.2:1, and 1:1.5, respectively. Thus, a similar stereochemical result was obtained for 1,5-diketone **34** and its corresponding 1,5-hydroxy ketone **6**, as well as for biacetyl and its corresponding 3-hydroxy-2-butanone (*2c*). Although there is alternation in the form of the major diol isomer, the product invariably possesses the *anti* geometry. This parallels the results in our studies, in which the *anti* 1,5-, 1,6-, and 1,7-diols are generally favored (with one exception for a 1,7 case). The relatively modest remote stereocontrol observed by Maier et al. for their reaction type, even with 1,2- and 1,3-diketones, serves to highlight the significance of our results for the 1,5- and 1,6-hydroxy amino ketones.

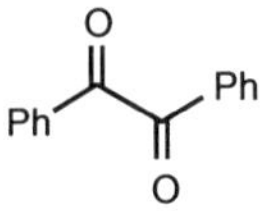

**31**, 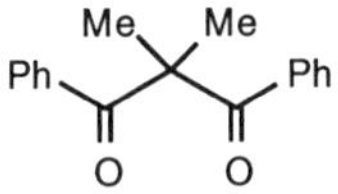 **32**, 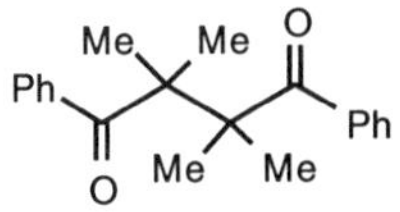 **33**, 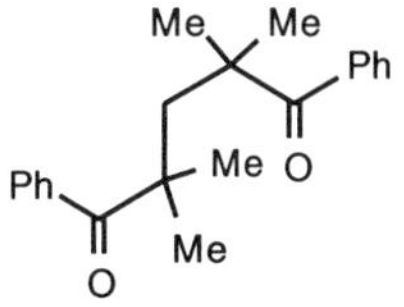 **34**

Maier et al. (8) proposed cyclic chelate models to rationalize the direction of hydride addition (Scheme 3). Since the ultimate stereochemistry is determined by the second reduction, the models show aluminum complexes of 1,n-hydroxy ketones. Details of their mechanistic discussion will not to be addressed here; the interested reader should refer to the original article.

**Scheme 3. Cyclic Aluminum Chelates (Adapted from Ref. 8).**

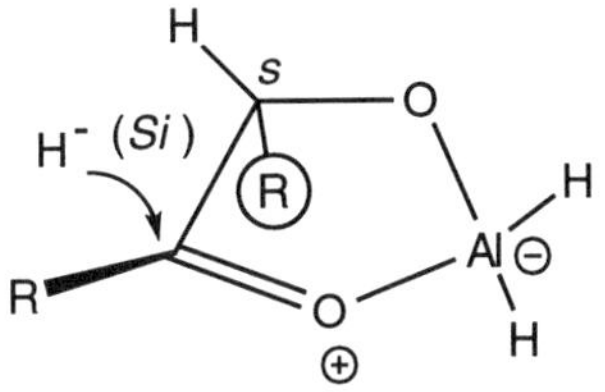

**1,2**, 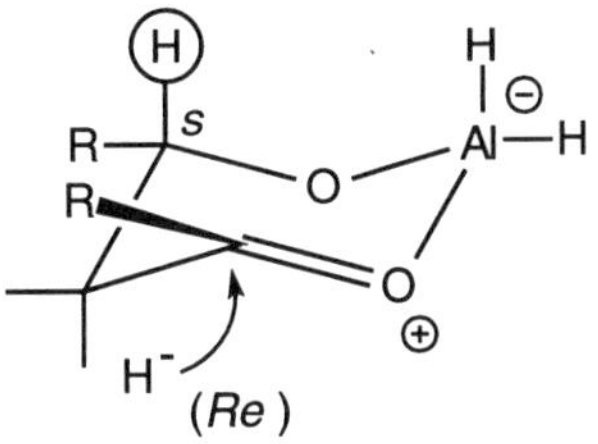 **1,3**, 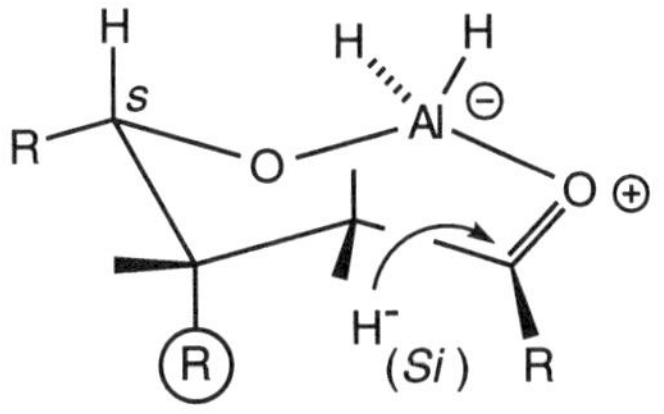 **1,4**, 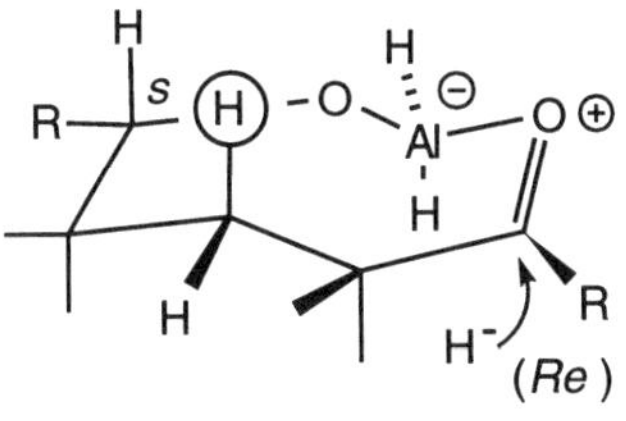 **1,5**

Very high diastereocontrol has been realized in the reduction of 1,3-hydroxy ketones with certain borohydride reagents (*17*, *18*). For stereochemically simple 1,3-hydroxy ketones, Evans et al. reported *anti* 1,3-diols of up to 92% diastereomeric excess (d.e.), i.e., 24:1 *anti:syn* ratio, by using $Me_4N^+BH(OAc)_3^-$ in THF at 40 °C (*17*). This high stereocontrol is connected with intramolecular hydride delivery via a cyclic transition state. Narasaka and Pai reported *syn* 1,3-diols of up to 96% d.e., i.e, 1:49 *anti:syn* ratio, by treatment with $Bu_3B$ in THF at 23 °C followed by $NaBH_4$ at -100 °C (*18*). The *syn* stereochemistry here is a consequence of chelate-controlled addition of external hydride.

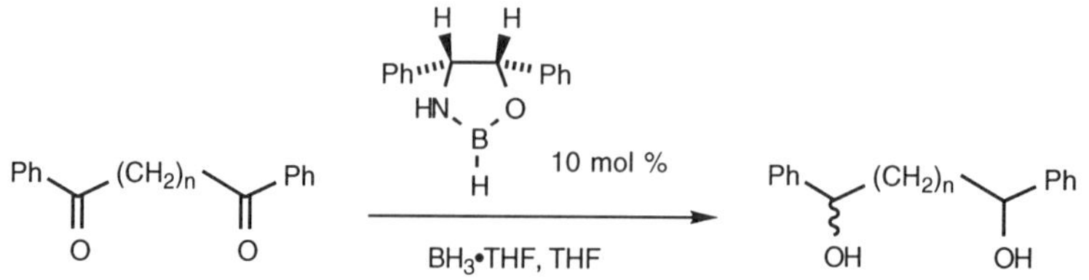

Recently, Quallich et al. (*19*) reported on the reduction of 1,2-, 1,4-, 1,5-, and 1,6-diketones to 1,n-diols in THF with borane-Me$_2$S alone and with a chiral oxaza-borolidine catalyst added (Eq 10). With borane-Me$_2$S *meso:dl* ratios were: 4.9:1 for 1,2, 1.6:1 for 1,4, 1.3:1 for 1,5, and 1:1 for 1,6. With 10 mol % catalyst added the *meso:dl* ratios were: 1:1.1 for 1,2, 1:3.8 for 1,4, 1:6.1 for 1,5, and 1:7.3 for 1,6. The high *dl* selectivity on increased separation of the reaction centers is reflective of the intrinsic, reagent-based enantioselectivity, which appears to be counteracted when the reaction centers are proximal. The magnitude and direction of the e.e. values were similar to those obtained for reduction of the monocarbonyl model, acetophenone (94% e.e., *S* configuration).

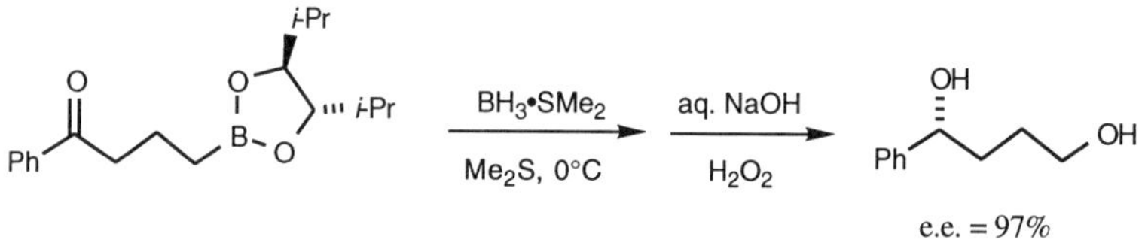

Molander and Bobbitt (*4a*) reported a most impressive 1,7 asymmetric reduction with borane-Me$_2$S that involves bicyclic chelation control (Eq 11). The dioxaborolane was suggested to complex with the carbonyl and then direct external hydride addition by the remote *i*-Pr substituents in the lower energy structure (Scheme 4).

**Scheme 4. Keto Boronate Bicyclic Transition State for 1,7 Diastereo-selectivity (Adapted from Ref. 4a).**

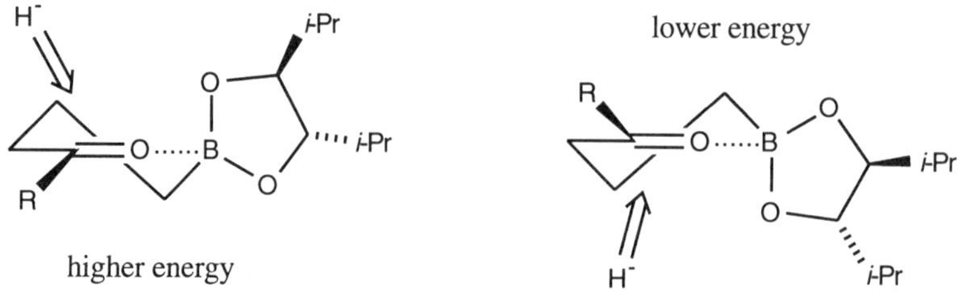

Another interesting example of a ketone reduction in an acyclic system that proceeds with high 1,5 diastereocontrol is given by the borohydride reduction in Eq 12, reported by Uemura et al. (*20*). The additional arene-Cr(CO)$_3$ stereogenic element relays stereochemical information into the conformationally flexible 1,5 system.

$$\text{(scheme 12)}$$

a: $R^1$ = Me, $R^2$ = H
b: $R^1$ = H, $R^2$ = Me

## Conclusion

High stereocontrol (>10:1) between remote 1,5 and 1,6 stereocenters was achieved in the reduction of certain hydroxy amino ketones, namely **4, 10, 12,** and **19,** by using *R*-Alpine-Hydride, $Zn(BH_4)_2$, or lithium triethylborohydride. *Anti* diastereoselectivity was optimized in a noncoordinating solvent, such as $CH_2Cl_2$, and at diminished temperatures, such as -78 °C. The distance between the hydroxyl and amino groups had a greater effect on the diastereoselectivity than the distance between the ketone and amino groups. It is reasonable to suggest that a bicyclic chelation model is responsible for the unusually high remote acyclic stereocontrol.

One of the major challenges in synthetic organic chemistry is the control of stereochemistry between remote sites in conformationally flexible systems. There are a limited number of examples of remote asymmetric induction in acyclic molecules with stereocenters at a distance of 1,5, 1,6, or greater, especially examples unperturbed by the presence of additional stereogenic features. To be sure, no general reactions for the construction of remote chiral relationships are yet known. We have discovered an entirely new reaction type that is capable of delivering high 1,5 and 1,6 *anti* asymmetric induction in an acyclic array, presumably on the basis of the broad concept of "bicyclic chelation control".

## Acknowledgments

We are grateful to Paul Lobben for technical assistance. The early part of this work was conducted in the Janssen Research Foundation, Spring House, PA.

## References and Notes

(1) For a preliminary report on our work, see: Zhang, H.-C.; Costanzo, M. J.; Maryanoff, B. E. *Tetrahedron Lett.*, **1994**, *35*, 4891.

(2) Reviews: (a) Oishi T.; Nakata, T. *Acc. Chem. Res.* **1984**, *17*, 338. (b) Reetz, M. T. *Angew. Chem., Int. Ed. Engl.* **1984**, *23*, 556. (c) Bartlett, P. A. *Tetrahedron* **1980**, *36*, 3. (d) *Asymmetric Synthesis*; Morrison, J. D., Ed.; Academic Press: New York, Vol. 2, 1983 and Vol. 3, 1984. (e) Hoveyda, A. H.; Evans, D. A.; Fu, G. C. *Chem. Rev.* **1993**, *93*, 1307.

(3) (a) Reissig, H.-U.; Angert, H. *J. Org. Chem.* **1993**, *58*, 6280. (b) Panek, J. S.; Xu, F. *Ibid.* **1992**, *57*, 5288. (c) Fujita, M.; Hiyama, T. *Ibid.* **1988**, *53*, 5405, 5415. (d) Evans, D. A.; Chapman, K. T.; Carreira, E. M. *J. Am. Chem. Soc.* **1988**, *110*, 3560. (e) Anwar, S.; Bradley, G.; Davis, A. P. *J. Chem. Soc. Perkin Trans. 1* **1991**, 1383. (f) Mori, Y.; Suzuki, M. *Tetrahedron Lett.* **1989**, *30*, 4383.

(4) (a) Molander, G. A.; Bobbitt, K. L. *J. Am. Chem. Soc.* **1993**, *115*, 7517. (b) Molander, G. A.; Haar, J. P., Jr. *Ibid.* **1993**, *115*, 40. (c) Panek, J.; Yang, Y.

*Ibid.* **1991**, *113*, 6594. (d) Erker, G.; Soana, F.; Betz, P.; Werner, S.; Kruger, C. *Ibid.* **1991**, *113*, 565. (e) Teerawutgulrag, A.; Thomas, E. *J. Chem. Soc., Perkin Trans. 1* **1993**, 2863. (f) Fang, J.-M.; Chang, C.-J. *Ibid.* **1989**, 1787. (g) Uemura, M.; Minami, T.; Hirotsu, K.; Hayashi, Y. *J. Org. Chem.* **1989**, *54*, 469. (h) Tamura, R.; Watabe, K-I.; Katayama, H.; Suzuki, H.; Yamamoto, Y. *Ibid.* **1990**, *55*, 408. (i) Denmark, S. E.; Marple, L. K. *Ibid.* **1990**, *55*, 1984. (j) Tamai, Y.; Koike, S.; Ogura, A.; Miyano, S. *J. Chem. Soc., Chem. Commun.* **1991**, 799.

(5) Van Lommen, G.; De Bruyn, M.; Schroven, M. *J. Pharm. Belg.* **1990**, *45*, 355. Also, see: *Drugs Future* **1991**, *16*, 964; *Drugs Future* **1995**, *20*, 1070.

(6) Burger, A.; Deinet, A. J. *J. Am. Chem. Soc.* **1945**, *67*, 566.

(7) Greeves, N. in *Comprehensive Organic Synthesis*, Vol. 8, Trost, B. M.; Fleming, I., Eds., Pergamon Press: Oxford, 1991, pp 1-24.

(8) Maier, G.; Roth, G.; Schmitt, R. K. *Chem. Ber.* **1985**, *118,* 704.

(9) Krishnamurthy, S.; Vogel, F.; Brown, H. C. *J. Org. Chem.* **1977**, *42*, 2534.

(10) Pirkle, W. H.; Hoover, D. J. *Top. Stereochem.* **1978**, *13*; 263.

(11) Midland, M. M.; Kazubski, A. *J. Org. Chem.* **1982**, *47*, 2495.

(12) Baker, R.; Cottrell, I. F.; Ravenscroft, P. D.; Swain, C. J. *J. Chem. Soc. Perkin Trans. 1* **1985**, 2463.

(13) We prepared **10** from 2-(benzylamino)-1-phenylethanol and 2-bromoacetophenone, according to: Brown, C. L.; Lutz, R. E. *J. Org. Chem.* **1952**, *17*, 1187.

(14) Huang, R. L.; Williams, P. J. *J. Chem. Soc.* **1958**, 2637.

(15) Partial reduction of the commercially available diketone with $NaBH_4$ provided a mixture of hydroxy ketone **16**, diol **17**, and starting material, from which **16** was separated by preparative TLC.

(16) Valle, G; Buso, M.; De Lucchi, O. *Zeit. Kristallogr.* **1987**, *181*, 95.

(17) Evans, D. A.; Chapman, K. T.; Carreira, E. M. *J. Am. Chem. Soc.* **1988**, *110*, 3560.

(18) Narasaka, K; Pai, H. C. *Chem. Lett.* **1980**, 1415.

(19) Quallich, G. J.; Keavey, K. N.; Woodall, T. M. *Tetrahedron Lett.*, **1995**, *36*, 4729.

(20) Uemura, M.; Minami, T.; Hirotsu, K.; Hayashi, Y. *J. Org. Chem.* **1989**, *54*, 469.

# Chapter 10

# Synthesis, Characterization, and Synthetic Utility of Lithium Aminoborohydrides
## A New Class of Powerful, Selective, Air-Stable Reducing Agents

Gayane Godjoian[1], Gary B. Fisher[1], Christian T. Goralski[2], and Bakthan Singaram[1,3]

[1]Department of Chemistry and Biochemistry, University of California, Santa Cruz, CA 95064
[2]Pharmaceuticals Process Research, Dow Chemical Company, Midland, MI 46874

> Lithium aminoborohydrides ($LiABH_3$) are a new class of powerful, selective, air-stable reducing agents. $LiABH_3$'s can be prepared as solids, as 1-2M THF solutions, or generated *in situ*, for immediate use. $LiABH_3$'s can be synthesized from any primary or secondary amine, thus allowing control of the steric and electronic environment of these reagents. Solid $LiABH_3$'s can be handled in dry air as easily as sodium borohydride and retain their chemical activity for at least 6 months when stored under nitrogen or dry air at 25 °C. THF solutions of $LiABH_3$'s retain their chemical activity for at least 9 months when stored under $N_2$ at 25 °C. $LiABH_3$'s are non-pyrophoric and only liberate hydrogen slowly in (or when using) protic solvents above pH 4. $LiABH_3$'s reduce aromatic and aliphatic esters in 30 minutes at 0 °C in air. Tertiary amides are selectively reduced to the corresponding amine or alcohol depending on the steric environment of the $LiABH_3$. $\alpha,\beta$-Unsaturated aldehydes and ketones undergo exclusive 1,2-reduction to the corresponding allylic alcohols. Aliphatic and aromatic azides are readily reduced to the corresponding primary amines using only 1.5 equivalents of $LiABH_3$.

Since the discovery of lithium aluminum hydride ($LiAlH_4$) in 1947 (1), enormous effort has been invested in the development of safe and convenient alternatives (2) to this useful but highly reactive reagent (3). Unfortunately, any increase in the stability of the new hydride reagents was accompanied by an even greater decrease in reactivity. Sodium bis(2-methoxy)aluminum hydride developed in 1969 and commercially available as Vitride® (4), is a significant improvement over $LiAlH_4$. Vitride® is non-pyrophoric, yet retains the reactivity of $LiAlH_4$. Unfortunately, the by-product of Vitride®, monomethoxyethanol, is a known teratogen (4d). Recently, the synthesis and characterization of the reducing properties of sodium aminoborohydrides was reported (5). Many of the functional groups reduced by $LiAlH_4$ were also reduced by sodium aminoborohydrides. However, no further work has appeared in the literature on the use of sodium aminoborohydride reducing agents and the goal of powerful, air-stable reducing agents seemed to be as elusive as ever.

[3]Corresponding author

However, in 1990, the serendipitous discovery (6, 7b) and subsequent characterization of the reducing properties of lithium aminoborohydrides (LiABH3) (6d), radically altered this picture.

## The Discovery of Lithium Aminoborohydrides

During our work on the hydroboration of $\beta,\beta$-disubstituted enamines, we were puzzled by the unexpected formation of dihydridoaminoboranes in the reaction product (eq. 1) (7a,b).

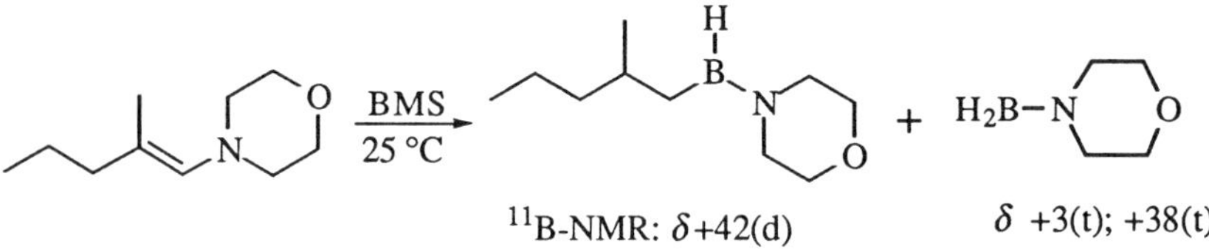

$$\text{enamine} \xrightarrow[25\,°C]{BMS} \text{alkyl-B(H)-morpholine} + H_2B\text{-morpholine} \quad (1)$$

$^{11}$B-NMR: $\delta$ +42(d) $\qquad \delta$ +3(t); +38(t)

In order to verify this result we needed authentic samples of these aminoboranes. Consequently, we developed a new method for the synthesis of dihydridoaminoboranes (eq. 2) (7).

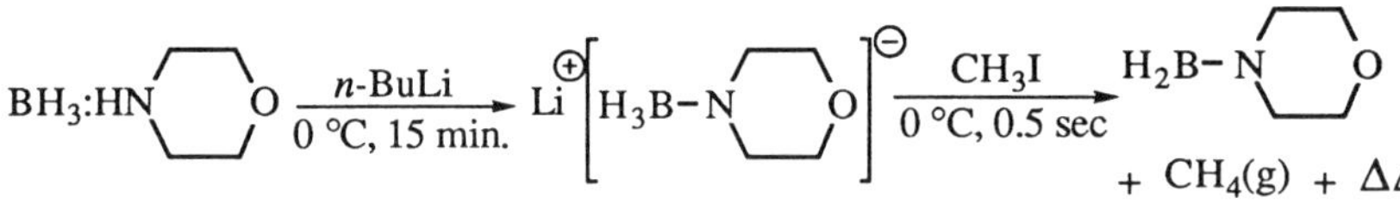

$$BH_3\!:\!HN\text{(morph)} \xrightarrow[0\,°C,\,15\,min.]{n\text{-BuLi}} Li^{\oplus}[H_3B\text{-N(morph)}]^{\ominus} \xrightarrow[0\,°C,\,0.5\,sec]{CH_3I} H_2B\text{-N(morph)} + CH_4(g) + \Delta\Delta \quad (2)$$

Reaction of $n$-butyllithium or methyllithium with amine-borane complexes, $H_3B\!:\!NHR_2$, in THF, readily afforded the corresponding LiABH3's in quantitative yields. When each of these LiABH3's were quenched with methyl iodide at 0 °C, a violent, exothermic reaction ensued and gave the corresponding aminoboranes in high purity as determined by $^{11}$B-NMR (eq. 2). Methyl iodide was known to react in such a vigorous fashion only with LiAlH4 and lithium triethylborohydride (LiEt3BH) (2e). This reaction suggested that the LiABH3's were a new type of powerful reducing agent, comparable in reducing power to LiAlH4 and LiEt3BH.

**Characterization of Spectral and Physical Properties.** The method used for synthesizing LiABH3's, shown in equation 2, is general. Following this procedure, several representative LiABH3's were prepared (Figure 1).

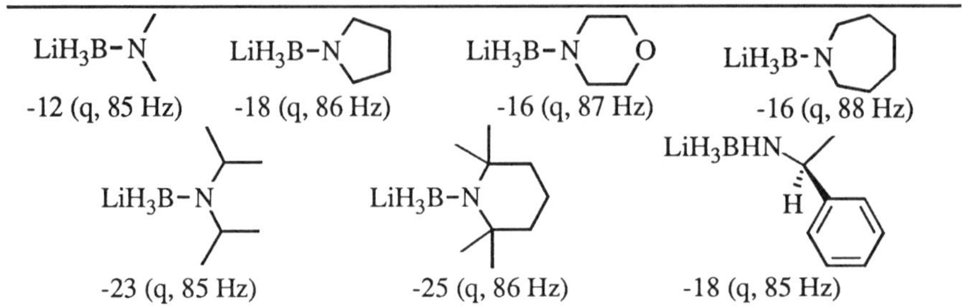

| LiH3B-N(Me)(iPr) | LiH3B-N(pyrrolidine) | LiH3B-N(morpholine) | LiH3B-N(azepane) |
|---|---|---|---|
| -12 (q, 85 Hz) | -18 (q, 86 Hz) | -16 (q, 87 Hz) | -16 (q, 88 Hz) |
| LiH3B-N(iPr)2 | LiH3B-N(piperidine) | LiH3BHN-CH(Me)Ph | |
| -23 (q, 85 Hz) | -25 (q, 86 Hz) | -18 (q, 85 Hz) | |

Figure 1. $^{11}$B-NMR shift values of representative LiABH3's. Although the chemical shifts of the corresponding amine-borane complexes are virtually identical to those of the LiABH3's, the $J$-values of the amine-borane complexes range from 95-98 Hz.

We have found two diagnostic criteria for determining the purity of LiABH$_3$'s: $^{11}$B-NMR coupling constants and the reaction of LiABH$_3$'s with methyl iodide. Although both LiABH$_3$'s and their corresponding amine-borane precursors appear as sharp quartets with virtually identical chemical shifts in their respective $^{11}$B-NMR spectra, the coupling constants of the LiABH$_3$'s and the corresponding amine-boranes are quite different. The LiABH$_3$'s exhibit $^{11}$B-NMR $J$-values of between 82 and 87 Hz. In contrast, amine-boranes have coupling constants ranging from 95-98 Hz. Additionally, LiABH$_3$'s react vigorously and exothermically with methyl iodide to liberate methane and the corresponding dihydridoaminoborane. Amine-borane complexes, however, are unreactive towards methyl iodide.

**Synthesis of THF Solutions of Lithium Aminoborohydrides.** THF solutions (1-2M) of the LiABH$_3$'s are quite stable and can be stored under nitrogen at 25 °C for at least nine months without undergoing any decomposition or loss of hydride activity.

**Synthesis of Lithium Piperidinoborohydride.** The following procedure for the synthesis of a 1M THF solution of lithium piperidinoborohydride is representative. A 1000-mL round-bottom flask equipped with a magnetic stirring bar and fitted with a rubber septum was charged by cannula with piperidine (34.1g, 39.6 mL, 400 mmol), THF (anhydrous, freshly distilled from benzophenone-ketyl, 160.4 mL), and cooled under positive nitrogen pressure to 0 °C. Borane methyl sulfide (10 M, 40 mL, 400 mmol) was added dropwise, with stirring, over a 30 min. period. The reaction mixture was stirred for an additional 1h at 0 °C. The reaction mixture was charged dropwise by cannula with $n$-butyllithium (2.5 M, 160 mL, 400 mmol) over a 90 minute period. The reaction mixture was stirred at 0 °C for 1h, then allowed to come to room temperature. The reaction mixture was stirred for an additional 1h at 25 °C to afford lithium piperidinoborohydride. $^{11}$B-NMR (THF): $\delta$ -16 (q, $J$ = 86 Hz); IR (THF): 2235 (B-H str), 1468 (B-N str), 1377 (B-N str), cm$^{-1}$.

**Synthesis of Solid Lithium Aminoborohydrides.** Solid LiABH$_3$'s were readily obtained by performing their synthesis in anhydrous diethyl ether (Et$_2$O) or hexanes followed by removal of the solvent under high vacuum (1 Torr, 40 °C). Solid LiABH$_3$'s are stable in dry air (6d,e) and can be stored under dry air in a closed container for over 6 months with no loss of hydride activity. LiABH$_3$'s are not pyrophoric either in solid form or as THF solutions (6d,e).

**Synthesis of Solid Lithium $N,N$-Diethylaminoborohydride.** The following procedure for the synthesis of solid lithium diethylamino-borohydride is representative. A 250-mL round-bottom flask equipped with a magnetic stirring bar and fitted with a rubber septum was charged with borane methyl sulfide (10 M, 5 mL, 50 mmol), and hexanes (anhydrous, 100 mL), and cooled under nitrogen to 0 °C. Diethylamine (3.6 g, 5.2 mL, 50 mmol) was added dropwise, with stirring, over a 2 min. period. The reaction mixture was stirred for an additional 1h at 0 °C. The reaction mixture was charged dropwise by cannula with $n$-butyllithium (2.5 M, 19 mL, 47.5 mmol, 0.95 equiv.) over a 15 minute period *(Note: The use of a substoichiometric amount of n-BuLi is essential to insure that the solid LiABH$_3$'s synthesized are non-pyrophoric! See ref. 6d)*. The reaction mixture was stirred at 0 °C for an additional 1h and allowed to come to room temperature. The reaction mixture was stirred for an additional 1h at 25 °C. The solvent was removed under reduce pressure (40 °C. 0.5 Torr) to yield $N,N$-diethylaminoborohydride as a white powder. $^{11}$B-NMR (THF): $\delta$ -18 (q, $J$ = 85 Hz); IR (THF): 2205 (B-H str), 1462 (B-N str), 1372 (B-N str), cm$^{-1}$.

**Safety Considerations** (6d,e). When synthesizing solid LiABH$_3$'s, *it is essential that no trace of n-BuLi remains in the LiABH$_3$!* Residual $n$-BuLi acts as a "fuse" and will cause the LiABH$_3$ to ignite in air. We have found that *use of a substoichiometric amount of n-BuLi (0.95 equiv.) ensures that the resulting solid LiABH$_3$ is completely non-pyrophoric*. However, when generating LiABH$_3$'s *in situ*, equimolar amounts of amine-borane and *n*-BuLi are routinely employed with no resulting safety problems. Acidic compounds which have a pK$_a$<4.0 react readily with LiABH$_3$'s, liberating hydrogen. However, unlike the violent and potentially dangerous reaction of LiAlH$_4$ with water or methanol, hydrogen was liberated only slowly during the reaction of LiABH$_3$'s with these solvents. Moreover, LiABH$_3$'s were converted to the unreactive amine-borane complex on addition to either methanol or water, as indicated by $^{11}$B-NMR spectra. LiABH$_3$'s that were converted to the corresponding amine-boranes by prolonged deliberate exposure to air-borne moisture were readily regenerated by deprotonation, under nitrogen, with *n*-BuLi.

**Characterization of Physical Properties.** Attempts to obtain melting point data on the solid LiABH$_3$'s were not successful. However, LiABH$_3$'s displayed remarkable thermal stability, with gradual decomposition occurring above 200 °C. The thermal stability of LiABH$_3$'s is particularly notable when compared to the explosive decomposition that results from heating LiAlH$_4$ above 100 °C (1e).

**Characterization of Reducing Properties.** We began a systematic study of the reducing properties of LiABH$_3$'s using lithium pyrrolidinoborohydride (LiPyrrBH$_3$) as the representative reagent (6d). Numerous types of functional groups were readily reduced and the products were easily isolable by a simple acidic work-up to eliminate contamination by boron-containing materials. Unlike other powerful reducing agents, once the LiABH$_3$ has been generated, either *in situ* or in solid form, reductions can be performed without any precautions to exclude air. However, the exclusion of adventitious moisture was essential to maximize the yields in reductions that require refluxing or more than one hour to complete. Thus, we recommend carrying out such reductions in tetrahydrofuran (THF) under nitrogen or an atmosphere of dry air. Additionally, recent work using solid LiABH$_3$'s has demonstrated that these reagents can be handled in dry air with the same ease as sodium borohydride (NaBH$_4$). Whether generated *in situ* or used in solid form, most reductions with LiABH$_3$'s were complete in 2-3 hours at ambient temperature, although some very hindered substrates required refluxing in THF for 2-3 hours to achieve a reasonable yield of the desired product.

**Reduction of Aldehydes and Ketones.** The reduction of aldehydes and ketones generally was complete in 15-30 minutes at 0 °C (6d). Thus, the reduction of 2-methylcyclohexanone with LiPyrrBH$_3$ gave a 40:60 ratio of *cis*- to *trans*-2-methylcyclohexanol in 88% isolated yield (eq. 3).

$$\text{2-methylcyclohexanone} \xrightarrow[\text{THF, 0 °C, 0.5 h}]{\text{LiPyrrBH}_3} \text{cis-2-methylcyclohexanol} + \text{trans-2-methylcyclohexanol} \quad (3)$$

(88% isolated yield)
40%    60%

Aromatic ketones were reduced with similar ease. Both aliphatic and aromatic ketones and aldehydes required only one equivalent of LiABH$_3$ to effect the complete reduction to the corresponding alcohol. The stereoselectivity of LiABH$_3$ reductions of 3-substituted cyclohexanones indicates that the LiABH$_3$'s behave like unhindered hydride reagents, regardless of the size of the amine moiety. Thus, the reduction of 3-methylcyclohexanone gave *cis*-3-methylcyclohexanol as the major product (eq. 4) (8).

$$\text{3-methylcyclohexanone} \xrightarrow[\text{THF, 0 °C, 0.5 h}]{\text{LiABH}_3} \text{cis-3-methylcyclohexanol (8%)} + \text{trans-3-methylcyclohexanol (92%)} \quad (4)$$

Additionally, the reduction of 4-*tert*-butylcyclohexanone using lithium di-*n*-propylamino-borohydride (Li(*n*-Pr)$_2$NBH$_3$) gave 99% *trans*-4-*tert*-butylcyclohexanol in 95% isolated yield (eq. 5).

$$\text{4-tert-butylcyclohexanone} \xrightarrow[\text{THF, 0 °C, 0.5 h}]{\text{Li}(n\text{-Pr})_2\text{NBH}_3} \text{trans-4-tert-butylcyclohexanol} \quad (95\% \text{ isolated yield}) \quad (5)$$

**Reduction of $\alpha,\beta$-Unsaturated Aldehydes and Ketones.** $\alpha,\beta$-Unsaturated aldehydes and ketones were reduced with remarkable regioselectivity to the corresponding allylic alcohols. Although other reagents are known that give high 1,2:1,4 reduction ratios (9), LiABH$_3$'s are the only reducing agents that give exclusive 1,2-reduction of both $\alpha,\beta$-unsaturated aldehydes and ketones, results that compliment those obtained using LiAlH$_4$ and the Luche reagent (9). For example, the reduction of cinnamaldehyde using LiAlH$_4$ gave exclusively the corresponding saturated alcohol (9). Although the Luche reagent (NaBH$_4$/CeCl$_3$) gives exclusive 1,2-reduction of $\alpha,\beta$-unsaturated ketones, it does not reduce $\alpha,\beta$-unsaturated aldehydes (9). In contrast, LiPyrrBH$_3$ reduces cinnamaldehyde exclusively to the 1,2-reduction product in 95% isolated yield (eq. 6).

$$\text{cinnamaldehyde} \xrightarrow[\text{THF, 25 °C, 3 h}]{\text{LiPyrrBH}_3} \text{cinnamyl alcohol} \quad (95\% \text{ isolated yield}) \quad (6)$$

(*R*)-(+)-Pulegone was reduced to the corresponding allylic alcohol, (*1R,3R*)-(-)-*cis*-pulegol, in 96% isolated yield (eq. 7).

$$\text{(R)-(+)-pulegone} \xrightarrow[\text{THF, 25 °C, 3 h}]{\text{LiPyrrBH}_3} \text{(1R,3R)-(-)-cis-pulegol} \quad (96\% \text{ isolated yield}) \quad (7)$$

Similarly, (*R*)-(-)-carvone was reduced to (*1R, 5R*)-(-)-*cis*-carveol (9). Reduction of Hagemann's ester (4-carbethoxy-3-methyl-2-cyclohexen-1-one) demonstrated the chemoselectivity of LiABH$_3$'s. Using one equivalent of LiPyrrBH$_3$, the $\alpha,\beta$-unsaturated ketone functionality was reduced leaving the ester moiety intact (eq. 8).

$$\text{3-methyl-4-oxo-cyclohex-2-ene-1-carboxylate} \xrightarrow[\text{THF, 25 °C, 3 h}]{\text{LiPyrrBH}_3} \text{3-methyl-4-hydroxy-cyclohex-2-ene-1-carboxylate} \quad (94\% \text{ isolated yield}) \quad (8)$$

**Reduction of Carboxylic Acid Esters.** Even though several reducing agents are available for the reduction of esters to the corresponding alcohols (10), all practical methodologies require the rigorous exclusion of air during the reduction. In contrast, the reduction of both aliphatic and aromatic esters with $LiABH_3$'s can be carried out rapidly in air (6d). Thus, ethyl octanoate was reduced, in air, to the corresponding alcohol in 0.5 h at 0 °C in >90% isolated yield (eq. 9).

$$\text{CH}_3(\text{CH}_2)_6\text{C(O)OEt} \xrightarrow[\text{THF, 0 °C, 0.5h}]{\text{LiPyrrBH}_3} \text{CH}_3(\text{CH}_2)_7\text{OH} \quad (92\% \text{ isolated yield}) \quad (9)$$

Further, reduction of ethyl cinnamate afforded exclusively the 1,2-reduction product, cinnamyl alcohol, in 95% isolated yield (eq. 10).

$$\text{PhCH=CHC(O)OEt} \xrightarrow[\text{THF, 0 °C, 0.5 h}]{\text{Li}(i\text{-Pr})_2\text{NBH}_3} \text{PhCH=CHCH}_2\text{OH} \quad (95\% \text{ isolated yield}) \quad (10)$$

Thus, for reduction of aliphatic or aromatic esters with $LiABH_3$'s, the mild reaction conditions, short reaction times, and chemoselectivity of $LiABH_3$'s make these reagents excellent alternatives to the reagents currently available for carrying out this transformation (6d, 10).

**General Procedure for Ester Reductions. Use of a 1M THF Solution of $LiABH_3$. In-Air Reduction of Ethyl Benzoate to Benzyl Alcohol.** The following procedure for the reduction of ethyl benzoate with a 1M THF solution of lithium pyrrolidinoborohydride ($LiPyrrBH_3$) is representative. A 125-mL Erlenmeyer flask equipped with a magnetic stirring bar and fitted with a rubber septum was cooled to room temperature under nitrogen. The flask was charged with ethyl benzoate (1.5 g, 1.4 mL, 10 mmol) and THF (anhydrous, 20 mL). The flask was cooled to 0 °C, opened to the air and a 1M THF solution of $LiPyrrBH_3$ (12 mL, 12 mmol) was added in a single portioin with stirring to the reaction mixture. The reaction was maintained at 0 °C and stirred for an additional 1 h. The reaction mixture was kept at 0 °C and quenched by the slow addition of 3 M HCl (17 mL, 51 mmol) *[Caution: Hydrogen evolution!]*. The aqueous and organic fractions were separated and the aqueous fraction extracted with $Et_2O$ (2 X 50 mL). The combined ethereal fractions were washed with water (3 X 10 mL) and dried over $MgSO_4$. The solvent was removed at 25 °C under reduced pressure (6 Torr). The residue was distilled to yield benzyl alcohol (bp: 47-48 °C, 2 Torr, 1.0 g, 95% isolated yield)

**Reduction of Tertiary Amides.** Primary and secondary amides were not reduced with $LiABH_3$'s even in refluxing THF and were recovered unchanged. However, a wide variety of aromatic and aliphatic tertiary amides were reduced in excellent yield in dry air with $LiABH_3$'s (6c, d). The reduction of unhindered tertiary amides, such as $N,N$-dimethylbenzamide, gave benzyl alcohol regardless of the $LiABH_3$ used. However, for more sterically demanding tertiary amides, selective C-O or C-N bond cleavage (6c, 11k) was achieved by varying the steric environment of the amine moiety of the LiABH$_3$. For example, reduction of 1-pyrrolidinooctanamide with LiPyrrBH$_3$ gave 1-octanol in 77% isolated yield. When the reduction of 1-pyrrolidino-octanamide was carried out with the significantly more sterically demanding LiABH$_3$, Li($i$-Pr)$_2$NBH$_3$, 1-octylpyrrolidine was obtained in 95% isolated yield. Similarly, reduction of $N,N$-diethyl-$m$-toluamide with Li($i$-Pr)$_2$NBH$_3$ gave 3-methyl-$N,N$-diethylbenzylamine in 95% isolated yield while reduction of this amide with LiPyrrBH$_3$ gave 3-methylbenzyl alcohol in 95% isolated yield (eq. 11).

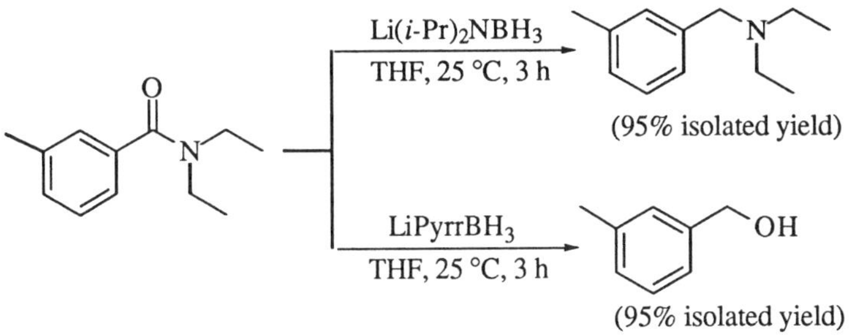

(11)

The selectivity of this reduction appears to involve a common intermediate, **1**, obtained as the initial reduction product of the amide (Figure 2) (6d, 11k).

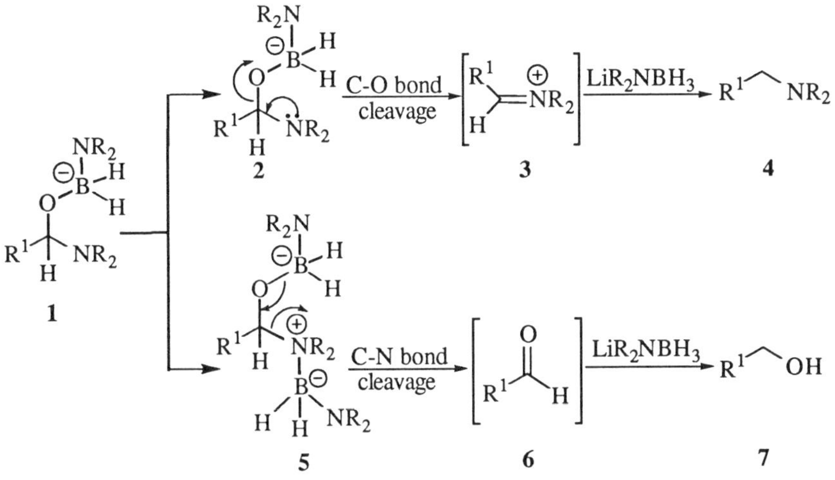

Figure 2. Mechanism to account for the selective C-O or C-N bond cleavage in LiABH$_3$ reductions of tertiary amides.

From the intermediate **1**, two different pathways lead to the corresponding amine or alcohol. In the first, the iminium species, **3**, is formed by the expulsion of the lithium dihydridoaminoborinate, **2**. LiABH$_3$ then rapidly reduces the iminium group to the corresponding amine, **4**. In the second pathway, the complexation of an aminoborane to the nitrogen of **1** converts the amine to an ammonium moiety, **5**, making it a better leaving group. Expulsion of the diaminodihydridoborohydride moiety results in the formation of an aldehyde, **6**, which is rapidly reduced to the corresponding alcohol, **7**. In all of the tertiary amide reductions that we have carried out, we have found that, as the groups bonded to the amide or LiABH$_3$ nitrogen are made more sterically demanding, amine formation through C-O bond cleavage is favored, apparently due to the unfavorable steric interactions between the LiABH$_3$

and the amide nitrogen. Whereas reductions performed with LiAlH$_4$ give mainly C-O bond cleavage and those carried out with LiEt$_3$BH (11k) give C-N bond cleavage, LiABH$_3$'s give selective C-O or C-N bond cleavage by simply altering the steric environment of the amine moiety of the LiABH$_3$.

**General Procedure for Tertiary Amide Reductions. Reduction of 1-Pyrrolidinooctanamide to 1-Octylpyrrolidine.** The following procedure is representative for all *in situ* reductions. A 50-mL round bottom flask fitted with a rubber septum and equipped with a magnetic stirring bar was cooled to room temperature under nitrogen. The flask was charged with a 1 M solution of Li(*i*-Pr)$_2$NBH$_3$ (12 mL, 12 mmol) followed by the dropwise addition, with stirring, of neat 1-pyrrolidinooctanamide (1.4 g, 6.9 mmol). The reaction mixture was stirred for an additional 2h at 25 °C. The reaction mixture was cooled to 0 °C and quenched by the *slow* addition of 3 M HCl (17 mL, 51 mmol) *[Caution: Hydrogen evolution!]*. The aqueous fraction was separated, layered with Et$_2$O (~40 mL), cooled to 0 °C and NaOH (s) was added until the reaction mixture was strongly basic to litmus. The ethereal fraction was separated, the aqueous fraction extracted with Et$_2$O (4 X 15 mL), and the ethereal fractions were combined and dried over MgSO$_4$. After filtration, the solvent was removed by aspirator vacuum (25 °C, 6mm) to yield 1-octylpyrrolidine (bp: 44-45 °C, 0.3 Torr; 1.7 g, 95% isolated yield).

**Reduction of Epoxides.** Epoxides were readily reduced to the corresponding alcohols with LiPyrrBH$_3$. No amino alcohols were formed in this reduction (5). Styrene oxide gave predominantly 1-phenylethanol. Additionally, cyclohexene oxide was reduced to cyclohexanol (eq. 12).

$$\text{cyclohexene oxide} \xrightarrow[\text{THF, 25 °C, 1 h}]{\text{LiPyrrBH}_3} \text{cyclohexanol} \quad (93\% \text{ isolated yield}) \quad (12)$$

**Reduction of Alkyl Halides.** The reduction of alkyl halides with LiABH$_3$'s gave the corresponding alkanes in excellent yields. Thus, 1-iododecane was cleanly reduced to decane (eq. 13).

$$\text{1-iododecane} \xrightarrow[\text{THF, 25 °C, 1 h}]{\text{LiPyrrBH}_3} \text{decane} \quad (92\% \text{ isolated yield}) \quad (13)$$

**Reducton of Azides.** Synthesis of primary amines by the reduction of azides is an important transformation in organic synthesis (12). Azides are commonly reduced using either an excess of LiAlH$_4$ (13) or catalytic hydrogenation (14), although other methods have been developed (15).

Lithium dimethylaminoborohydride (Li(Me)$_2$NBH$_3$) was employed for our azide reductions to facilitate separation of the amine generated during workup from the primary amine reduction product. Reductions of aliphatic and aromatic azides to the corresponding primary amines with Li(Me)$_2$NBH$_3$ (16a) were run under mild conditions: 1.5 equivalents of Li(Me)$_2$NBH$_3$ in THF at 25 °C. The reductions were complete within 2 to 5 hours, depending on the electronic and steric environment of the azide. Thus, the reduction of benzyl azide gave benzylamine in 85% isolated yield (eq. 14).

$$\text{PhCH}_2\text{N}_3 \xrightarrow[\text{THF, 25 °C, 2 h}]{\text{LiMe}_2\text{NBH}_3} \text{PhCH}_2\text{NH}_2 \quad (85\% \text{ isolated yield}) \quad (14)$$

Further, during the course of our studies of the nicotinic acetycholine receptor (AchR) (16a, b), a cell membrane protein (16b), we required 3α-aminocholest-5-ene and 3β-aminocholest-5-ene for use as non-radio-labeled cell membrane photo-affinity labels (16b). Previous syntheses of these cholestene amines utilized the reduction of the corresponding azides (16c). However, the reduction of these azides required the use of 12-25 equivalents of $LiAlH_4$ under an inert atmosphere (16c). When 3α-azidocholest-5-ene and 3β-azidocholest-5-ene were treated with $LiMe_2NBH_3$, we obtained the corresponding cholestene amines in 98% isolated yield (eqs. 15 and 16) (16a).

$$3\alpha\text{-azidocholest-5-ene} \xrightarrow[\text{THF, 65 °C, 3h}]{\text{LiMe}_2\text{NBH}_3 \ (1.5 \text{ eq.})} 3\alpha\text{-aminocholest-5-ene} \quad (98\% \text{ isolated yield}) \quad (15)$$

$$3\beta\text{-azidocholest-5-ene} \xrightarrow[\text{THF, 65 °C, 3h}]{\text{LiMe}_2\text{NBH}_3 \ (1.5 \text{ eq.})} 3\beta\text{-aminocholest-5-ene} \quad (98\% \text{ isolated yield}) \quad (16)$$

Particularly noteworthy aspects of these reductions were: (1) only 1.5 equivalents of $LiMe_2NBH_3$ were required to obtain a 98% isolated yield of both the 3α-aminocholest-5-ene and 3β-aminocholest-5-ene, respectively; (2) the reductions were complete in 2-3 hours; and (3) the reductions of both 3α-azidocholest-5-ene and 3β-azidocholest-5-ene were carried out *in air*.

**General Procedure for the Reduction of Aliphatic and Aromatic Azides. Reduction of Cyclohexylazide to Cyclohexylamine.** The following procedure for the reduction of cyclohexylazide to cyclohexylamine using lithium dimethylaminoborohydride ($LiMe_2NBH_3$) is representative. A 50-mL round bottom flask equipped with a magnetic stirring bar and fitted with a rubber septum was cooled to 0 °C under nitrogen and charged by syringe with cyclohexylazide (1.3 g, 10 mmol) and THF (anhydrous, 10 mL). The reaction mixture was charged dropwise by syringe with a 1M THF solution of $LiMe_2NBH_3$. The reaction mixture was stirred for 30 min at 0 °C, then for an additional 2h at 25 °C. Reaction progress was monitored by FT-IR: disappearance of the very strong azide peak (~2100 cm$^{-1}$) indicated completion of the reduction. The reaction was quenched by the *slow* addition of 12M HCl (60 mmol, 5mL) *[Caution!! Hydrogen evolution!]*. The aqueous fraction was separated, washed with diethyl ether ($Et_2O$; 3 X 10 mL), and the ethereal fractions discarded. The aqueous fraction was layered with fresh $Et_2O$ (~30 mL). The reaction mixture was basified with NaOH (s) and 1-2 mL of 3M NaOH (aq) until strongly basic to litmus and all color had transferred to the ethereal layer. The organic layer was separated and the intractable solid that remained was washed with $Et_2O$ (10 mL aliquots) until colorless. The ethereal fractions were combined and dried over MgSO$_4$. The solvent was removed by distillation at 25 °C to yield cyclohexylamine (1.0 g, 98% isolated yield, BP: 128-130 °C, 760 Torr).

**Transfer Reactions of LiABH$_3$'s: A General Synthesis of Alkyl-Substituted Borohydrides** Lithium trialkylborohyrides are powerful and selective reducing agents (17). Because of their synthetic utility as reducing agents, several synthetic routes to lithium trialkylborohydrides have been explored (18). Trialkylborohydrides can be synthesized by addition of lithium hydride (LiH) to unhindered trialkyborane (19). However, this procedure is not suitable for the synthesis of hindered trialkylborohydrides. This has led to the development of a number of reactive LiH equivalents, such as *tert*-butyllithium and LiAlH$_4$ (20, 21).

We envisioned LiABH$_3$'s to be an ether-soluble source of activated LiH, much like LiAlH$_4$. This in turn suggested that LiABH$_3$'s would be an excellent source of LiH for transfer of these elements to alkylboranes. Recent results from our laboratories have confirmed that LiABH$_3$'s readily add LiH to borane, monoalkylboranes, dialkylboranes, and hindered trialkylboranes to form the corresponding lithium alkylborohydrides (eq. 17) (22).

$$\begin{array}{c} R_1 \\ | \\ R_3 \diagdown B \diagup R_2 \end{array} + LiH_3BN\begin{array}{c} C_3H_7 \\ \diagdown \\ C_3H_7 \end{array} \xrightarrow{0\ °C,\ THF,\ 1h} \left[ \begin{array}{c} H \\ | \ominus \\ R_1 \diagdown B \diagup R_3 \\ | \\ R_2 \end{array} \right] Li^{\oplus} + H_2B-N\begin{array}{c} C_3H_7 \\ \diagdown \\ C_3H_7 \end{array} \quad (17)$$

$R_1, R_2, R_3$ = Alkyl, H

The exchange of LiH from LiABH$_3$'s to the trialkylboranes indicates that the aminoboranes are less Lewis acidic than these alkylboranes. Even highly hindered trialkylboranes, such as trisiamylborane and *B*-isopinocampheyl-9-borabicyclo (3.3.1)nonane, reacted readily with lithium di-*n*-propylaminoborohydride in THF to afford the corresponding lithium trialkylborohydrides (eqs. 18 and 19) (22).

$$[\text{trisiamylborane}] \xrightarrow{Li(n-Pr)_2NBH_3} \left[ \text{trisiamyl-B-H} \right] Li^{\oplus} \quad (18)$$

$^{11}$B-NMR: $\delta$ -13.5 (d, $J$ = 78 Hz)

$$[\text{B-isopinocampheyl-9-BBN}] \xrightarrow{Li(n-Pr)_2NBH_3} \left[ \text{product} \right] Li^{\oplus} \quad (19)$$

$^{11}$B-NMR: $\delta$ -6.6 (d, $J$ = 75 Hz)

Generally, due to incomplete reaction between LiH and trialkylboranes, lithium trialkylborohydrides give broad undefined peaks in their $^{11}$B-NMR spectrum (23). However, the $^{11}$B-NMR spectrum of the trialkylborohydrides obtained by the above transfer reaction showed sharp splitting patterns, indicative of a complete transfer of LiH from LiABH$_3$ to the trialkylboranes.

The transfer of LiH converts the LiABH$_3$ to an aminoborane (R$_2$NBH$_2$) which remains in the reaction mixture (eq. 17). Since R$_2$NBH$_2$'s are relatively unreactive and are poor reducing agents (7b), the presence of the R$_2$NBH$_2$ in the final product does not influence the reactivity of the lithium alkylborohydride.

Lithium trisiamylborohydride is known to react with 4-*tert*-butylcyclohexanone to yield the corresponding *cis* alcohol (24). Lithium trisiamylborohydride produced by the exchange reaction also reduces 4-*tert*-butylcyclohexanone in the presence of the aminoborane byproduct to give only the *cis* alcohol (eq. 20).

$$\text{4-}t\text{Bu-cyclohexanone} + \left[\left(\text{siamyl}\right)_3\text{B-H}\right]^\ominus \text{Li}^\oplus \xrightarrow[-78\,°\text{C}]{\text{H}_2\text{BN}(n\text{-Pr})_2} \text{4-}t\text{Bu-cyclohexanol (}cis\text{)} \quad (20)$$

We next sought to determine the generality of this reaction. Further studies of the exchange reaction showed that LiABH$_3$'s transfer LiH to dialkylboranes, such as diisopinocampheylborane (eq. 21).

$$(\text{Ipc})_2\text{BH} \xrightarrow[\text{THF, 0 °C, 1h}]{\text{LiH}_3\text{BN}(n\text{-Pr})_2} \left[(\text{Ipc})_2\text{BH}_2\right]^\ominus \text{Li}^\oplus \quad (21)$$

$$\delta = -5.8 \ (t, J = 67 \text{ Hz})$$

In addition to forming di- and trialkylborohydrides, we also tested the exchange reaction on monoisopinocampheylborane, a monoalkylborane. The exchange reaction afforded the corresponding monoalkylborohydride (eq. 22).

$$\text{IpcBH}_2 \xrightarrow[\text{THF, 0 °C, 1 h}]{\text{LiH}_3\text{BN}(n\text{-Pr})_2} \left[\text{IpcBH}_3\right]^\ominus \text{Li}^\oplus \quad (22)$$

$$^{11}\text{B-NMR: } \delta - 21.1 \ (q, J = 75 \text{ Hz})$$

The exchange reaction also works with borane and affords lithium borohydride in essentially quantitative yield (eq. 23).

$$\text{H}_3\text{B:SMe}_2 \xrightarrow[\text{THF, 0 °C, 1h}]{\text{LiH}_3\text{BN}(n\text{-Pr})_2} \text{LiBH}_4 \quad (23)$$

$$^{11}\text{B-NMR: } \delta -41 \ (qnt, J = 84 \text{ Hz})$$

The aminoborane byproduct is much less Lewis acidic than borane or alkylboranes. Consequently, the exchange reaction is the result of the thermodynamically favorable exchange of lithium hydride. Thus, trialkyl-, dialkyl-, monoalkylboranes can be converted into the corresponding lithium alkylborohydrides by this simple exchange reaction. The mild reaction conditions and the generality of the reaction makes this procedure attractive for the synthesis of a wide variety of substituted borohydrides.

## Conclusion

The future use of lithium aminoborohydride reducing agents in both industry and academia appears to be quite promising. Lithium aminoborohydrides are already finding useful applications in organic synthesis. Recently, lithium pyrrolidinoborohydride was utilized to selectively cleave the amide C-N bond in pseudoephedrine amides to yield highly enantiomerically enriched primary alcohols in 81-88% yield (eq 24) (25).

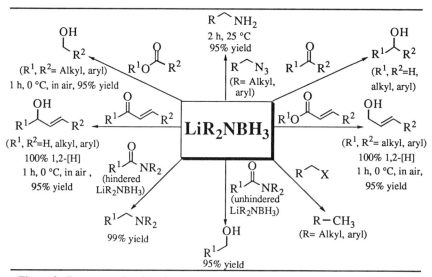

Although the reduction was problematic for R = phenyl and R' = ethyl, the use of LiH$_2$NBH$_3$ in THF gave the desired primary alcohol in 80% yield and 88% ee.

Lithium aminoborohydrides are a new class of powerful yet selective reducing agents that reproduce, in air, virtually all of the transformations for which LiAlH$_4$ is now used (Figure 3).

Figure 3. Representative functional groups reduced by lithium aminoborohydrides.

The reactivity of LiABH$_3$'s is comparable to that of both LiAlH$_4$ and Vitride. LiABH$_3$'s are air-stable, non-pyrophoric, thermally stable, and liberate hydrogen only slowly with protic solvents above pH 4. LiABH$_3$'s, whether solid or as THF solutions, retain their chemical activity for at least 6 months when stored under nitrogen at 25 °C. LiABH$_3$'s can be synthesized from any primary or secondary amine, thus allowing precise control of the steric and electronic environment of these reagents.

In undergraduate teaching laboratories, transformations that would seldom be attempted due to the need to use LiAlH$_4$ or borane, such as the reduction of tertiary amides or esters, may become routine experiments with the use of LiABH$_3$'s. For example, for the past four years, students in the U. C. Santa Cruz introductory organic chemistry laboratory class have employed 1 M THF solutions of LiABH$_3$'s to reduce aliphatic, aromatic, and $\alpha,\beta$-unsaturated esters to the corresponding aliphatic, aromatic, and allylic alcohols, in air, in 70-98% isolated yields without incident or difficulty. In academic research laboratories, the short reaction time, ease of generation and handling, and simple work-up procedure for performing reductions with LiABH$_3$'s makes these new reagents attractive alternatives to LiAlH$_4$ or Lithium triethylborohydride ("SuperHydride"®) reductions.

Currently, our research group is conducting a comprehensive investigation of the reducing properties of LiABH$_3$'s containing an optically active amine moiety.

**References**

(1) (a) Nystrom, R. F.; Brown, W. G. *J. Am. Chem. Soc.* **1947**, *69*, 1197. (b) Finholt, A. E.; Bond, Jr., A. C.; Schlesinger, H. I. *J. Am. Chem. Soc.* **1947**, *69*, 1199. (c) Nystrom, R. F.; Brown, W. G. *J. Am. Chem. Soc.* **1947**, *69*, 2548. (d) Nystrom, R. F.; Brown, W. G. *J. Am. Chem. Soc.* **1948**, *70*, 3738. (e) Block, J.; Gray, H. P. *Inorg. Chem.* **1965**, *4*, 304.

(2) (a) Brown, H. C. *Hydroboration* Benjamin-Cummings: Reading, 1980, p. 296-297. (b) Brown, H. C.; Mead, E. J. *J. Am. Chem. Soc.* **1953**, *73*, 6263. (c) Brown, H. C.; Subba-Rao, B. C. *J. Am. Chem. Soc.* **1956**, *78*, 2582. (d) Brown, H. C.; Mead, E. J.; Schoaf, C. J. *J. Am. Chem. Soc.* **1956**, *78*, 3616. (e) Brown, H. C.; Krishnamurthy, S. *Tetrahedron* **1979**, *35*, 567. (f) Walker, E. R. H. *Chem. Soc. Rev.* **1976**, *5*, 23. (g) Brown, H.C.; Singaram, S.; Singaram, B. *J. Organomet. Chem.* **1982**, *239*, 43. (h) Golden, J. H.; Schreier, C.; Singaram, B.; Williamson, S. M. *Inorg. Chem.* **1992**, *31*, 1533.

(3) (a) Hudlicky, M. *Reductions in Organic Chemistry* Ellis Horwood: Chichester, 1984, p. 20 and references cited therein. (b) Hajos, A. *Complex Hydrides* Elsevier: New York, 1979; p. 83-86 and references cited therein.

(4) (a) Capka, M.; Chvalovsky, V; Kochloefl, K.; Kraus, M. *Coll. Czech. Chem. Comm.* **1969**, *34*, 118. (b) Cerny, V.; Malek, J.; Capka, M.; Chvalovsky, V. *Coll. Czech. Chem. Comm.* **1969**, *34*, 1025. (c) Vit, J. *Eastman Organic Chemical Bulletin* **1970**, *42*, 1. (d) Lerga, R. E., Ed. *The Sigma-Aldrich Library of Chemical Safety Data, Edition 1* Aldrich Chemical Co.: Milwaukee, 1985; p. 1182

(5) (a) The first synthesis of *sodium* aminoborohydrides was reported in 1961. Aftandilian, V. D.; Miller, H. C.; Muetterties, E. L. *J. Am. Chem. Soc.* **1961**, *83*, 2471. (b) Hutchins, R. O.; Learn, K.; El-Telbany, F.; Stercho, Y. P. *J. Org. Chem.* **1984**, *49*, 2438. (c) Keller, P. C. *J. Am. Chem. Soc.* **1969**, *91*, 1231.

(6) (a) Fisher, G. B.; Harrison, J.; Fuller, J. C.; Goralski, C. T.; Singaram, B. *Tetrahedron Lett.* **1992**, *33*, 4533. (b) Fuller, J. C.; Stangeland, E. L.; Goralski, C. T.; Singaram, B. *Tetrahedron Lett.* **1993**, *34*, 257. (c) Fisher, G. B.; Fuller, J. C.; Harrison, J.; Goralski, C. T.; Singaram, B. *Tetrahedron Lett.* **1993**, *34*, 1091. (d) Fisher, G. B.; Fuller, J. C.; Harrison, J.; Alvarez, S. G.; Burkhardt, E. R.; Goralski, C. T.; Singaram, B. *J. Org. Chem.* **1994**, *59*, 6378. (e) Singaram, B.; Fisher, G. B.; Fuller, J. C.; Harrison, J.; Goralski, C. T. U. S. Patent 5 466 798, 1995. (f) Under conditions of extremely high humidity (90-100%), lower molecular weight LiABH$_3$'s were found to smoulder and, occasionally, catch fire. However, the pyrophoricity is due to residual *n*-BuLi present in the LiABH$_3$ acting as a "fuse" on exposure to moist air rather than to any inherent pyrophoricity of the LiABH$_3$.

(7) (a) Singaram, B.; Goralski, C. T.; Fisher, G. B. *J. Org. Chem.* **1991**, *56*, 5691. (b) Fisher, G. B.; Juarez-Brambila, J. J.; Goralski, C. T.; Wipke, W. T.; Singaram, B. *J. Am. Chem. Soc.* **1993**, *115*, 440. (c) Fisher, G. B.; Nicholson, L. W.; Goralski, C. T.; Singaram, B. *Tetrahedron Lett.* **1993**, *34*, 7693.

(8) For a review of the factors affecting the stereochemistry of hydride reductions of

substituted cyclohexanones, see Wigfield, D. C. *Tetrahedron* **1979**, *35*, 449.
(9) Please see ref. 6d, footnote 8.
(10) (a) Brown, H. C.; Choi, Y. M.; Narasimhan, S. *J. Org. Chem.* **1982**, *47*, 3153. (b) Brown, H. C.; Narasimhan, S. *J. Org. Chem.* **1982**, *47*, 1604. (c) Brown, H. C.; Mathew, C. P.; Pyun, C.; Son. J. C.; Yoon, N. M. *J. Org. Chem.* **1984**, *49*, 3091. (d) Yoon, N. M.; Brown, H. C. *J. Am. Chem. Soc.* **1968**, *90*, 2927. (e) Nystrom, R. F.; Brown, W. G. *J. Am. Chem. Soc.* **1947**, *69*, 1197.
(11) LiAlH$_4$: (a) Uffer, H.; Schlitter, E. *Helv. Chim Acta* **1948**, *31*, 1397. (b) Gaylord, N. G. *Reductions With Complex Metal Hydrides* Wiley-Interscience: New York, 1956; p.544-592. (c) Zabicky, J., Ed. *The Chemistry of Amides*, Wiley-Interscience: New York, 1970, p.795-801. Borane: (d) Brown, H. C.; Heim, P. *J. Am. Chem. Soc.* **1964**, *86*, 3566. (e) Brown, H. C.; Heim, P. *J. Org. Chem.* **1973**, *38*, 912. (f) Brown, H. C.; Narasimhan, S.; Choi, Y. M. *Synthesis* **1981**, 441. (g) Brown, H. C.; Narasimhan, S.; Choi, Y. M. *Synthesis* **1981**, 996. NaBH$_4$: (h) Borch, R. F. *Tetrahcdron Lett.* **1968**, 61. (i) Satoh, T.; Suzuki, S. *Tetrahedron Lett.* **1969**, 4555. (j) Rahman, A. U.; Basha, A.; Waheed, N. *Tetrahedron Lett.* **1969**, 219. LiEt$_3$BH: (k) Brown, H. C.; Kim, S. C. *Synthesis* **1977**, 635. This paper also reported the first observation of competing C-O vs. C-N bond cleavage in the reduction of tertiary amides. However, LiEt$_3$BH gives predominant C-N bond cleavage only for aromatic *N,N*-dimethylamides.
(12) (a) Sheradsky, T. *The Chemistry of the Azido Group* Patai, S., Ed.; Wiley-Interscience: New York, 1971; Chapter 6.
(13) (a) Boyer, J. H. *J. Am. Chem. Soc.* **1951**, *73*, 5865. (b) Boyer, J. H.; Canter, F. C. *Chem. Rev.* **1954**, *54*, 1. (c) Hojo, H.; Kobayashi, S. ; Soai, J.; Ikeda, S.; Mukaiyama, T. *Chem. Lett.* **1977**, 635. (d) Kyba, E.P.; John, A. M. *Tetrahedron Lett.* **1977**, 2737.
(14) Corey, E. J.; Nicolaou, K. C.; Balanson, R. D.; Machida, Y. *Synthesis* **1975**, 590.
(15) (a) Pei, Y.; Wickham, B. O. S. *Tetrahedron Lett.* **1993**, *34*, 7509. (b) Vaultier, M.; Knouzi, N.; Carrie, R. *Tetrahedron Lett.* **1983**, *24*, 763. (c) Rolla, F. *J. Org. Chem.* **1982**, *47*, 4327. (d) Samano, M.C.; Robins, M. J. *Tetrahedron Lett.* **1991**, *32*, 6293. (e)Sarma, J. C.; Sharma, R. P. *Chem. Ind.* **1987**, *21*, 764. (f) Bayley, H.; Standring, D. N.; Knowles, J. R. *Tetrahedron Lett.* **1978**, 3633.
(16) (a) Alvarez, S. G.; Fisher, G. B. Singaram, B. *Tetrahedron Lett.* **1995**, *36*, 2567. (b) Please see ref. 15a, footnotes 1 and 2. (c) Please see ref. 15a, footnote 3.
(17) (a) Krishnamurthy, S. *Aldrichimica* **1974**, *7*, 55. (b) Krishnamurthy, S.; Brown, H. C. *J. Org. Chem.* **1976**, *41*, 3064. (c) Brown, H. C.; Kim, S. C. *Synthesis* **1977**, 635.
(18) (a) Brown, H. C.; Krishnamurthy, S.; Hubbard, J. L. *J. Am. Chem. Soc.* **1978**, *100*, 3343. (b) Brown, C. A.; Krishnamurthy, S. *J. Organomet. Chem.* **1978**, *156*, 111. (c) Brown, H. C.; Krishnamurthy, S.; Hubbard, J. L. *J. Organomet. Chem.* **1979**, *166*, 271.
(19) Brown, H. C.; Moerikofer, A. W. *J. Am. Chem. Soc.* **1962**, *84*, 1478.
(20) Corey, E. J.; Albonico, S. M.; Koelliker, U.; Schaaf, T. K.; Varma, R. K. *J. Am. Chem. Soc.* **1971**, *93*, 1491.
(21) (a) Brown, H. C.; Hubbard, J. L.; Singaram, B. *J. Org. Chem.* **1979**, *44*, 5004. (b) Brown, H. C.; Hubbard, J. L.; Singaram, B. *Tetrahedron* **1981**, *37*, 2359.
(22) (a) Harrison, J.; Alvarez, S. G.; Godjoian, G.; Singaram, B. *J. Org. Chem.* **1994**, *59*, 7193.
(23) Brown, H. C.; Khuri, A.; Krishnamurthy, S. *J. Am. Chem. Soc.* **1977**, *99*, 6237.
(24) Krishnamurthy, S.; Brown, H. C. *J. Am. Chem. Soc.* **1976**, *98*, 3383.
(25) Myers, A.G.; Yang, B.H.; Chen, H; Gleason, J.L. *J. Am. Chem. Soc.* **1994**, *116*, 9361.

## Chapter 11

# Sodium Borohydride and Carboxylic Acids: A Novel Reagent Combination

Gordon W. Gribble

Department of Chemistry, Dartmouth College, Hanover, NH 03755

The combination of sodium borohydride ($NaBH_4$) and carboxylic acids – sodium acyloxyborohydrides – represents a remarkably versatile and powerfully efficient synthetic tool. This reagent manifold, the reactivity of which can be controlled depending on the nature and number of acyloxy groups, reduces and $N$-alkylates indoles, quinolines, isoquinolines, related heterocycles, imines, enamines, oximes, enamides, and similar functional groups. It reduces amides and nitriles to amines in the presence of esters, aryl alcohols and ketones to hydrocarbons, aldehydes to alcohols in the presence of ketones, and β-hydroxyketones to 1,3-diols stereoselectively. This reagent is also an extraordinarily useful methodology for the $N$-alkylation of primary and secondary amines, in a reaction sequence that is believed to involve sequential reduction of the carboxylic acid to the corresponding aldehyde followed by a standard reductive amination process. Frequently, the monoalkylation of primary amines can be achieved. The use of sodium cyanoborohydride ($NaBH_3CN$) militates against $N$-alkylation, and, for example, the union of $NaBH_3CN$/HOAc cleanly reduces indoles to indolines *sans* alkylation. Depending on the circumstances and conditions, alkenes can be hydroborated, esters and carboxylic acids can be reduced to alcohols, and arenes can be induced to undergo the Baeyer condensation. No other chemical system can boast of such amazing flexibility!

More than 20 years ago, as part of an undergraduate research project at Dartmouth, we decided to attempt the reduction of 1,2,3,4-tetrahydrocarbazole with neat formic acid. A general indole double-bond reduction method was lacking at that time (*1*) and we felt that a Leuckart-type reaction (*2*) on the protonated indole might occur (equation 1). However, the only product of this reaction was the $N$-formyl derivative of **1** (95%) (Gribble, G. W.; Strickman, D., Dartmouth College, unpublished result).
Nevertheless, since indoles are well known to undergo C-3 protonation in mineral acids (*1, 3*), we felt that a better hydride source than formate might succeed in ambushing the presumed indolenium ion (*e.g.*, **2**), if indeed carboxylic acids are capable of protonating the indole double bond. We chose to study sodium borohydride

[Equation (1): Compound **1** + HCO₂H, Δ → [**2**] → (with H-C(=O)-O⁻) → **3**]

(NaBH₄) and acetic acid (HOAc) since Marshall and Johnson had established the compatibility of these unlikely partners a few years earlier in the reduction of steroidal dienamines and enamines (4, 5). Ironically, at about the same time, it was claimed that reductions involving NaBH₄ in glacial acetic acid could *not* be performed (6). In the event, treatment of indole (**4**) in glacial HOAc with NaBH₄ did not give indoline (**5**) (2,3-dihydroindole) but, rather, *N*-ethylindoline (**6**) in high yield (equation 2) (7). This very surprising result spawned the chemistry in this chapter.

[Equation (2): **4** → (NaBH₄, HOAc, 86%) → [**5**] → **6** (N-CH₂CH₃)]

Since several reviews of this area have appeared (*8-10*), the present chapter focuses mainly on recent developments and the author's own work. The interesting history of the discovery of acyloxyborohydrides has been previously discussed (*8, 9*).

**Reduction of Indoles.**

The remarkable tandem reduction and *N*-alkylation of indole (equation 2) is reasonably general for indoles and carboxylic acids (equation 3) (7). Only pivalic acid gives the corresponding product (*N*-neopentylindoline) in low yield.

[Equation (3): Indole with R₂, R₃, R₄ substituents + NaBH₄, R₁CO₂H, 50-60 °C, 49-90% → indoline product with N-CH₂R₁]

$R_1$ = H, Me, Et, *i*-Pr
$R_2$, $R_3$ = H, Me, –(CH₂)₄–
$R_4$ = 5-Br, 7-Me

The reaction of indole with NaBH₄/formic acid, which is exceptionally vigorous, leads to by-product **7**, the structure of which was confirmed by independent synthesis from the well-known indole dimer **8** (*11*).

[Scheme: Compound **7** (N-Me indoline with CH₂CH₂-aryl-NMe₂ side chain) → 1. MeI; 2. KOH; 3. H₂ Pd/C; 4. KOH, DMSO, MeI; 5. NaBH₃CN, HOAc → Compound **8**]

In an effort to suppress the *N*-alkylation reaction, we attempted the reduction of indole with NaBH₄ and the stronger trifluoroacetic acid (TFA). Indeed, this reaction affords indoline (**5**) and a small amount of *N*-trifluoroethylindoline (**9**), in addition to the interesting Baeyer condensation product (**10**) (equation 4) (*12*). The structures of

**9** and **10** were confirmed by independent syntheses and conversion to known compounds. For example, reaction of **9** under these conditions gives **10** in 57% yield.

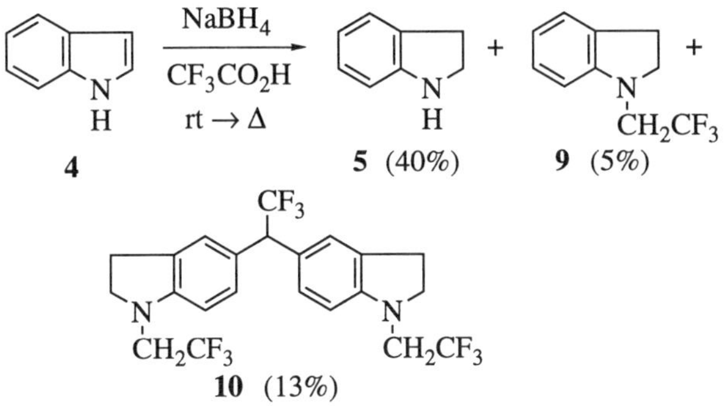

$$\text{indole (4)} \xrightarrow[\text{rt} \to \Delta]{\text{NaBH}_4 / \text{CF}_3\text{CO}_2\text{H}} \text{indoline 5 (40\%)} + \text{N-CH}_2\text{CF}_3\text{ indole 9 (5\%)} + \text{10 (13\%)} \quad (4)$$

where **10** is bis(N-trifluoroethylindolin-5-yl)(trifluoromethyl)methane.

The formation of **10** and the N-alkylation of indoles with NaBH$_4$/RCO$_2$H suggested to us that aldehydes, or their synthetic equivalent, were the source of the alkyl group. This pathway is discussed in the next section. As expected, treatment of the N-alkylindoles with NaBH$_4$/HOAc affords the corresponding N-alkylindolines (equation 5) (7).

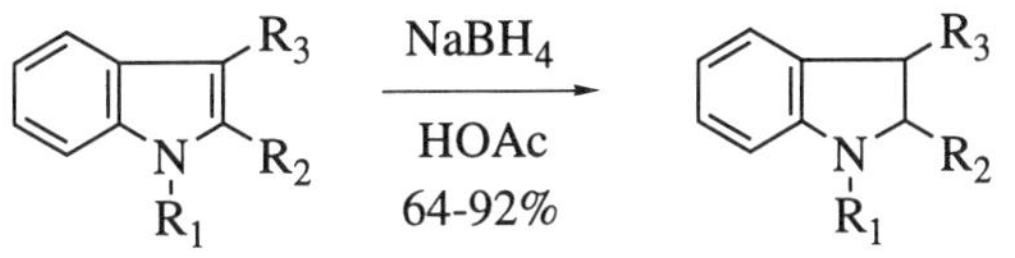

$$\text{N-alkylindole} \xrightarrow[\text{HOAc}]{\text{NaBH}_4} \text{N-alkylindoline} \quad 64\text{-}92\% \quad (5)$$

$R_1$ = Me, Et
$R_2$ = H, Me
$R_3$ = H
$R_2, R_3$ = –(CH$_2$)$_4$–

Since the original goal of this research program was to discover a new indole to indoline reduction method, *sans* N-alkylation, we examined the reaction of indole with sodium cyanoborohydride (NaBH$_3$CN) in HOAc. Much to our delight, this reaction afforded indoline (**5**) in 91% distilled yield, with no trace of N-ethylindoline (*7, 13*). At higher temperatures, N-ethylation is observed (*13, 14*). The reaction is rapid, general, and efficient (equation 6), failing only with electron-withdrawing substituents on the indole ring (e.g., 5-nitro- and 2,3-diphenylindole.

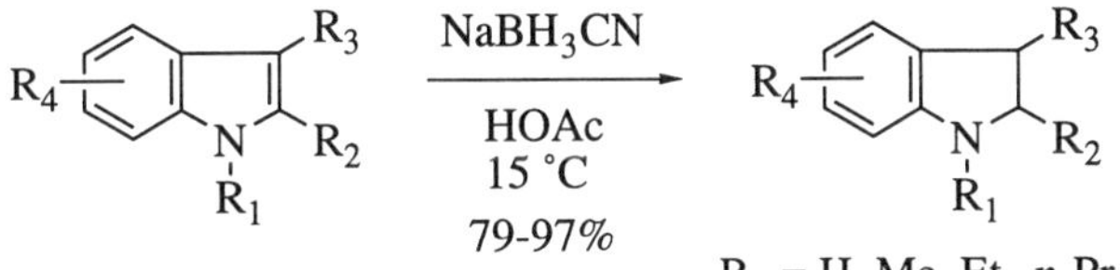

$$\text{indole} \xrightarrow[\substack{\text{HOAc} \\ 15\,°\text{C} \\ 79\text{-}97\%}]{\text{NaBH}_3\text{CN}} \text{indoline} \quad (6)$$

$R_1$ = H, Me, Et, n-Pr, CH$_2$Ph
$R_2, R_3$ = H, Me, –(CH$_2$)$_4$–
$R_4$ = 5-Br, 7-Me

This very useful indole reduction method has been employed by many groups and some of the indolines thusly prepared are shown in Scheme 1.

**Scheme 1**

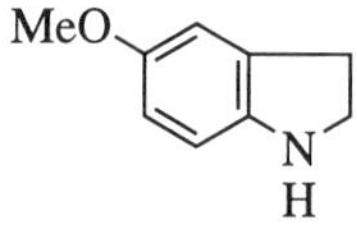

74% (*15*) 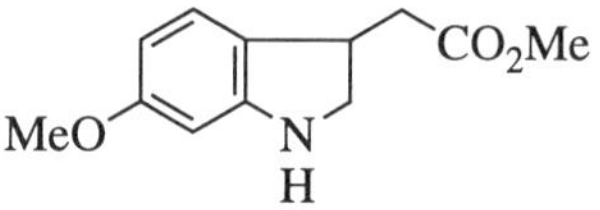   98% (*16*) 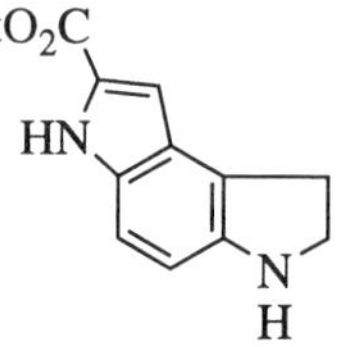   **11** 96% (*17*)

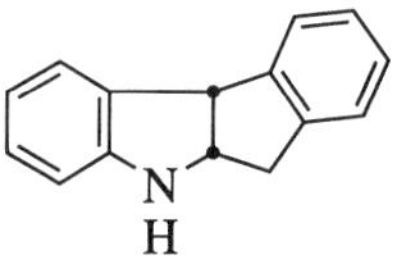

98% (*18*) 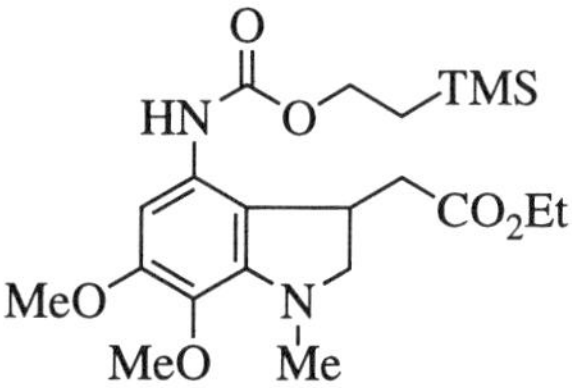   (*19*) 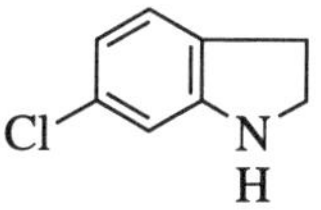   92% (*20*)

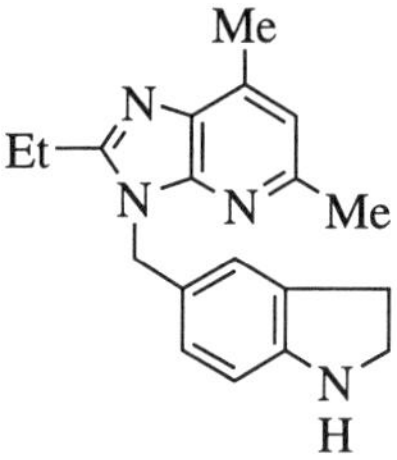

75% (*21*) 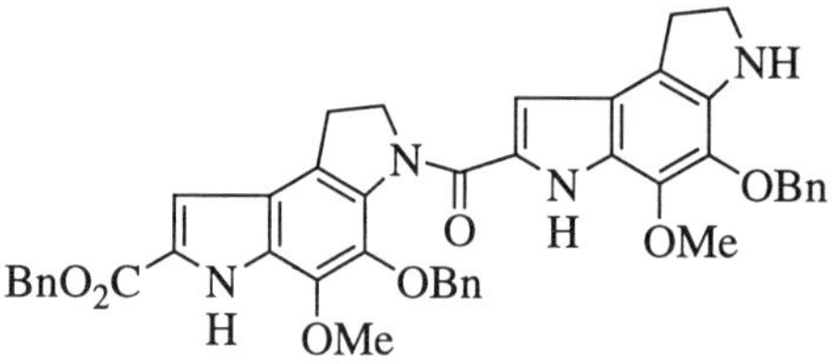   **12** 61% (*22*)

In particular, this reagent combination has been very successful in the selective reduction of the more basic indole double bond in CC-1065 and PDE precursors (*17, 22-27*), such as **11** and **12** in Scheme 1. Not surprisingly, treatment of indole with NaBH$_3$CN/HOAc in the presence of added acetaldehyde affords *N*-ethylindoline in 87% yield (Gribble, G.W. Dartmouth College, unpublished result). Thus, it would appear that NaBH$_3$CN is less effective than NaBH$_4$ in generating aldehydes (or their equivalents) from carboxylic acids. Ironically, the first report of the treatment of an indole with NaBH$_4$/HOAc did *not* result in the reduction of the indole double bond (equation 7) (*28*). We believe that this lack of reduction in this and related systems that contain a basic nitrogen (*29-31*) is due to protonation of the basic nitrogen which prevents a second protonation of the indole double bond.

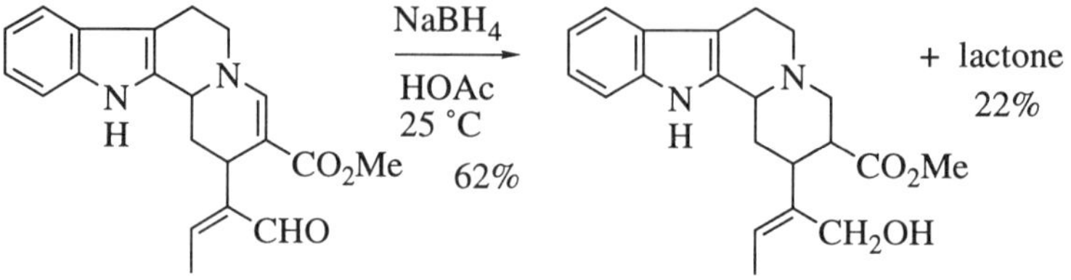

$$\xrightarrow[\text{HOAc} \atop 25\,°\text{C} \quad 62\%]{\text{NaBH}_4}$$ + lactone 22% (7)

Indeed, we found that TFA in combination with NaBH$_4$ smoothly reduces basic indoles such as indolo[2,3-*a*]quinolizidines (equation 8) (*32*).

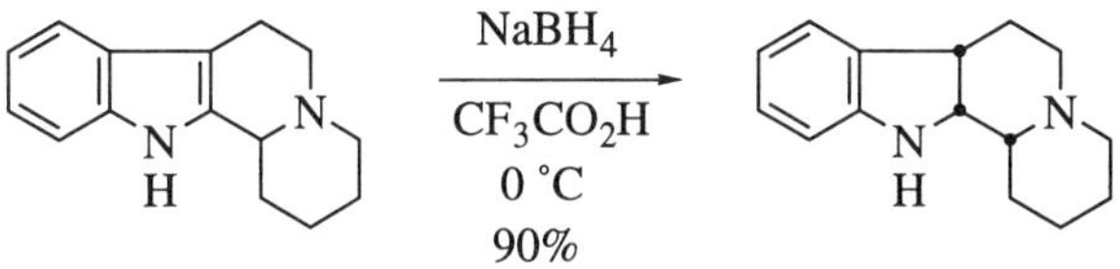

(8)

Several other workers have used the combination of NaBH$_4$/TFA to reduce the indole double bond of yohimbine (33), carbolines (34), and other amino indoles (35, 36). In some cases, NaBH$_3$CN/TFA is superior in this regard (37, 38). Indeed, this latter combination efficiently reduces N-(phenylsulfonyl)indoles to the corresponding indolines and 2- and 3-acyl-N-(phenylsulfonyl)indoles to the corresponding 2- and 3-alkyl-N-(phenylsulfonyl)indolines (39). This ketone to hydrocarbon reduction is discussed later.

### N-Alkylation of Amines.

The unexpected propensity of NaBH$_4$ in carboxylic acids to effect the N-alkylation of primary and secondary amines led us to explore the scope of this novel and potentially useful method. Depending on the temperature, aniline can be mono- or dialkylated (Scheme 2) (7, Gribble, G.W. Dartmouth College, unpublished results). Moreover, the incorporation of an aldehyde or ketone in this protocol allows for the formation of unsymmetrical tertiary amines in one pot, including the introduction of the bulky neopentyl group (from pivalic acid).

**Scheme 2**

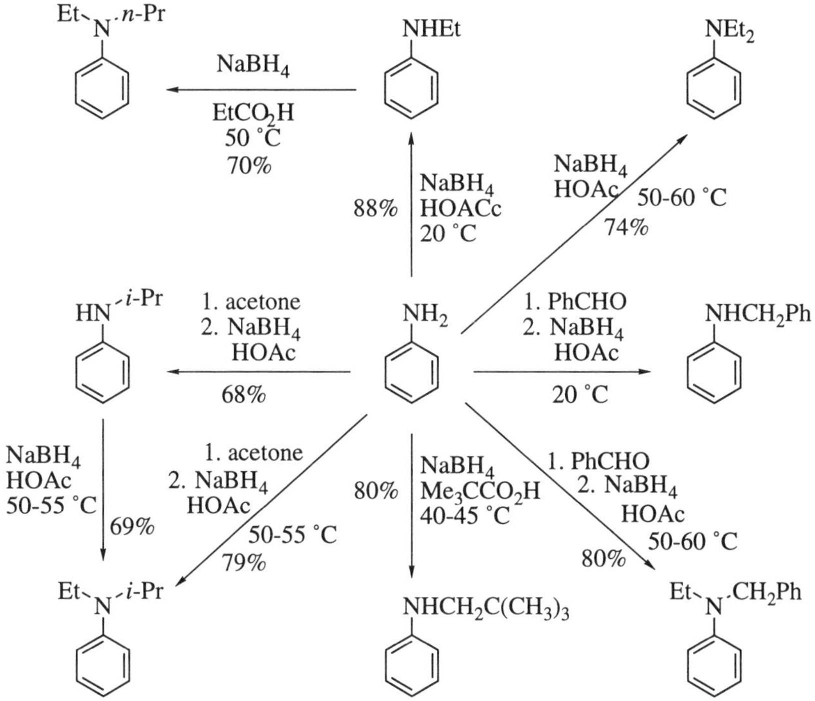

This N-alkylation is general for a range of secondary to tertiary anilines as well as for different carboxylic acids (equation 9) (7, Gribble, G.W. Dartmouth College, unpublished results), including solid carboxylic acids in a cosolvent (*40*). Even the weakly basic carbazole can be *N*-ethylated with NaBH$_4$/HOAc (92% yield) (*7*).

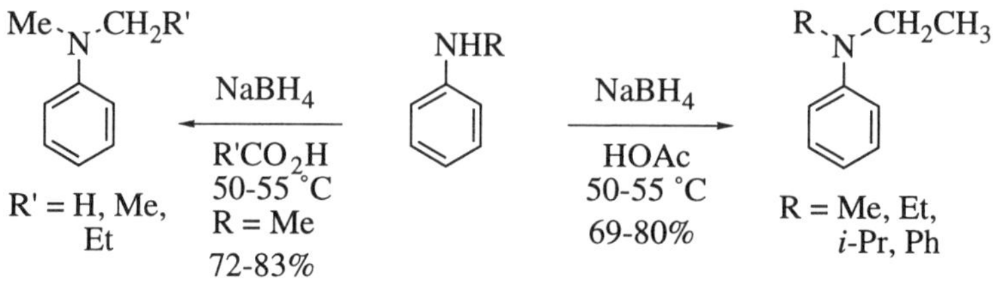

$$\underset{\substack{R' = H, Me, \\ Et}}{\text{Me-N-CH}_2\text{R'} \text{ (Ph)}} \xleftarrow[\substack{R'CO_2H \\ 50\text{-}55\,°C \\ R = Me \\ 72\text{-}83\%}]{\text{NaBH}_4} \underset{}{\text{PhNHR}} \xrightarrow[\substack{HOAc \\ 50\text{-}55\,°C \\ 69\text{-}80\%}]{\text{NaBH}_4} \underset{R = Me, Et, \\ i\text{-}Pr, Ph}{\text{R-N-CH}_2\text{CH}_3 \text{ (Ph)}} \quad (9)$$

Although it is generally sluggish, in some cases, the combination of NaBH$_4$/TFA leads to *N*-trifluoroethylation of aromatic amines (e.g., equations 10-12) (*12, 34, 41*).

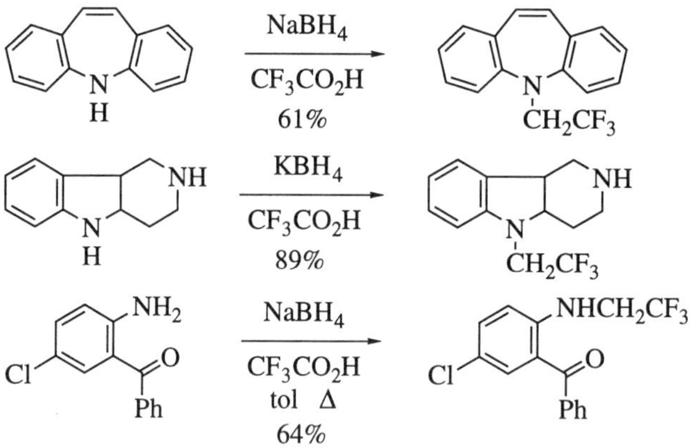

(10) iminostilbene + NaBH$_4$/CF$_3$CO$_2$H, 61% → *N*-CH$_2$CF$_3$ iminostilbene

(11) tetrahydro-β-carboline + KBH$_4$/CF$_3$CO$_2$H, 89% → *N*-CH$_2$CF$_3$ product

(12) 2-amino-5-chlorobenzophenone + NaBH$_4$/CF$_3$CO$_2$H, tol Δ, 64% → NHCH$_2$CF$_3$ product

This very useful aromatic amine alkylation has been employed many times in recent years and a few of the resulting compounds are shown in Scheme 3. In some cases, the alkyl group is derived from added aldehyde or ketone.

**Scheme 3**

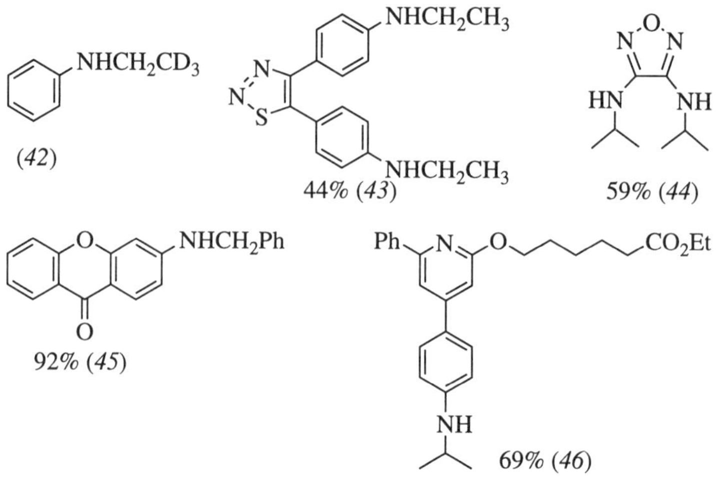

PhNHCH$_2$CD$_3$ (*42*)

44% (*43*) — thiadiazole bis(*p*-NHCH$_2$CH$_3$-phenyl)

59% (*44*) — furazan bis(NH-*i*-Pr)

92% (*45*) — xanthone-NHCH$_2$Ph

69% (*46*) — Ph-pyridine-O-(CH$_2$)$_4$-CO$_2$Et with *p*-(NH-*i*-Pr)phenyl

We also discovered that the *N*-alkylation of more basic aliphatic amines could be accomplished with $NaBH_4/RCO_2H$ (*47*). Thus, *N*-alkylbenzylamines can be *N*-ethylated with $NaBH_4$/HOAc and *N*-methylbenzylamine can be *N*-alkylated with $NaBH_4/RCO_2H$ (equation 13).

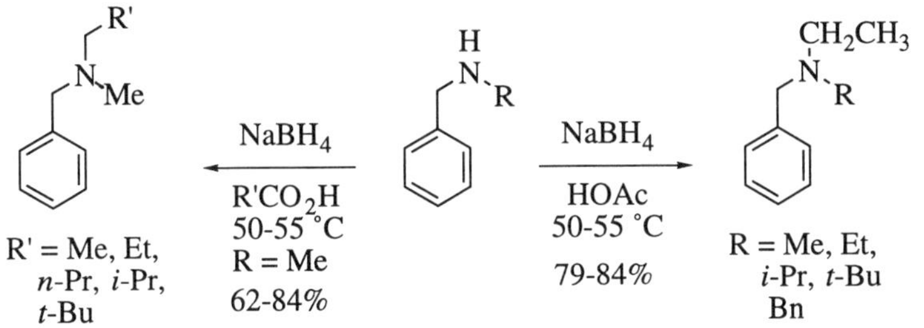

NaBH4
R'CO2H
50-55 °C
R = Me
62-84%
R' = Me, Et, n-Pr, i-Pr, t-Bu
NaBH4
HOAc
50-55 °C
79-84%
R = Me, Et, i-Pr, t-Bu Bn


(13)

Furthermore, *N*-benzylamine can be manipulated in the same way that aniline can be (Scheme 2), as summarized in equation 14 (*47*). The preparation of the unsymmetrical tertiary amine, *N*-ethyl-*N*-*i*-propylbenzylamine in one pot is noteworthy.

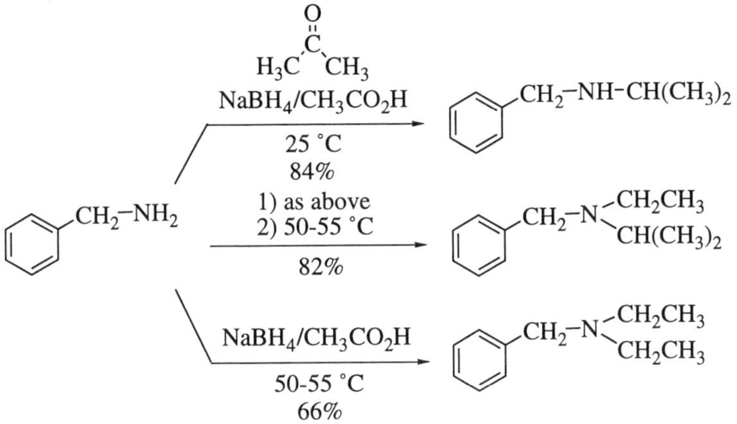

NaBH4/CH3CO2H
25 °C
84%
1) as above
2) 50-55 °C
82%
NaBH4/CH3CO2H
50-55 °C
66%


(14)

Some additional aliphatic amines that we have *N*-ethylated using $NaBH_4$/HOAc are shown in Scheme 4 (*47*), Gribble, G.W. Dartmouth College, unpublished results). Only in the case of highly hindered secondary amines or with hindered carboxylic acids does this alkylation proceed poorly.

**Scheme 4**

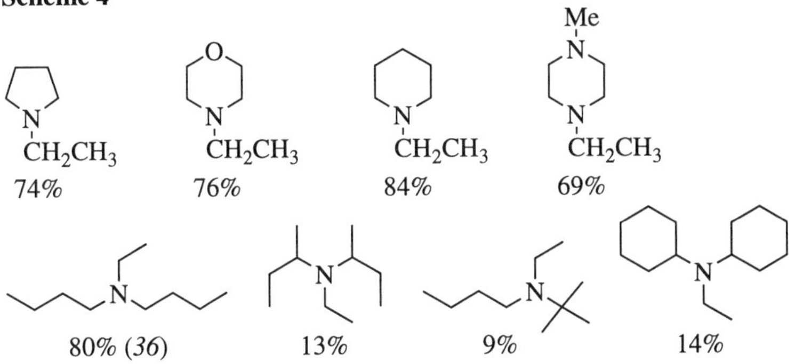

74%
76%
84%
69%
80% (36)
13%
9%
14%

Longer chain carboxylic acids can be used to alkylate diethylamine and dimethylamine (equations 15, 16) (*47*).

$$\begin{array}{c}CH_3CH_2\\ \phantom{CH_3CH_2}NH\\ CH_3CH_2\end{array} \xrightarrow[\substack{50\text{-}55\,°C\\70\%}]{\substack{NaBH_4\\CH_3(CH_2)_6CO_2H}} \begin{array}{c}CH_3CH_2\\ \phantom{CH_3CH_2}N\text{-}(CH_2)_7\text{-}CH_3\\ CH_3CH_2\end{array} \quad (15)$$

$$\begin{array}{c}CH_3\\ \phantom{CH_3}NH\cdot HCl\\ CH_3\end{array} \xrightarrow[\substack{THF\\ CH_3(CH_2)_7CO_2H\\50\text{-}55\,°C\\78\%}]{\substack{NaBH_4\\NaOAc}} \begin{array}{c}CH_3\\ \phantom{CH_3}N\text{-}(CH_2)_8\text{-}CH_3\\ CH_3\end{array} \quad (16)$$

Several other examples of the *N*-alkylation of aliphatic amines have been described in recent years and a few of these are summarized in Scheme 5. In most examples, either an ethyl or *n*-propyl group has been introduced, and, in some cases, multiple *N*-alkylation is the objective.

**Scheme 5**

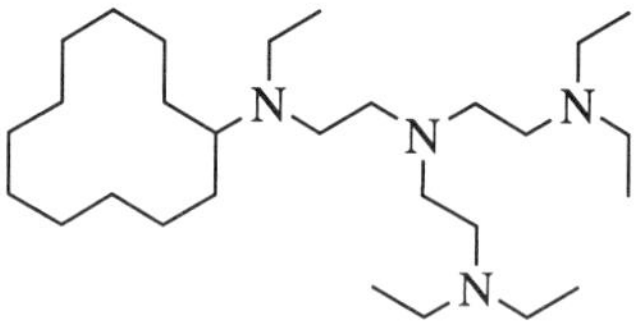

51% (*48*) 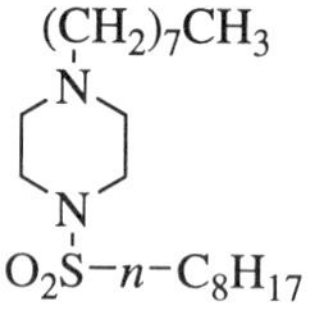       (*49*)

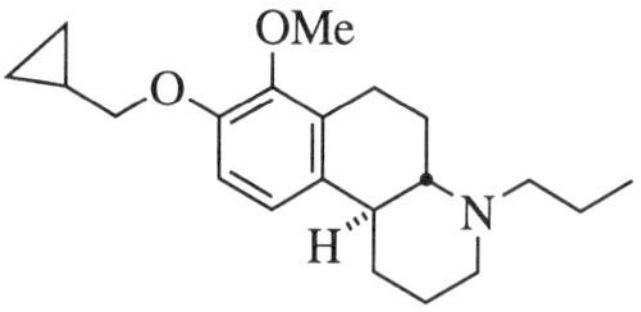

75% (*50*) 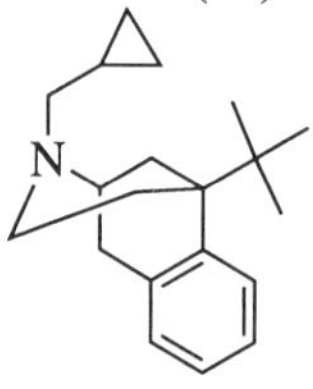       61% (*51*) 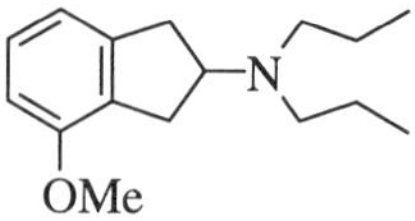       88% (*52*)

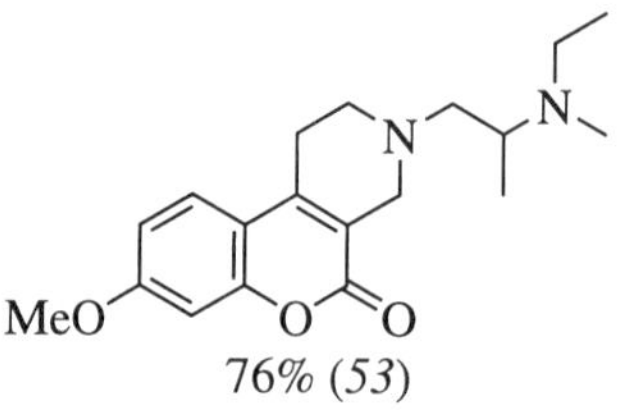

76% (*53*) 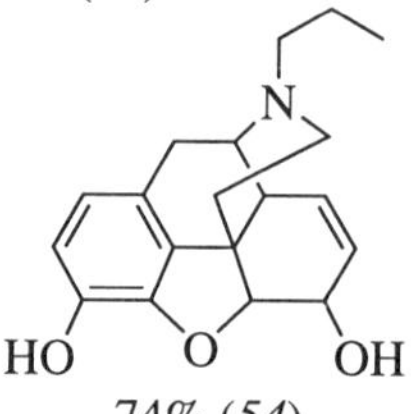       74% (*54*)

Because the reaction of $NaBH_4$ with formic acid ($HCO_2H$) is exceptionally vigorous, we have developed an alternative *N*-methylation protocol that utilizes paraformaldehyde in conjunction with $NaBH_4$/TFA or $NaBH_3CN$/HOAc (*55*) (equations 17, 18).

$(PhCH_2CH_2)_2NH \xrightarrow[\text{THF}]{\underset{\text{NaBH}_4 \text{ TFA}}{(HCHO)_n}} (PhCH_2CH_2)_2NCH_3$  (17)

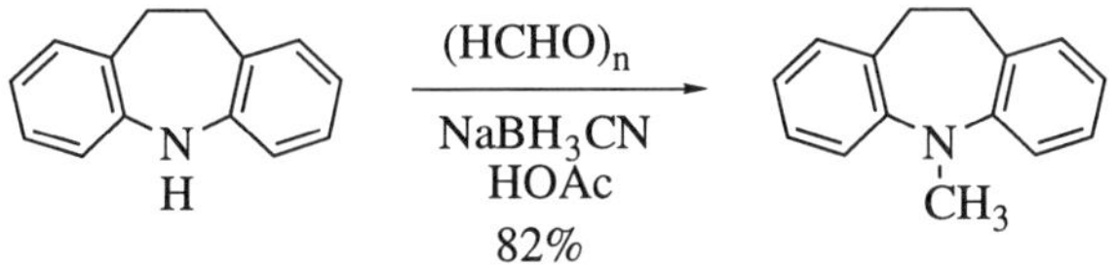

87%

[iminodibenzyl] $\xrightarrow[\text{HOAc}]{\underset{\text{NaBH}_3CN}{(HCHO)_n}}$ [N-methyl iminodibenzyl]  (18)

82%

An important variation on this reductive amination methodology has been developed by Abdel-Magid and is presented in a separate Chapter in this volume.

We believe that the mechanism of the $N$-alkylation of amines with $NaBH_4/RCO_2H$ involves the *in situ* generation of a triacyloxyborohydride species, which forms in the presence of excess carboxylic acid and which has been isolated and characterized (*7, 9, 40, 41, 56, 57*). This material then suffers self-reduction to generate either the free aldehyde or a boron species at the aldehyde oxidation level. Reductive amination of this aldehyde species completes the $N$-alkylation sequence (Scheme 6). Control experiments reveal that $N$-acylation is not involved since, for example, neither $N$-acetylindoline nor $N$-acetylindole are reduced to $N$-ethylindoline under the reaction conditions (*7*). However, as will be seen, amides are reduced to amines with the more reactive monotriacyloxyborohydride (*vide infra*). Support for the intermediacy of aldehydes is the fact that added aldehydes readily undergo reductive amination under the reaction conditions (Gribble, G.W. Dartmouth College, unpublished results), and that acetaldehyde can be trapped as its 2,4-DNP derivative from the evolved gases when $NaBH_4$ reacts with excess HOAc (*7*).

**Scheme 6**

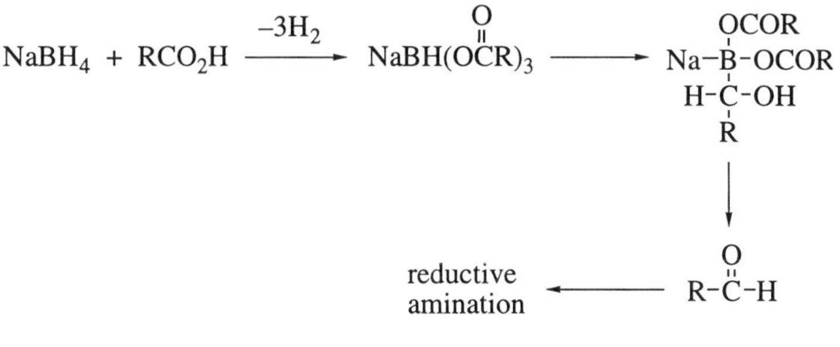

$NaBH_4 + RCO_2H \xrightarrow{-3H_2} NaBH(OCR)_3 \longrightarrow \underset{\underset{R}{\overset{|}{H-C-OH}}}{\overset{\overset{OCOR}{|}}{Na-B-OCOR}}$

$\downarrow$

reductive amination $\longleftarrow$ $R-\overset{O}{\overset{\|}{C}}-H$

**Reduction of Other Heterocycles.**

Not surprisingly, a variety of other nitrogen and oxygen heterocycles are reduced and $N$-alkylated by the action of $NaBH_4$ and $RCO_2H$.

Following an earlier report on the partial reduction of nitroquinolines with $NaBH_4$/HOAc (*58*), we examined this reaction with quinoline and isoquinoline in some detail (*59*). These two heterocycles exhibit a similar reaction manifold as summarized for quinoline in Scheme 7. Once again the reaction conditions may be varied to allow or to avoid $N$-alkylation, and to introduce secondary alkyl groups using a ketone additive.

**Scheme 7**

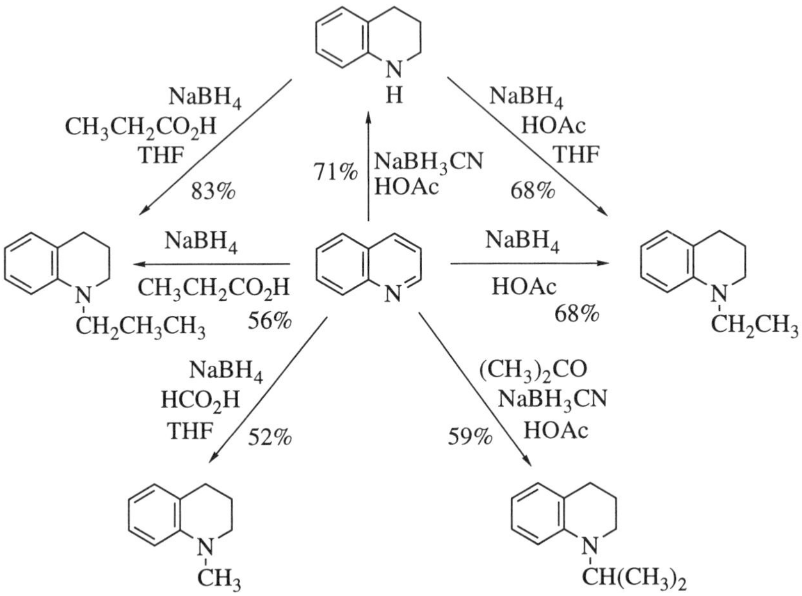

Interestingly, the isolated double bond in dihydroquinolines is not reduced under the typical reaction conditions (equations 19, 20) (Gribble, G.W. Dartmouth College, unpublished results; *60*).

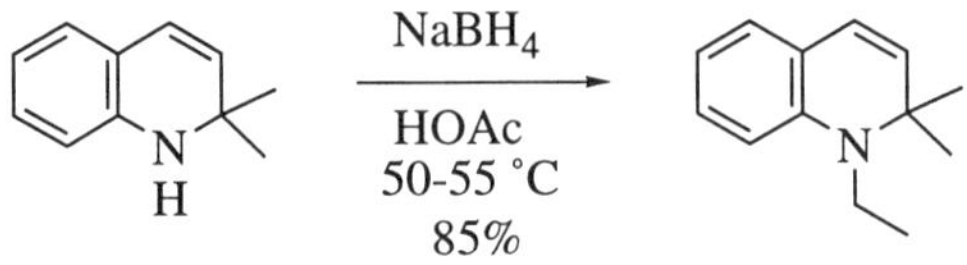

(19)

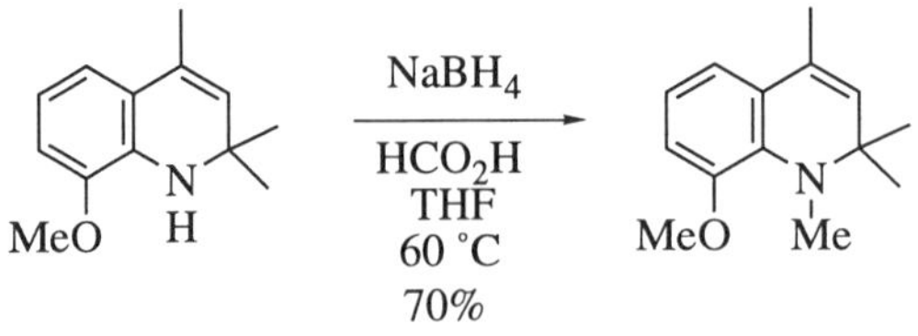

(20)

Acridine is reduced to acridan at 20 °C and converted to *N*-ethylacridan at higher temperature (equation 21) (Gribble, G.W. Dartmouth College, unpublished results).

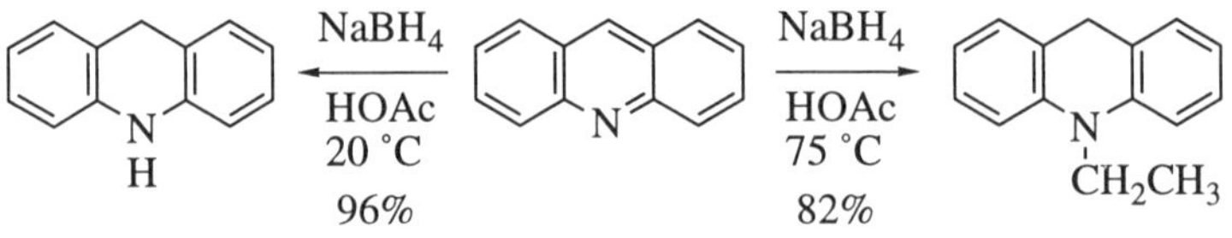

(21)

Several other examples of the reduction of quinolines and related heterocycles are summarized in equations 22-27 (*61-66*). The reaction shown in equation 24 involves an interesting migration of the acetyl group (*63*), and the reduction of the ketone in tryptanthrin (equation 26) (*65*) may involve amine-directed hydride transfer since ketones are normally not reduced under these conditions. The selective N-ethylation in equation 27 is noteworthy (66).

Other examples of imine reduction in nitrogen heterocycles have been uncovered. For example, benzoxazoles undergo reductive cleavage (equation 28) (67) and the synthesis of lennoxamine involves imine reduction and subsequent lactamization of a benzazepine (equation 29) (*68*). The imine group in a benzodiazepine is selectively reduced in the presence of an indole double bond (equation 30) (*69*). Presumably, the (protonated) basic nitrogen protects the indole ring towards protonation. A recent study of the reduction of indeno[1,2-*b*]quinoxalines and benzo[*b*]phenazines has been reported (e.g., equation 31) (*70*).

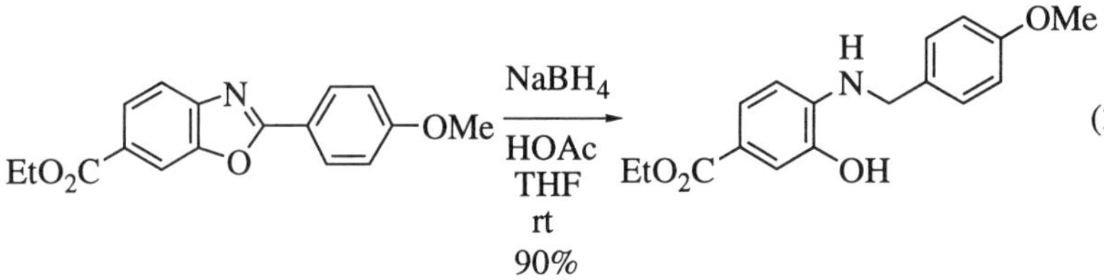

(28)

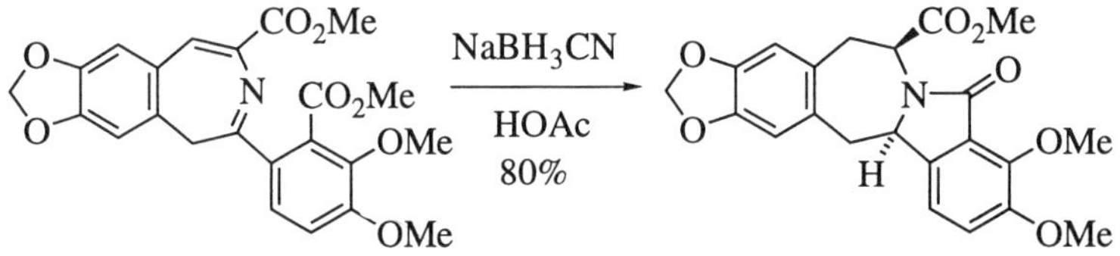

(29)

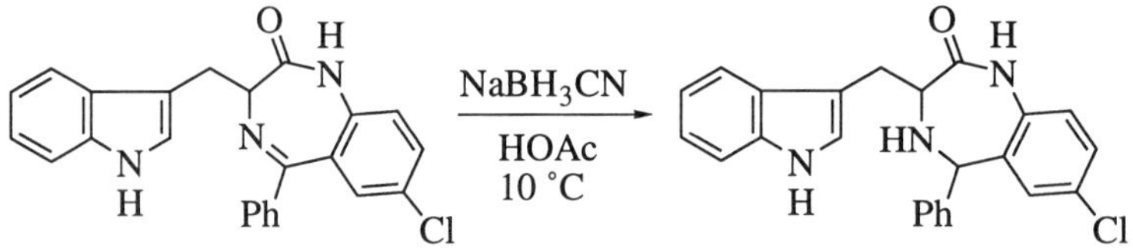

(30)

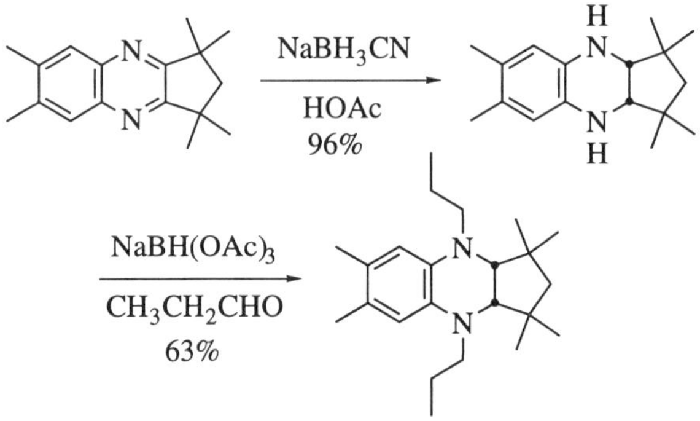

(31)

In a series of papers, Balaban and his colleagues have utilized NaBH$_4$/HOAc to reduce pyrylium salts (equation 32) (*71-74*).

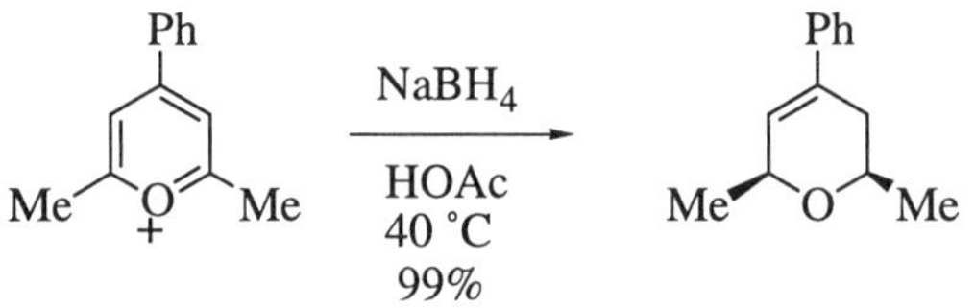

(32)

Although our early attempts to reduce simple pyrroles were unsuccessful (Gribble, G.W. Dartmouth College, unpublished results), Ketcha has succeeded in reducing *N*-(phenylsulfonyl)pyrroles with NaBH$_3$CN/TFA (equation 33) (*75*). The fully reduced compounds are minor products.

$$\text{pyrrole-SO}_2\text{Ph} \xrightarrow[\text{75\%}]{\text{NaBH}_3\text{CN, TFA}} \text{2,5-dihydropyrrole-SO}_2\text{Ph} \qquad (33)$$

Saturated heterocycles are reductively cleaved under these conditions, analogous to the reaction of acetals (*vide infra*). Two examples are shown (equations 34, 35) (*76, 77*).

$$\text{bicyclic N-Bn, O} \xrightarrow[\text{100\%}]{\text{NaBH}_3\text{CN, HOAc}} \text{piperidine with OH, Bn} \qquad (34)$$

trans/cis = 6:1

$$\text{(imidazolidine with S-S)} \xrightarrow{\text{NaBH}_4, \text{HOAc}} \text{(open-chain NH)} \quad 61\% \;\; + \;\; \text{(open-chain N-Et)} \quad 34\% \qquad (35)$$

### Reduction of Oximes, Imines, Enamines and Related Compounds.

Not surprisingly, in view of the facile reductions of indole, quinoline, and related heterocycles with $\text{NaBH}_4/\text{RCO}_2\text{H}$ (*vide supra*), a wide range of C=N compounds are also transformed with these reagents. Moreover, compounds, such as enamines and enamides that can be protonated to form iminium ions are also readily reduced. An earlier review documents many examples of this type (*9*).

Oximes can be reduced and *N*-alkylated under the influence of $\text{NaBH}_4/\text{RCO}_2\text{H}$ and this procedure represents an excellent way to prepare unsymmetrical *N,N*-dialkylhydroxylamines (*78*). Three examples are shown (equations 36-38). Once again, alkylation can be suppressed by using $\text{NaBH}_3\text{CN}$ (equation 38).

$$(\text{CH}_3)_2\text{C=N-OH} \xrightarrow[\substack{50\,°\text{C}\\68\%}]{\text{NaBH}_4,\;\text{CH}_3\text{CH}_2\text{CO}_2\text{H}} (\text{CH}_3)_2\text{CH-N(OH)-CH}_2\text{CH}_2\text{CH}_3 \qquad (36)$$

$$\text{PhCH=N-OH} \xrightarrow[\substack{40\,°\text{C}\\65\%}]{\text{NaBH}_4,\;(\text{CH}_3)_2\text{CHCO}_2\text{H}} \text{PhCH}_2\text{-N(OH)-CH(CH}_3)_2 \qquad (37)$$

$$\text{cyclohexanone oxime} \xrightarrow[\substack{\text{HOAc}\\25\,°\text{C}\\81\%}]{\text{NaBH}_3\text{CN}} \text{cyclohexyl-NHOH} \qquad (38)$$

In some cases, over-reduction results (equations 39, 40) (*78, 79*).

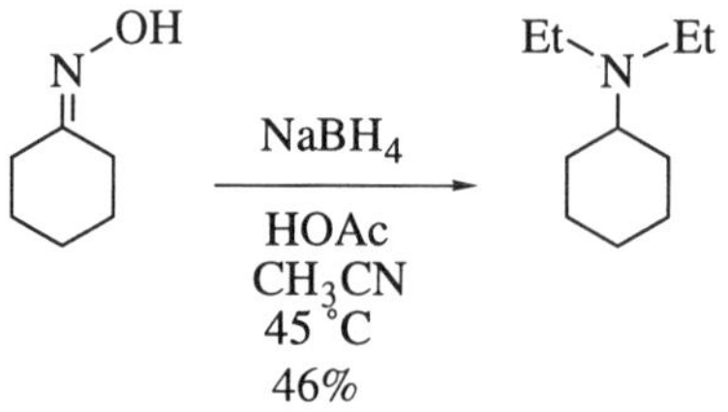

$$\text{(cyclohexanone oxime)} \xrightarrow[\substack{\text{HOAc} \\ \text{CH}_3\text{CN} \\ 45\,°\text{C} \\ 46\%}]{\text{NaBH}_4} \text{(N,N-diethylcyclohexylamine)} \quad (39)$$

$$\xrightarrow[\substack{\text{HOAc} \\ \text{THF} \\ 0\,°\text{C}}]{\text{NaBH}_3\text{CN}} \quad (40)$$

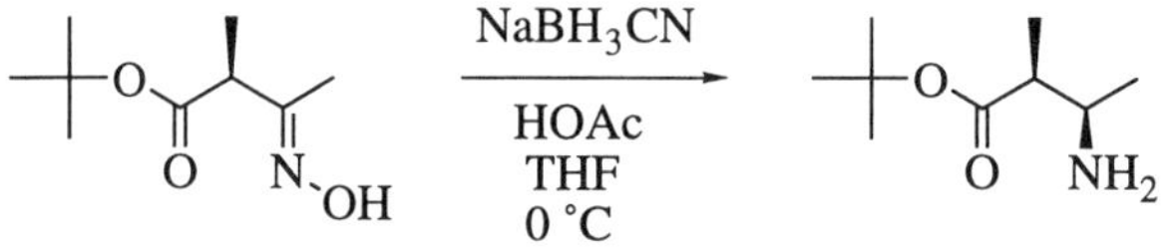

Several workers have exploited this methodology to reduce oximes or oxime ethers and the products of these reactions are summarized in Scheme 8.

**Scheme 8**

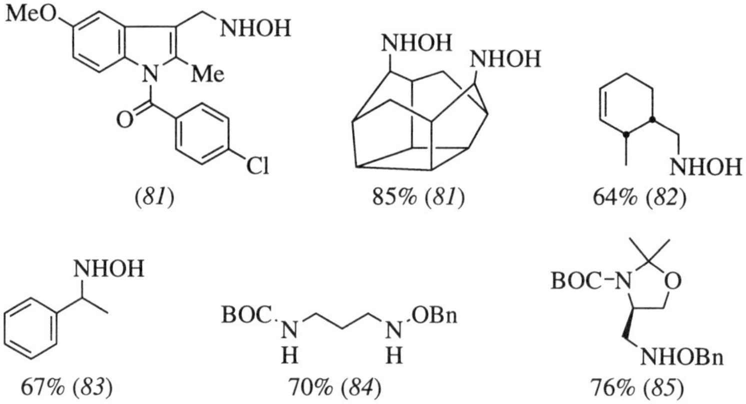

(*81*)   85% (*81*)   64% (*82*)

67% (*83*)   70% (*84*)   76% (*85*)

Furthermore, Williams has described the hydroxy-directed reduction of oxime ethers using the isolated reagent $NaBH(OAc)_3$ (equation 41) (*86*), a powerful tactic that will be discussed again in the reduction of β-hydroxy ketones.

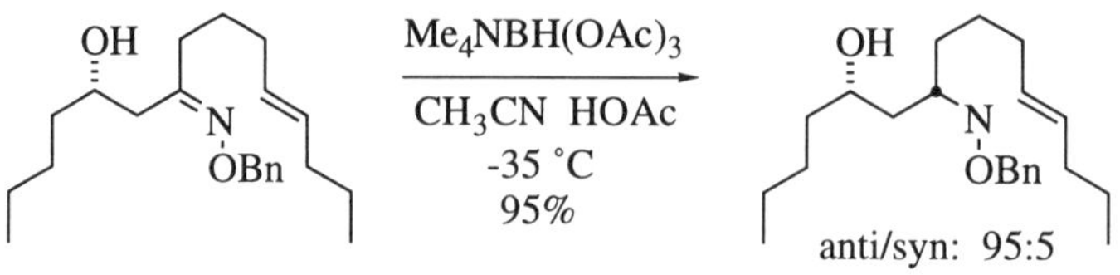

$$\xrightarrow[\substack{\text{CH}_3\text{CN HOAc} \\ -35\,°\text{C} \\ 95\%}]{\text{Me}_4\text{NBH(OAc)}_3} \quad (41)$$

anti/syn: 95:5

A number of simple imines are reduced with $NaBH_4$ or $NaBH_3CN$ in carboxylic acids, as tabulated in equations 42-44 (*87-89*).

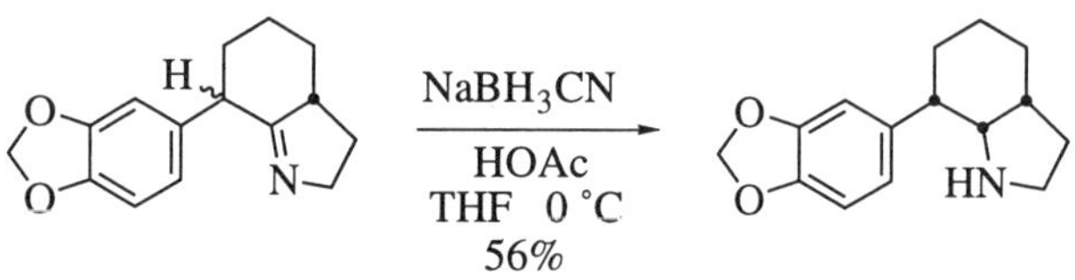

$$\xrightarrow[\substack{\text{HOAc} \\ \text{THF  0\,°C} \\ 56\%}]{\text{NaBH}_3\text{CN}} \quad (42)$$

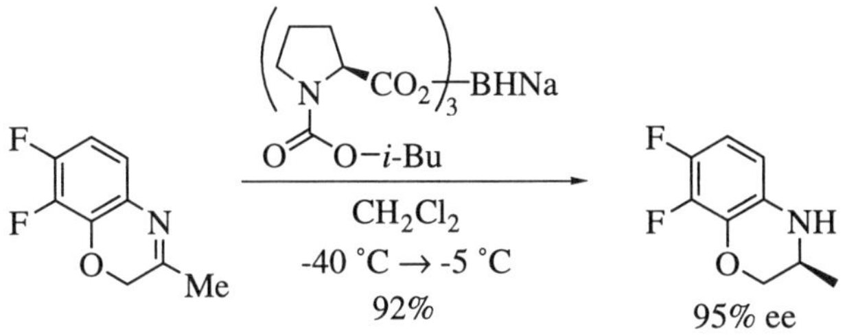

$$\text{[structure]} \xrightarrow[\substack{\text{CH}_2\text{Cl}_2 \\ -40\,°\text{C} \to -5\,°\text{C} \\ 92\%}]{\left(\substack{\text{pyrrolidine-CO}_2}\right)_{\!3}\text{-BHNa},\ \text{O=C(O-}i\text{-Bu)}} \text{[product]} \quad 95\%\ ee \tag{43}$$

$$\text{[structure]} \xrightarrow[\substack{\text{TFA} \\ \text{rt} \\ 45\%}]{\text{NaBH}_3\text{CN}} \text{[product]} \quad (+\text{ aziridine}) \tag{44}$$

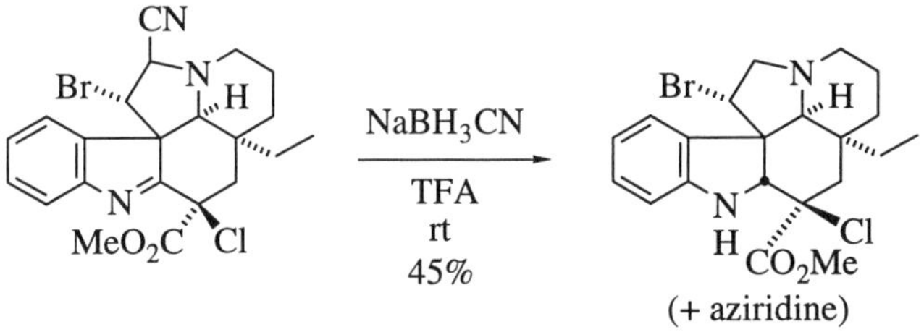

The imine-immonium ion can be generated *in situ* as illustrated by the examples in equations 45, 46 (*90, 91*).

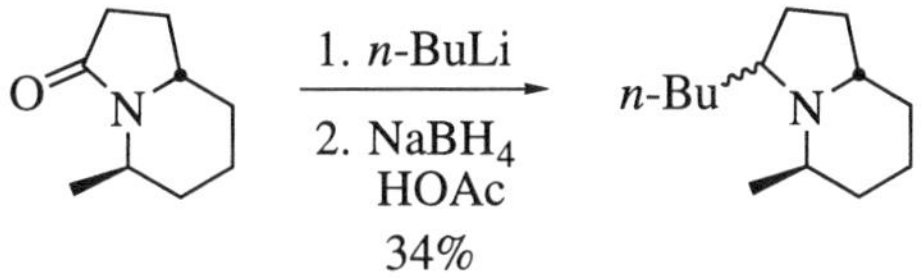

$$\text{[structure]} \xrightarrow[\substack{\text{2. NaBH}_4 \\ \text{HOAc} \\ 34\%}]{1.\ n\text{-BuLi}} n\text{-Bu-[product]} \tag{45}$$

$$\text{[structure]} \xrightarrow[\substack{\text{2. NaBH}_3\text{CN} \\ \text{HOAc} \\ 71\%}]{1.\ \text{MeMgCl}} \text{[product]} \tag{46}$$

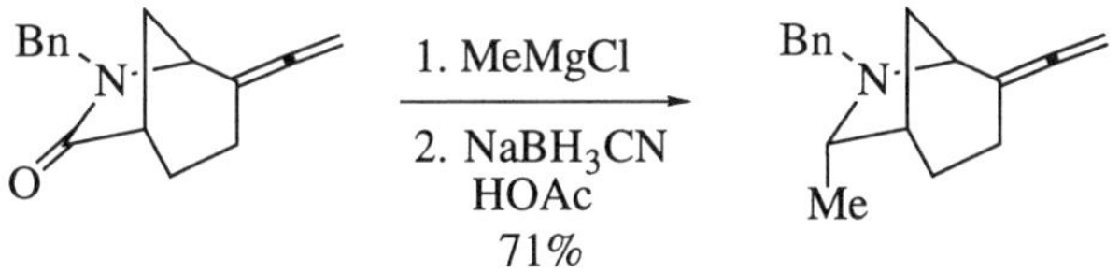

The combination of NaBH$_3$CN/TFA reduces all eight imine units in a four-fold bridged double-decker prophyrin (*92*).

Following the early studies of Marshall and Johnson (*4-6*) and the subsequent exhaustive study by Hutchins (*93*), several recent examples of enamine reduction using NaBH$_4$ or NaBH$_3$CN in carboxylic acids have been described (equations 47-50) (*94-98*). Normalindine (equation 49) can also be reached via the same reduction from the exocyclic enamine (*97*).

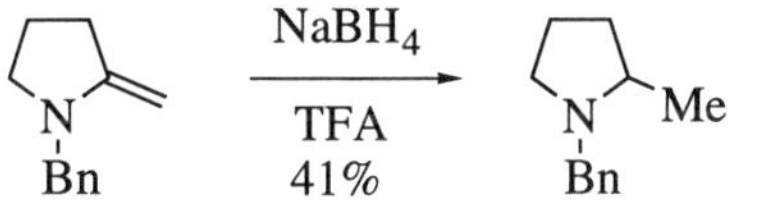

$$\text{[N-Bn pyrrolidine with exocyclic CH}_2\text{]} \xrightarrow[\substack{\text{TFA} \\ 41\%}]{\text{NaBH}_4} \text{[N-Bn 2-methylpyrrolidine]} \tag{47}$$

$$\text{[OMe,OMe tetrahydro N-Bn structure]} \xrightarrow[\substack{\text{HOAc} \\ \text{MeCN rt} \\ 56\%}]{\text{NaBH}_3\text{CN}} \text{[product]} \quad (+3\%\ \textit{trans}) \tag{48}$$

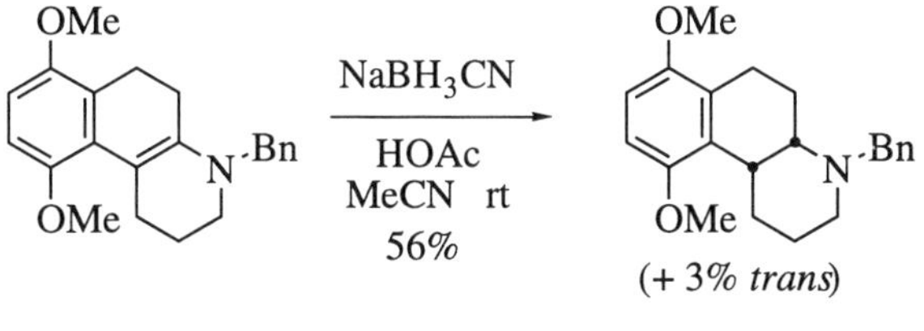

[Equation (49): NaBH₃CN, HOAc, 10 °C, 55% → normalindine (+ 6% cis)]

[Equation (50): NaBH₄, HOAc, THF 20 °C, 62%]

Similarly, enamides, vinylogous amides, and related compounds are generally smoothly reduced to their saturated analogues, and some recent cases are depicted in equations 51-55 (*99-103*). The Bartoli chemistry (equation 51) is highly diastereoselective (*99*), and the reaction shown in equation 52 was actually performed on a homochiral substrate to give diastereomers (*100*).

[Equation (51): NaBH(OAc)₃, HOAc, 10 °C, 89%]

[Equation (52): NaBH₃CN, HCO₂H, 33%]

[Equation (53): NaBH₃CN, HCO₂H, 90%]

[Equation (54): NaBH₃CN, CH₂Cl₂, TFA, −42 °C, 55%]

[Equation (55): NaBH₃CN, CH₂Cl₂, TFA, −45 °C, 76% (+22% cis)]

Several such reductions are known from the indole field (equations 56-59) (*104-107*), including Djerassi's original observation (equation 7) (*28*). As expected, the indole double bond is impervious to these reduction conditions.

(56) NaBH$_4$, HOAc, 10 °C, 90% (+ *trans*, 1:1)

(57) NaBH$_3$CN, HOAc, 10-28 °C, 82% (+ 12% ketone)

(58) NaBH$_3$CN, HOAc, MeCN, 20 → 60 °C, 80%

(59) NaBH$_4$, HOAc, 90 °C, 55% (+ 15% epimeric at ester)

**Reduction of Amides.**

It has been seen that amides survive unscathed in the reaction medium presented thus far, which generates NaBH(OCOR)$_3$ species (excess RCO$_2$H). Umino and his colleagues discovered that amides are, in fact, reduced to amines under conditions that generate the more reactive NaBH$_3$OCOR species (*108*). This important extension of the NaBH$_4$/RCO$_2$H technology has been utilized by several groups to reduce amides and lactams (*9*). A few recent examples are tabulated in equations 60-63 (*109-112*). The selectivity in equations 61 and 63 is noteworthy, and the latter reduction is thought to involve hydroxyl participation since the OTBS ether is not reduced (*112*).

(60) NaBH$_4$, HOAc, dioxane, 97%

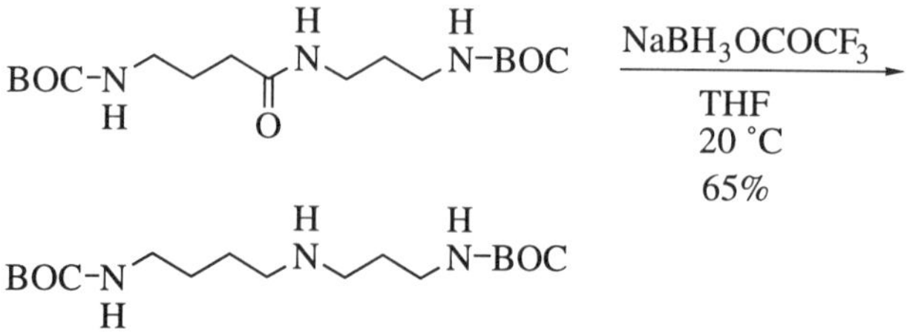

$$\text{BOC-NH-(CH}_2)_3\text{-C(O)-NH-(CH}_2)_3\text{-NH-BOC} \xrightarrow[\substack{\text{THF} \\ 20\,^\circ\text{C} \\ 65\%}]{\text{NaBH}_3\text{OCOCF}_3} \text{BOC-NH-(CH}_2)_4\text{-NH-(CH}_2)_3\text{-NH-BOC} \quad (61)$$

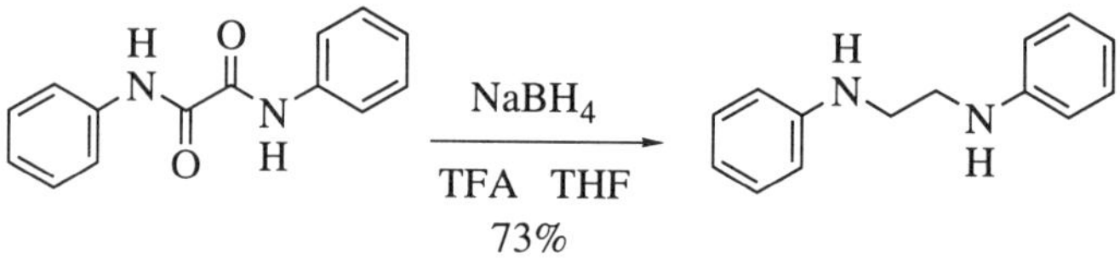

$$\text{PhNH-C(O)-C(O)-NHPh} \xrightarrow[\substack{\text{TFA THF} \\ 73\%}]{\text{NaBH}_4} \text{PhNH-CH}_2\text{-CH}_2\text{-NHPh} \quad (62)$$

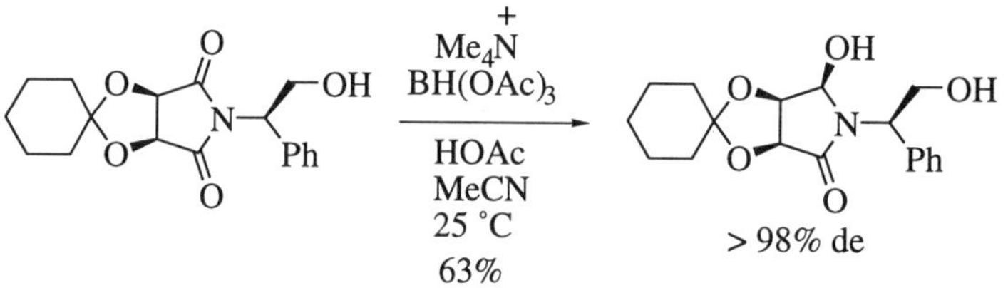

(63) — reduction of bicyclic imide with Me$_4$N$^+$BH(OAc)$_3$, HOAc, MeCN, 25 °C, 63%, giving hydroxy lactam, >98% de.

## Reduction of Nitriles.

Umino also discovered that nitriles can be reduced to primary amines with NaBH$_3$OCOCF$_3$, but poorly with NaBH$_3$OAc (*113*). A few recent cases are shown here (equations 64-68) (*114-117*, Gribble, G.W. Dartmouth College, unpublished results). The lack of nitro group reduction is particularly noteworthy since conventional reduction methods would invariably reduce the nitro group before the cyano group.

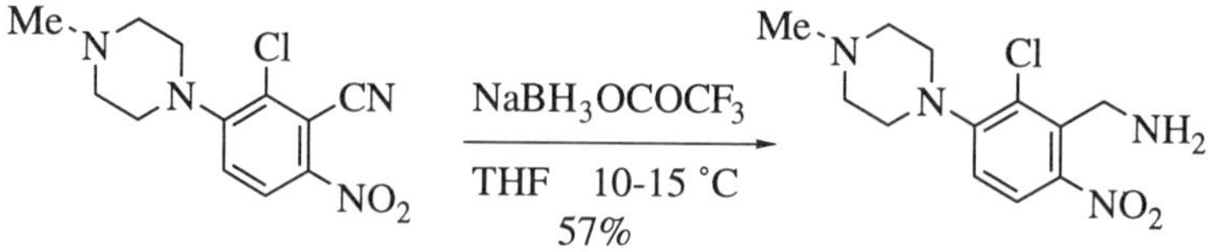

(64) Aryl-CN (with Me-piperazine, Cl, NO$_2$ substituents) $\xrightarrow{\text{NaBH}_3\text{OCOCF}_3,\ \text{THF},\ 10\text{-}15\,^\circ\text{C},\ 57\%}$ Aryl-CH$_2$NH$_2$

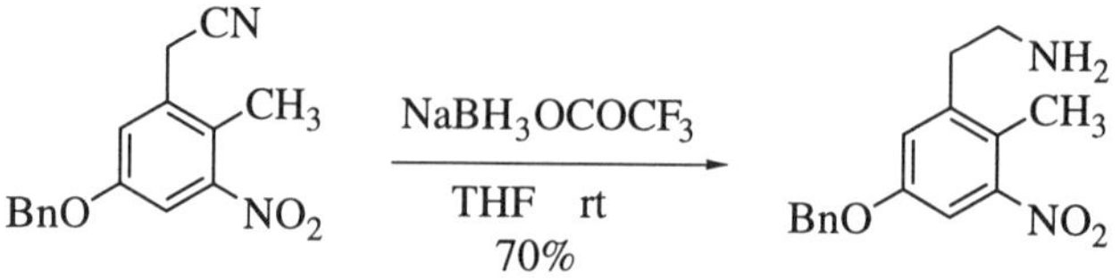

(65) ArCH$_2$CN (BnO, CH$_3$, NO$_2$ substituted) $\xrightarrow{\text{NaBH}_3\text{OCOCF}_3,\ \text{THF rt},\ 70\%}$ ArCH$_2$CH$_2$NH$_2$

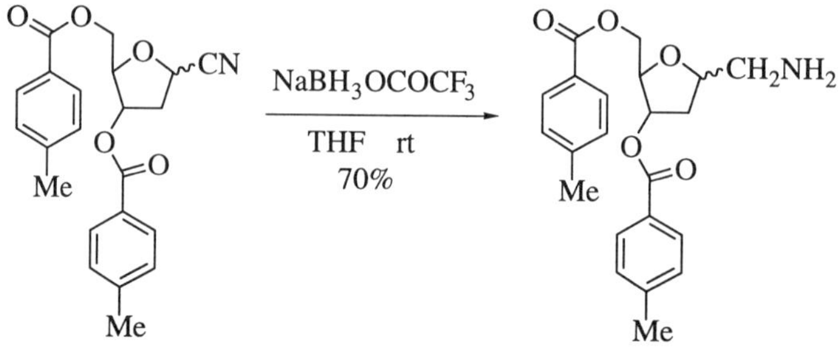

(66) Sugar-CN (bis-toluoyl ester) $\xrightarrow{\text{NaBH}_3\text{OCOCF}_3,\ \text{THF rt},\ 70\%}$ Sugar-CH$_2$NH$_2$

$$\text{(67)}$$

(CN compound with N-CO₂Bn) → NaBH₃OCOCF₃, THF rt, 70% → (CH₂NH₂ compound with N-CO₂Bn)

$$\text{(68)}$$

(Pentafluorobenzyl cyanide) → NaBH₃OCOCF₃, THF rt, 67% → (Pentafluorophenethylamine)

Itsuno has been able to trap the *N*-boryl imines, generated from nitriles and NaBH₃OCOR, with alkyllithium reagents to give amines (e.g., equation 69) (*118*).

$$\text{(69)}$$

PhCN → 1. NaBH₃OCO-*i*-Pr; 2. *n*-BuLi, 68% → 1-phenyl-1-aminobutane

## Hydroboration of Alkenes.

Marshall and Johnson also described the use of NaBH₄/HOAc to hydroborate alkenes (*119*), and several recent examples and variations have been reported (*9*). A few recent examples are illustrated in equations 70-72 (*120-122*). The NaBH₃OAc in equations 70 and 71 can also be generated from NaBH₄ and Hg(OAc)₂ (*120, 121*). In related chemistry, organomercurials can be reduced with NaBH(OAc)₃ (*123*).

$$\text{(70)}$$

Styrene → 1. NaBH₄, HOAc; 2. H₂O₂, OH⁻, 89% → 2-phenylethanol

$$\text{(71)}$$

(OTMS diene) → 1. NaBH₃OAc; 2. H₂O₂, OH⁻, 83% → (OTMS alcohol)

$$\text{(72)}$$

Cyclohexene → 1. NaBH₃OAc; 2. NaOMe; 3. H₂O₂, OH⁻, 67% → dicyclohexyl ketone

## Reduction of Alkenes.

In addition to hydroboration, alkenes can be reduced to alkanes in a few cases. The first such example was our observation that 1,1-diphenylethylene was reduced to 1,1-diphenylethane with NaBH₄/TFA in 93% yield, undoubtedly via the highly stabilized carbocation (*124*). However, only a few other examples of alkene reductions with NaBH₄/RCO₂H have been reported (e.g., equations 73, 74) (*125, 126*).

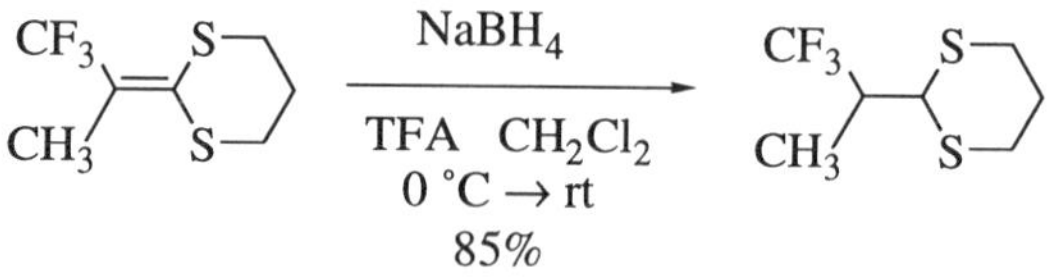

$$\text{(73)}$$

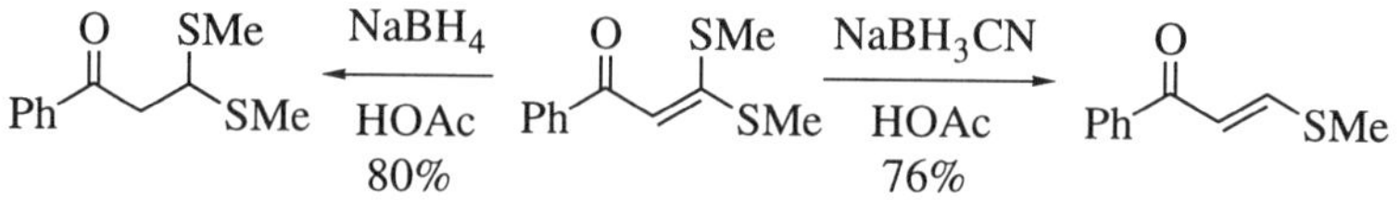

[Equation 73: CF₃/CH₃ ketene dithioacetal + NaBH₄, TFA/CH₂Cl₂, 0 °C → rt, 85% → CF₃CH(CH₃)-dithiane]

$$\text{(74)}$$

[Equation 74: Ph-CO-CH=C(SMe)₂ ← NaBH₄/HOAc, 80% — Ph-CO-CH(SMe)₂ from Ph-CO-CH=CH-SMe via NaBH₃CN/HOAc, 76%]

### Reduction of Alcohols.

Early in our research program, when we realized that trifluoroacetic acid (TFA) and NaBH$_4$ were reasonably compatible, we thought that benzylic alcohols would be reduced to hydrocarbons under these conditions, since TFA is an excellent solvent for solvolysis and other S$_N$1 reactions (ionizing power Y value = 1.84). Indeed, diphenylmethanol and triphenylmethanol are reduced to diphenylmethane and triphenylmethane in 93% and 99% yields, respectively (*124*). The reaction is very general for di- and triarylcarbinols but is poorer for monobenzylic alcohols (*9, 124*) where the more reactive intermediate carbocations undergo side reactions (*9*). However, Nutaitis has found that the more reactive NaBH$_3$OCOCF$_3$ can reduce certain monobenzylic alcohols (equation 75) (*127*). Secondary and tertiary monobenzylic alcohols give higher yields.

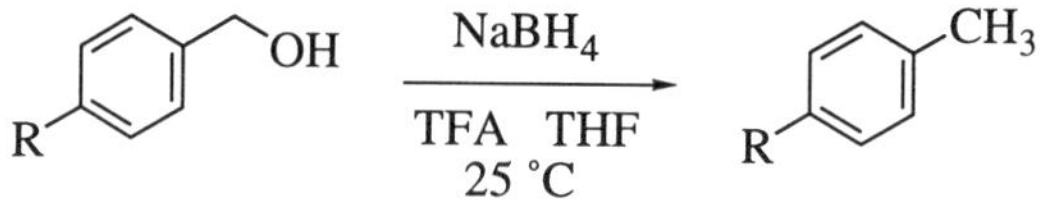

$$\text{(75)}$$

[Equation 75: 4-R-C₆H₄-CH₂OH + NaBH₄, TFA/THF, 25 °C → 4-R-C₆H₄-CH₃]

| R | % |
|---|---|
| Me$_2$N | 78% |
| OH | 45% |
| SMe | 41% |
| OMe | 19% |
| Me | 0% |
| Ph | 0% |

Some other recent examples of the reduction of monobenzylic alcohols are cited in equations 76-78 (*128-130*). Interestingly, Olah has found that NaBH$_4$/CF$_3$SO$_3$H is even more effective (98%) in reducing the alcohol shown in equation 77 (*129*). The selective deoxygenation shown in equation 78 is remarkable indeed (*130*).

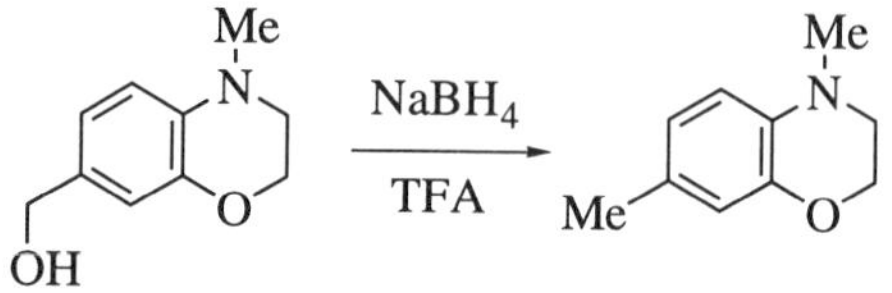

$$\text{(76)}$$

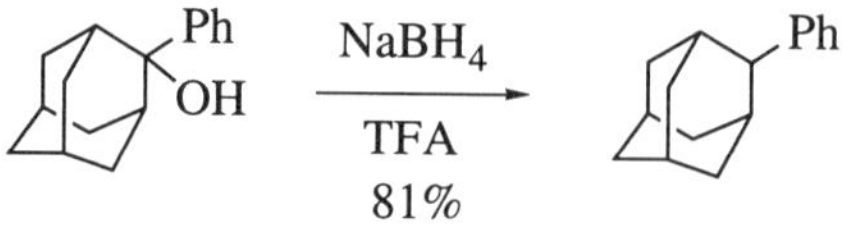

[Equation 76: N-methyl benzomorpholine with CH₂OH substituent + NaBH₄/TFA → N-methyl benzomorpholine with CH₃ substituent]

$$\text{(77)}$$

[Equation 77: 2-phenyl-2-adamantanol + NaBH₄/TFA, 81% → 2-phenyladamantane]

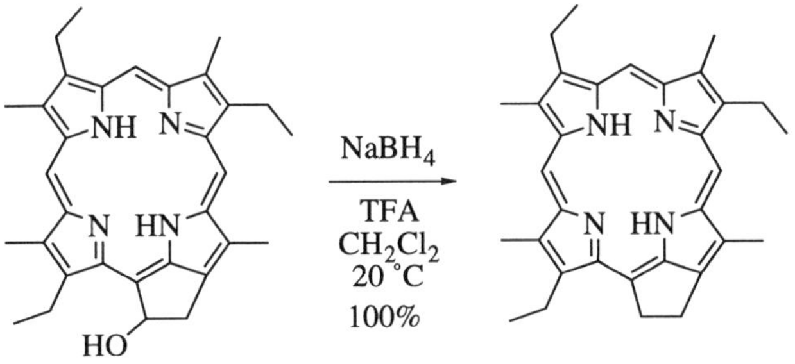

11. GRIBBLE  *Sodium Borohydride and Carboxylic Acids*  187

(78)

Following an initial result by Kabalka (*131*), Nutaitis and his coworkers have shown that a wide range of doubly-benzylic heterocyclic alcohols are reduced by NaBH$_4$/TFA to the corresponding hydrocarbons (equations 79-81) (*131*, *132*).

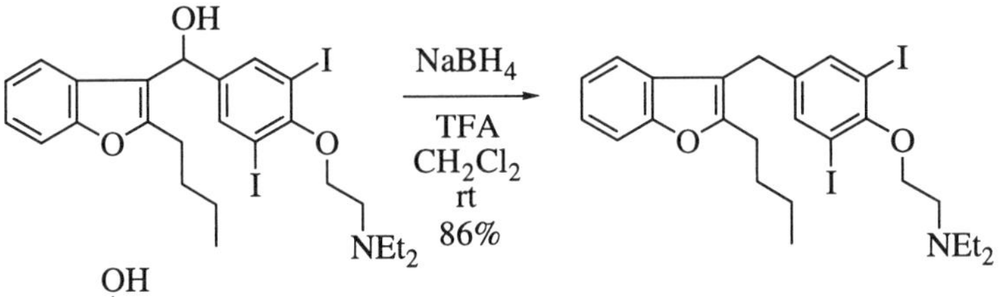

(79)

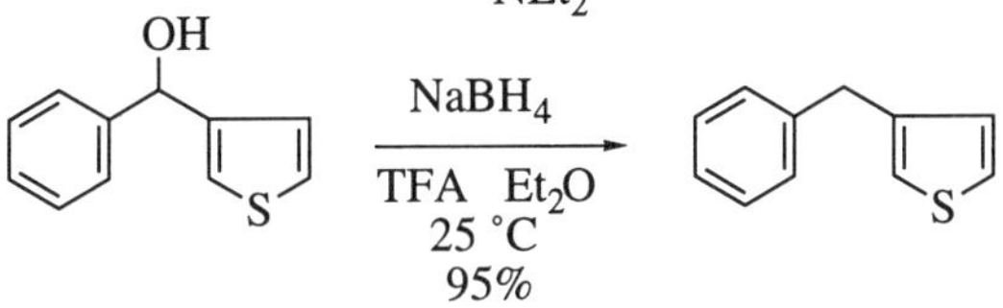

(80)

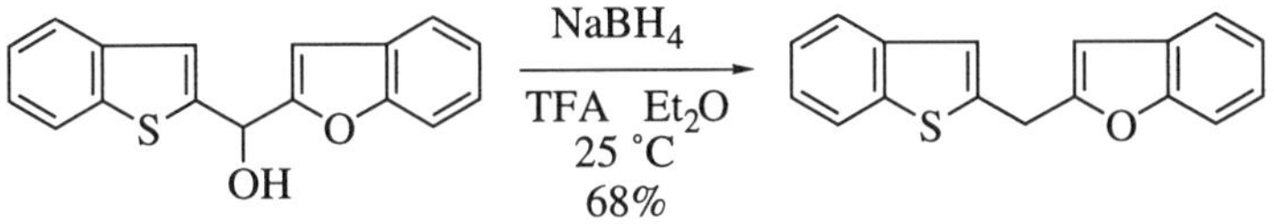

(81)

Other substrates that have been recently synthesized from the corresponding benzyl alcohols and NaBH$_4$/TFA are listed in Scheme 9 (see also *137*).

**Scheme 9**

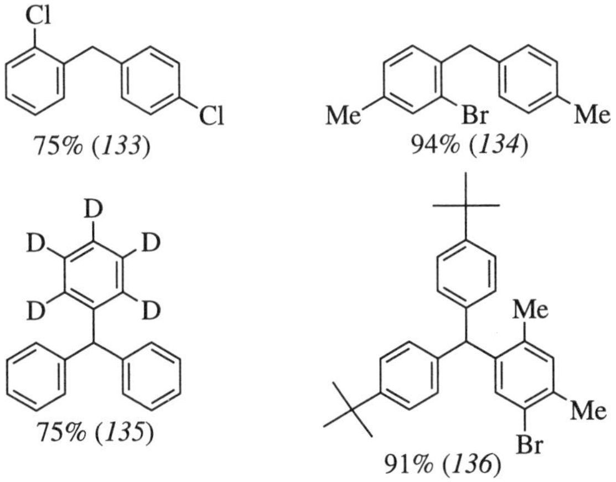

75% (*133*)

94% (*134*)

75% (*135*)

91% (*136*)

Nicholas has shown that acetylenic diol cobalt complexes are smoothly deoxygenated with NaBH$_4$/TFA (equations 82, 83) (*138, 139*).

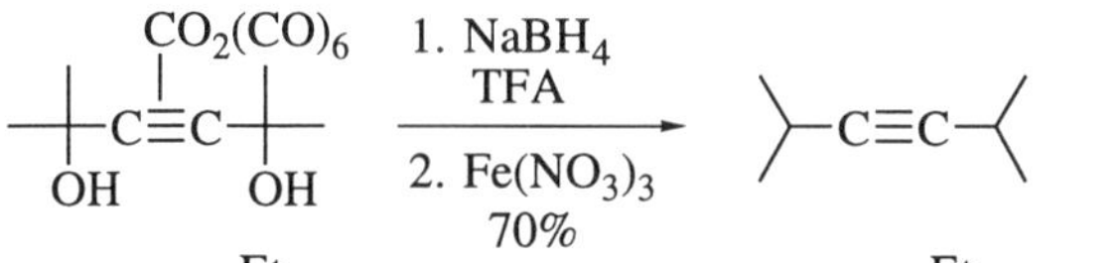

$$\underset{\underset{OH}{|}}{-}\overset{\overset{CO_2(CO)_6}{|}}{C}\equiv C\underset{\underset{OH}{|}}{-} \xrightarrow[\text{2. Fe(NO}_3)_3]{\text{1. NaBH}_4 \text{ TFA}} \text{>-C≡C-<} \quad (82)$$

$$70\%$$

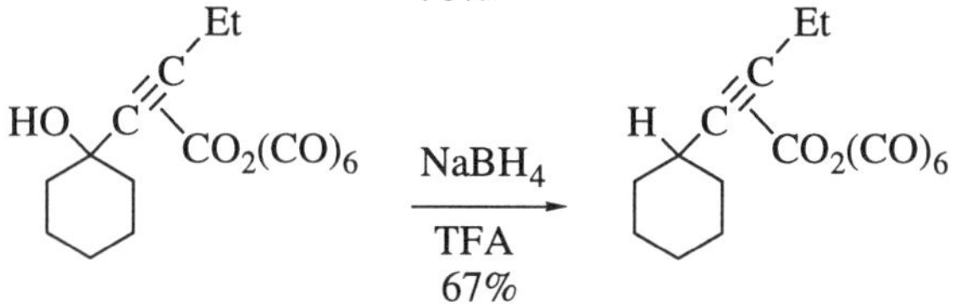

$$\text{(83)}$$

$$67\%$$

The enormous power and versatility of the NaBH$_4$/TFA reducing system is beautifully revealed by the companion reactions discovered by Maryanoff and his colleagues (equations 84, 85) (*140*). In the first reaction, the alcohol is reduced by NaBH(OCOCF$_3$)$_3$ in the presence of excess TFA, but, in the second reaction, the more reactive NaBH$_3$OCOCF$_3$ reduces only the lactam since excess TFA is not present.

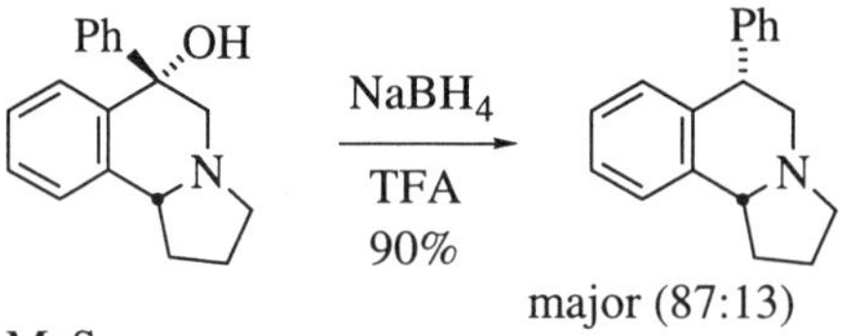

$$\text{(84)}$$
90%
major (87:13)

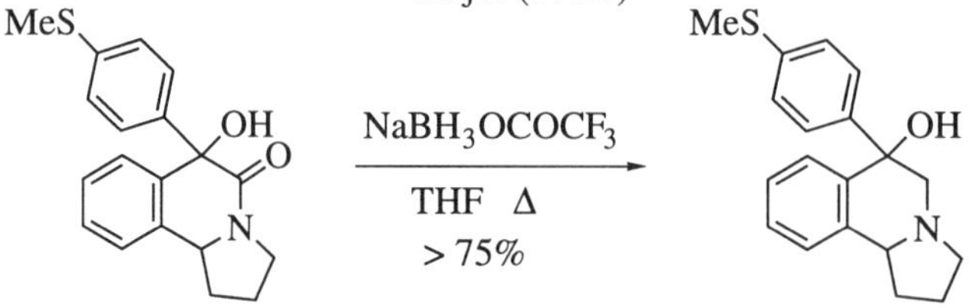

$$\text{(85)}$$
> 75%

**Reduction of Ketones to Hydrocarbons.**

During our early research on the reduction of diarylmethanols to diarylmethanes (*124*) (*vide supra*), we also discovered that benzophenone is reduced to diphenylmethane with NaBH$_4$/TFA in 92% distilled yield (*124*). Subsequent studies in our laboratory revealed the generality of this novel and efficient reduction method (equation 86) (*141*). A range of functional groups tolerates these reaction conditions and the reaction only fails or fares poorly when the ketone is highly hindered (dimesityl ketone) or contains a nitro group (*141, 9*).

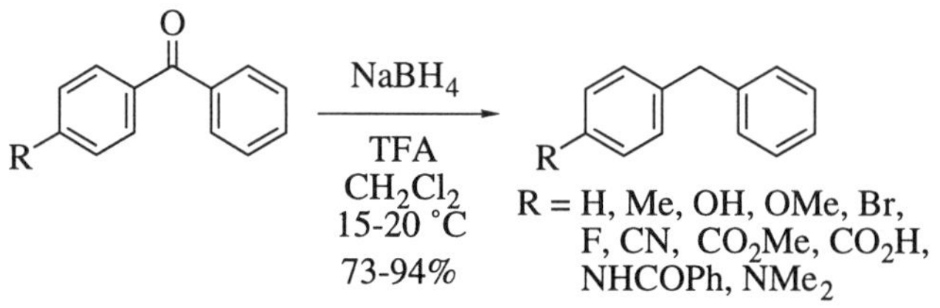

$$\text{(86)}$$

NaBH$_4$
TFA
CH$_2$Cl$_2$
15-20 °C
73-94%

R = H, Me, OH, OMe, Br, F, CN, CO$_2$Me, CO$_2$H, NHCOPh, NMe$_2$

## 11. GRIBBLE  *Sodium Borohydride and Carboxylic Acids*

Some recent compounds that have been synthesized by the action of NaBH$_4$/TFA on the corresponding ketone (indicated by an arrow) are listed in Scheme 10. The NaBH$_3$CN/TFA reaction leading to the pyrrole imide (*147*) appears to be the first reduction of a formyl group to a methyl group using this methodology.

**Scheme 10**

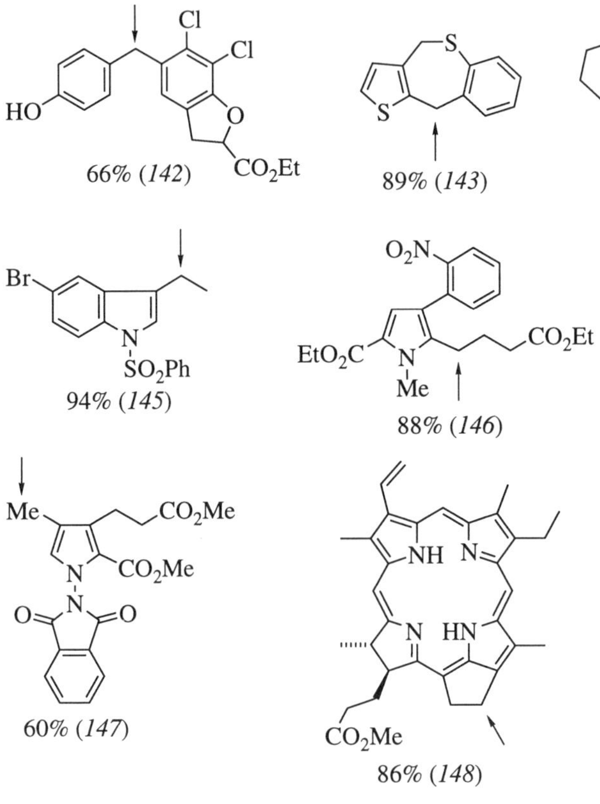

66% (*142*)   89% (*143*)   74% (*144*)

94% (*145*)   88% (*146*)

60% (*147*)   86% (*148*)

The reaction of 2-iodo-3-acetyl-*N*-(phenylsulfonyl)indole with NaBH$_4$/TFA gives 3-ethyl-*N*-(phenylsulfonyl)indole (75% yield) (*145*). This appears to be the first example of dehalogenation with NaBH$_4$/RCO$_2$H.

In related chemistry discovered sometime ago, Hutchins utilized NaBH$_4$/HOAc to reduce the tosylhydrazones of aldehydes and ketones to hydrocarbons (*149*). A recent example of this reaction is shown in equation 87 (*150*).

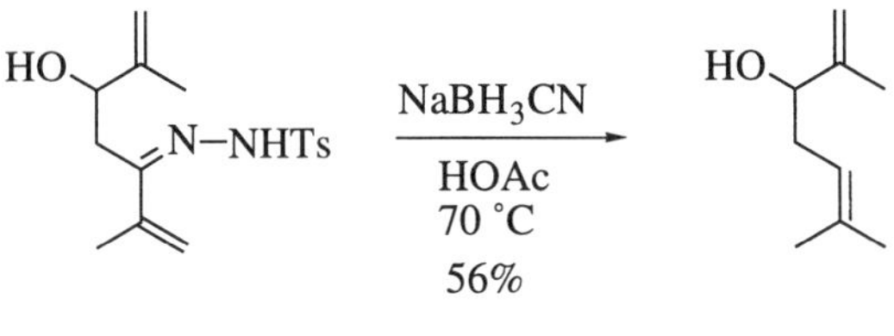

(87)

## Reduction of Carboxylic Acids and Esters.

In view of the facile reduction of carboxylic acids to aldehydes with $NaBH_4$, via acyloxyborohydrides, it is not surprising that complete reduction to primary alcohols has been observed by two groups (equations 88, 89) (*151, 152*), following the pioneering work by Liberatore and colleagues (*40*).

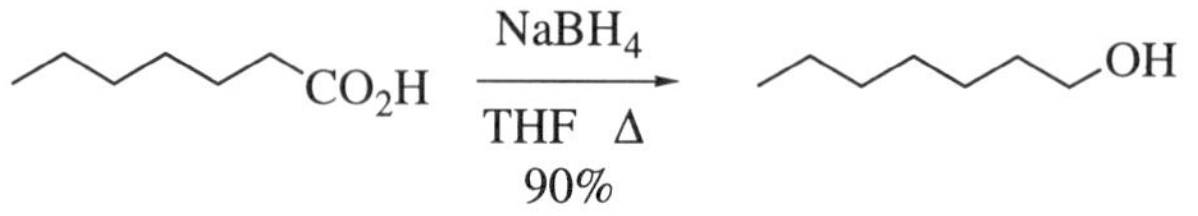

$$CH_3(CH_2)_4CH_2CO_2H \xrightarrow[THF\ \Delta]{NaBH_4} CH_3(CH_2)_4CH_2CH_2OH \quad 90\% \tag{88}$$

$$MeO-CO-(CH_2)_8CO_2H \xrightarrow[\substack{TFA\ THF\\25\ ^\circ C}]{NaBH_4} MeO-CO-(CH_2)_9OH \quad 78\% \tag{89}$$

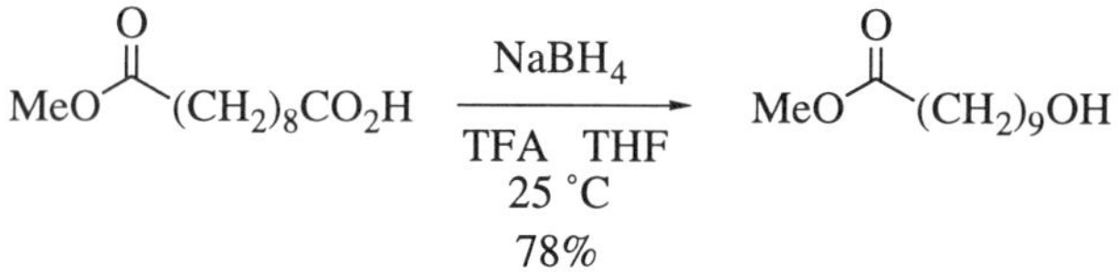

The only report of the $NaBH_4/RCO_2H$ reduction of an ester group appears to be that shown in equation 90 (and related examples) (*153*).

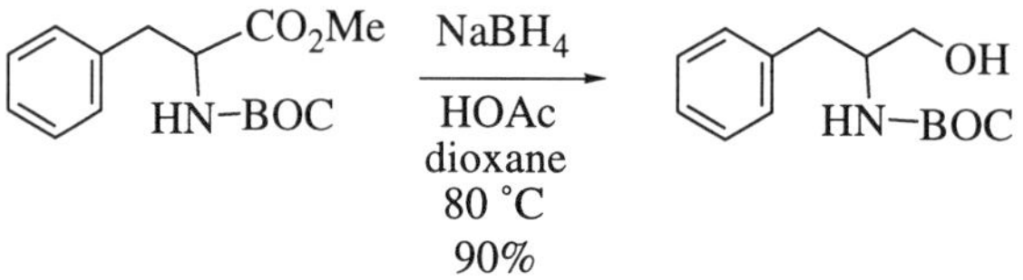

$$PhCH_2CH(NHBOC)CO_2Me \xrightarrow[\substack{HOAc\\dioxane\\80\ ^\circ C}]{NaBH_4} PhCH_2CH(NHBOC)CH_2OH \quad 90\% \tag{90}$$

## Reductive Cleavage of Acetals, Ketals and Ethers.

As might be anticipated, the action of $NaBH_4/RCO_2H$ effects the cleavage of acetals, ketals, ethers and related compounds (*9*). We have reported both the reductive cleavage of cyclic acetals and ketals (e.g., equation 91) (*154*) and the reductive deoxygenation of 1,4-epoxy-1,4-dihydronaphthalenes (e.g., equation 92) (*155*). We utilized the latter reaction in the synthesis of a dimethylbenzo[*b*]carbazole (equation 93) (*156*).

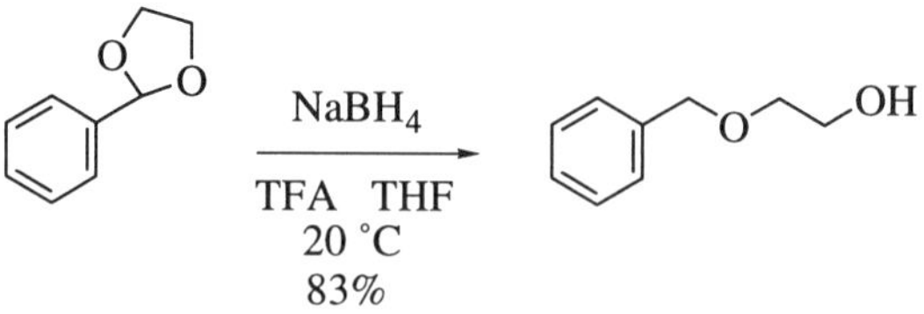

$$Ph\text{-}(1,3\text{-dioxolane}) \xrightarrow[\substack{TFA\ THF\\20\ ^\circ C}]{NaBH_4} PhCH_2\text{-}O\text{-}CH_2CH_2OH \quad 83\% \tag{91}$$

$$\text{(1,4-dimethyl-1,4-epoxy-1,4-dihydronaphthalene)} \xrightarrow[\substack{TFA\ THF\\0\text{-}5\ ^\circ C}]{NaBH_4} \text{(naphthalene)} \quad 90\% \tag{92}$$

$$\text{(2,3-dimethyl-N-PhO}_2\text{S-furo-indole)} \xrightarrow[\substack{1.\ benzyne\\2.\ NaBH_4,\ TFA\\3.\ HO^-}]{} \text{(5,11-dimethylbenzo[}b\text{]carbazole)} \quad 91\% \tag{93}$$

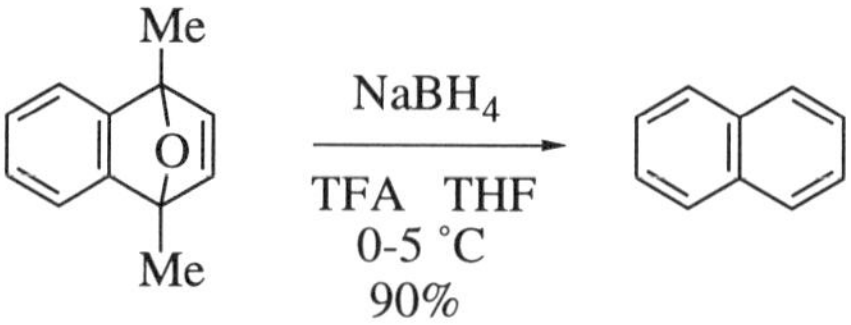

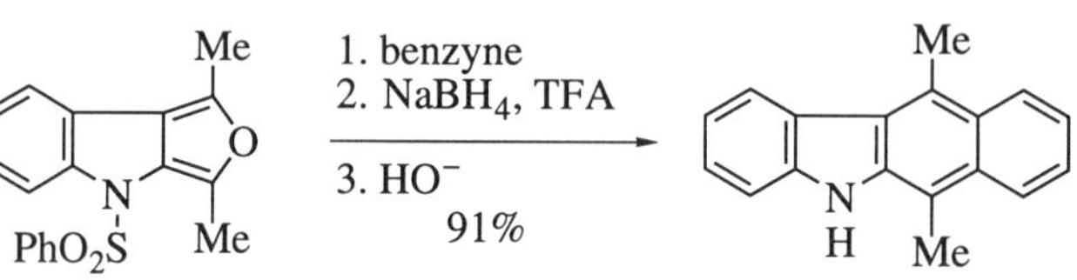

The reductive cleavage of isochromanes with NaBH$_4$/TFA has recently been described (equation 94) (*157*).

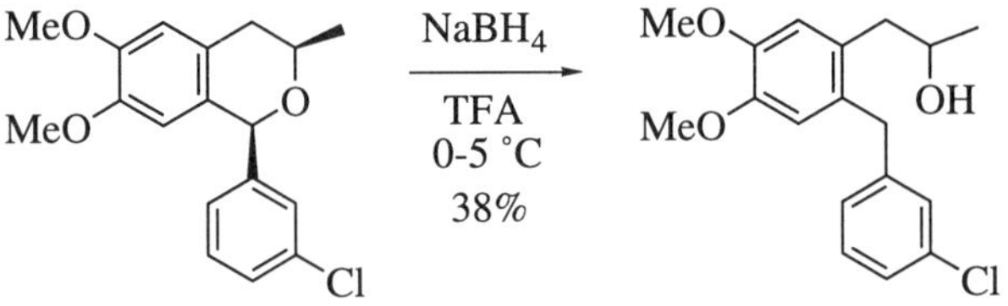

$$\text{MeO-isochromane-Cl} \xrightarrow[\text{0-5 °C}]{\substack{\text{NaBH}_4 \\ \text{TFA} \\ 38\%}} \text{MeO-diarylmethane-OH-Cl} \quad (94)$$

**Friedel-Crafts Alkylation of Arenes.**

During our investigation of the reaction of indole with NaBH$_4$/TFA, we observed the formation of the interesting "Baeyer condensation" product **10** (*9, 12*). This compound presumably arises from the reaction between indoline and trifluoroacetaldehyde, analogous to the synthesis of DDT from chlorobenzene, trichloroacetaldehyde, and H$_2$SO$_4$.

This reaction is reasonably general and the products shown in Scheme 11 have been prepared using this method (arene/NaBH$_4$/TFA) (*158*). With sterically congested arenes (e.g., mesitylene, durene), the reaction stops at the carbinol stage (34% and 15%, respectively).

**Scheme 11**

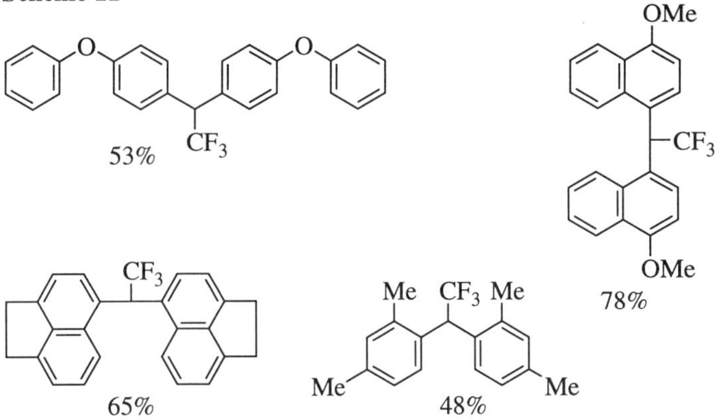

53%   78%   65%   48%

**Selective Aldehyde Reduction**

Early in our work with NaBH$_4$/RCO$_2$H, we observed that aldehydes and, especially, ketones are reduced much more slowly to alcohols by NaBH$_4$/HOAc than in conventional alcoholic or aqueous media. Indeed, this is why the *N*-alkylation of amines in this medium is successful! For example, although benzaldehyde is completely reduced to benzyl alcohol after 1 hr at 15 °C with a large excess of NaBH$_4$ in glacial HOAc, acetophenone is only reduced to the extent of 60% at 25 °C after 40 hr (*9*, Gribble, G.W. Dartmouth College, unpublished results).

These and related observations paved the way for the chemoselective reduction of aldehydes, in the presence of ketones. We found that the isolated reagents (NaBH(OAc)$_3$ (*56*) or *n*-Bu$_4$NBH(OAc)$_3$ (*159*) in benzene worked extremely well in this regard (equations 95-97). For other examples, see ref. *9*.

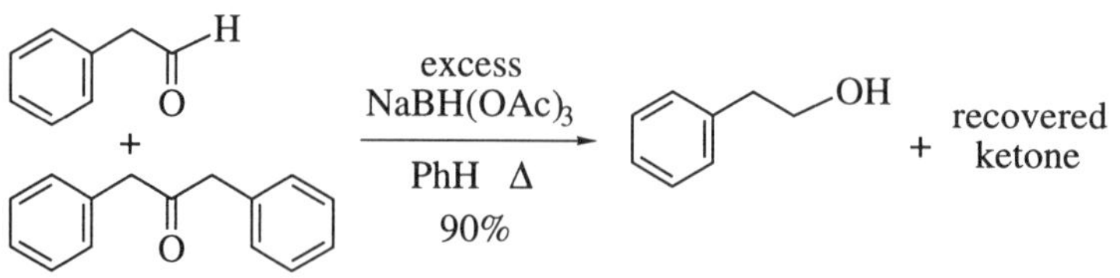

# REDUCTIONS IN ORGANIC SYNTHESIS

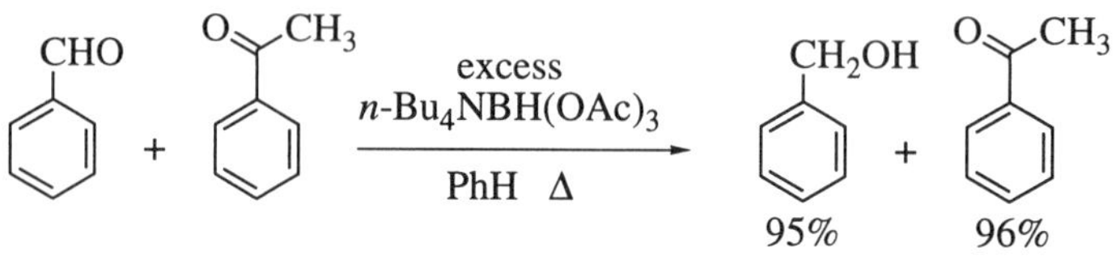

(95) PhCH₂CHO + PhCH₂C(O)CH₂Ph →[excess NaBH(OAc)₃, PhH Δ, 90%] PhCH₂CH₂OH + recovered ketone

(96) PhCHO + PhC(O)CH₃ →[excess n-Bu₄NBH(OAc)₃, PhH Δ] PhCH₂OH (95%) + PhC(O)CH₃ (96%)

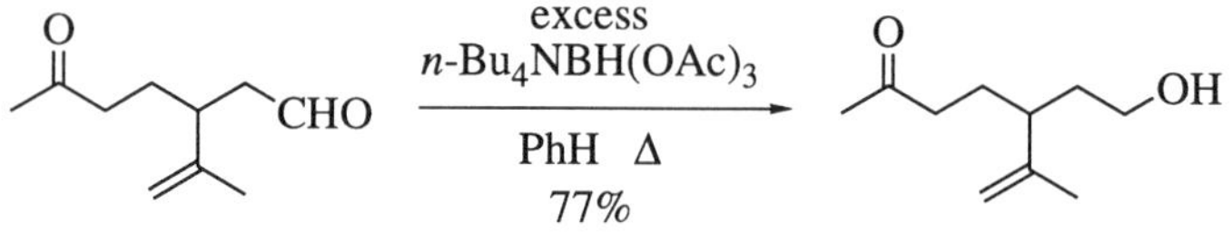

(97) [methyl ketone with pendant CHO and isopropenyl] →[excess n-Bu₄NBH(OAc)₃, PhH Δ, 77%] [methyl ketone with pendant CH₂OH and isopropenyl]

More recently, other workers have exploited this method for the selective reduction of aldehydes in the presence of ketones (Scheme 12). In each case, the primary alcohol was derived from the corresponding ketoaldehyde using NaBH(OAc)₃ in benzene.

**Scheme 12**

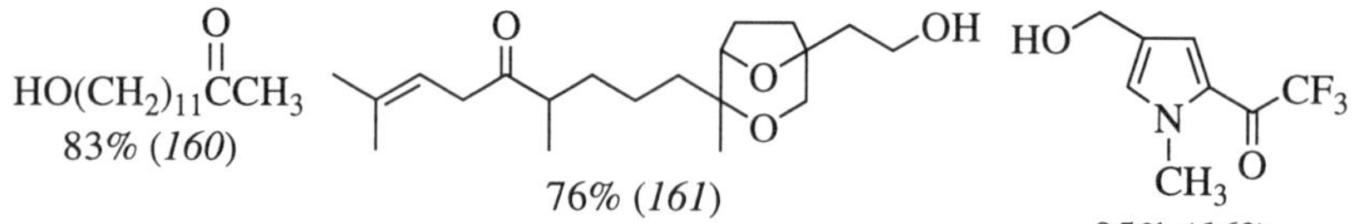

$HO(CH_2)_{11}CCH_3$ 83% (160)

[bicyclic ketal with pendant CH₂CH₂OH] 76% (161)

[N-methylpyrrole with CH₂OH and COCF₃] 95% (162)

Intrinsically more reactive ketones (cyclic, α- and β-keto) are reduced by NaBH₄/RCO₂H and some recent examples are shown in equations 98-100 (163-165).

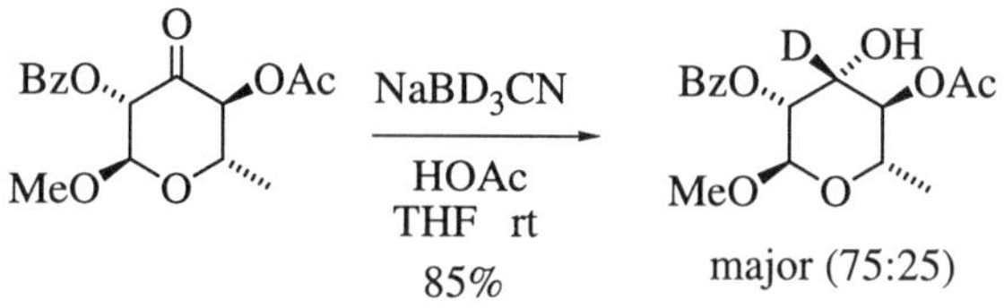

(98) [BzO, OAc, MeO pyranone] →[NaBD₃CN, HOAc, THF rt, 85%] [BzO, D, OH, OAc, MeO pyranose] major (75:25)

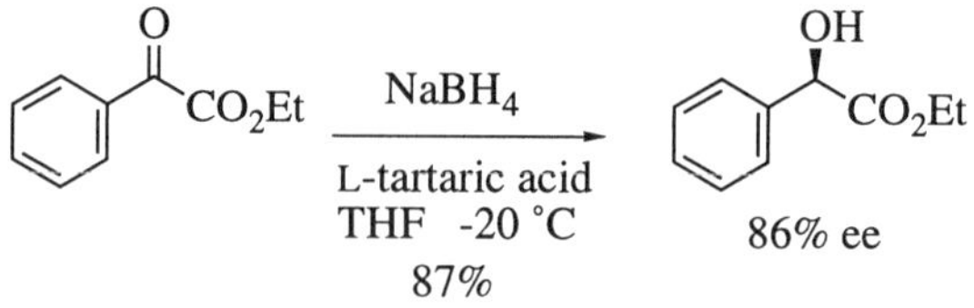

(99) PhC(O)CO₂Et →[NaBH₄, L-tartaric acid, THF -20 °C, 87%] PhCH(OH)CO₂Et 86% ee

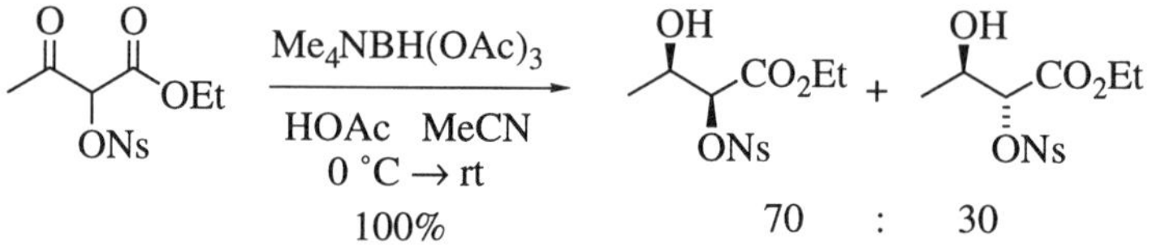

(100) CH₃C(O)CH(ONs)CO₂Et →[Me₄NBH(OAc)₃, HOAc MeCN, 0 °C → rt, 100%] CH₃CH(OH)CH(ONs)CO₂Et + CH₃CH(OH)CH(ONs)CO₂Et, 70 : 30

Nutaitis has studied the reduction of enones with $NaBH_4/HOAc$, conditions which give 1,2-reduction (equations 101-102) (*166*).

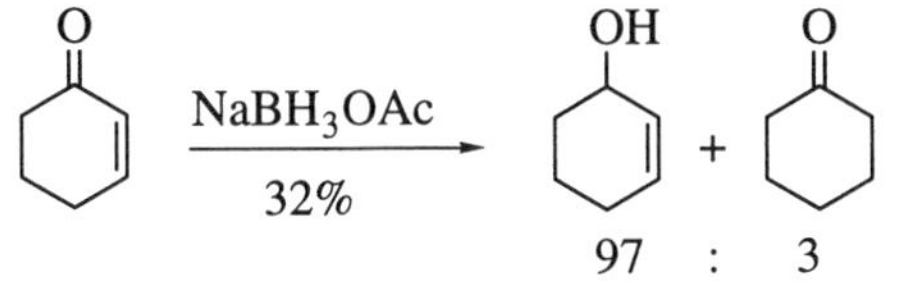

$$\text{cyclohexenone} \xrightarrow[32\%]{NaBH_3OAc} \text{cyclohexenol} + \text{cyclohexanone} \quad 97:3 \quad (101)$$

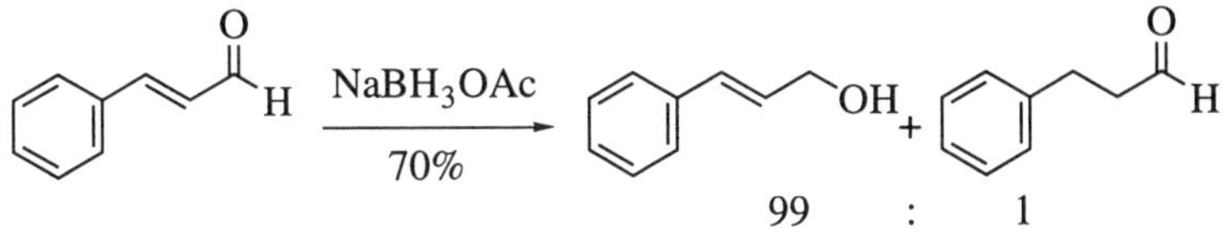

$$\text{PhCH=CHCHO} \xrightarrow[70\%]{NaBH_3OAc} \text{PhCH=CHCH}_2\text{OH} + \text{PhCH}_2\text{CH}_2\text{CHO} \quad 99:1 \quad (102)$$

**Hydroxyl-Directed Carbonyl Reduction.**

During our studies on the chemoselective reduction of aldehydes, we reported the reduction of the ketoaldehyde shown in equation 103 with $n$-$Bu_4NBH(OAc)_3$ and postulated an intramolecular hydride delivery as illustrated (*159*).

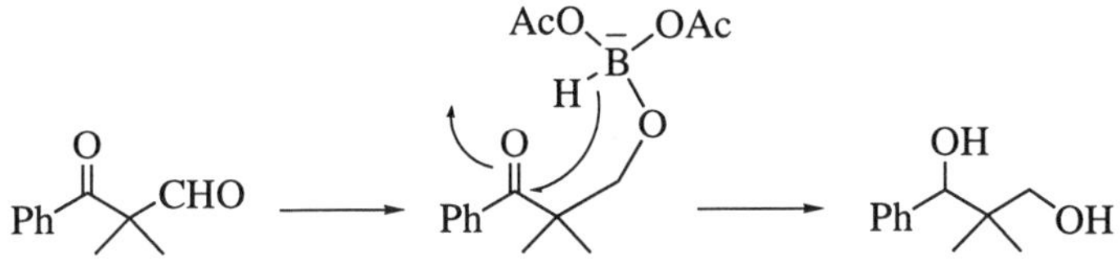

$$(103)$$

During our studies, Saksena reported the same reaction of $NaBH(OAc)_3$ with β-hydroxyketones in steroidal systems (*167*). In particular, he observed excellent stereoselectivities of this (intramolecular) hydride reduction. Subsequently, Evans developed this novel reduction into an extraordinarily useful stereoselective reduction method of β-hydroxyketones (*168, 57*), and he fully characterized several of the $MBH(OAc)_3$ reagents for the first time. Examples of this reduction procedure are listed in equations 104-110 (*168-174*).

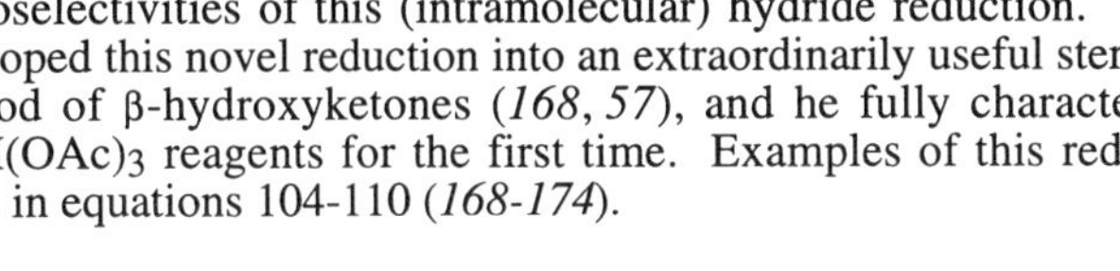

$$\xrightarrow[\substack{HOAc\ MeCN \\ -20\,°C \\ 92\%}]{Me_4NBH(OAc)_3} \quad 98:2\ \textit{anti/syn} \quad (104)$$

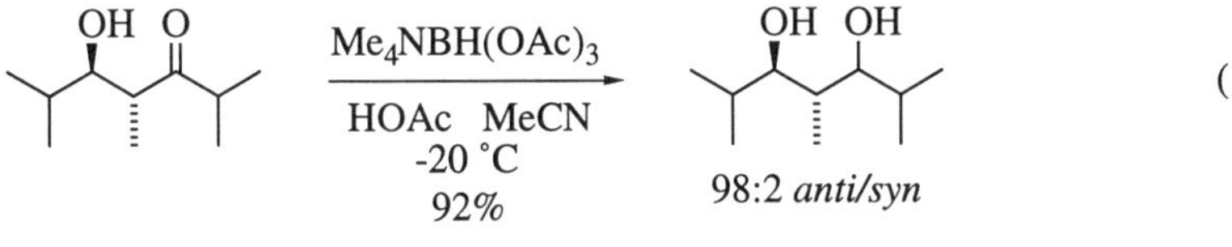

$$\xrightarrow[\substack{HOAc\ MeCN \\ -40\,°C \\ 93\%}]{Me_4NBH(OAc)_3} \quad 13:1\ \textit{anti/syn} \quad (105)$$

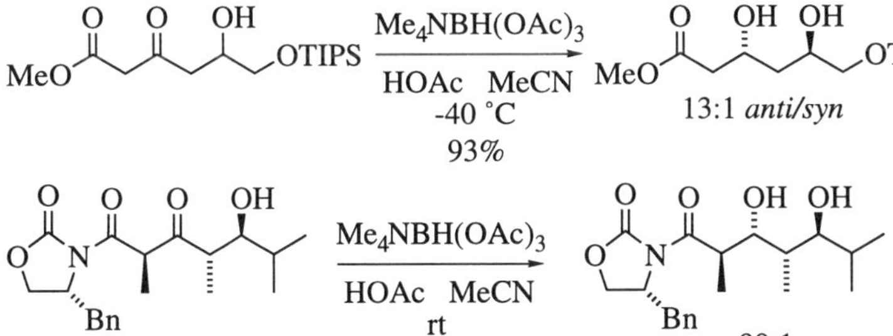

$$\xrightarrow[\substack{HOAc\ MeCN \\ rt \\ 81\%}]{Me_4NBH(OAc)_3} \quad >99:1 \quad (106)$$

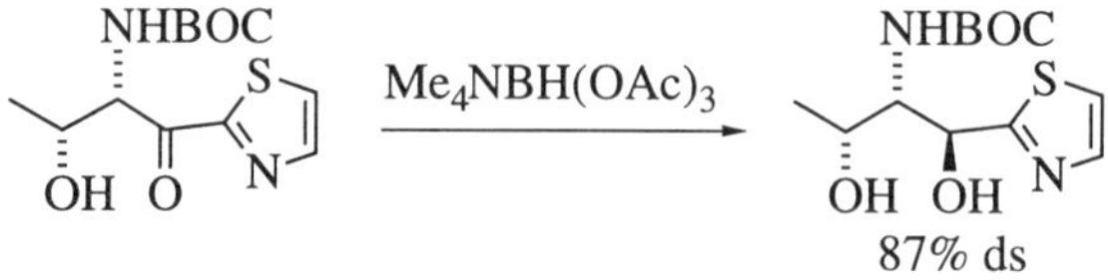

$$\text{(107)}$$

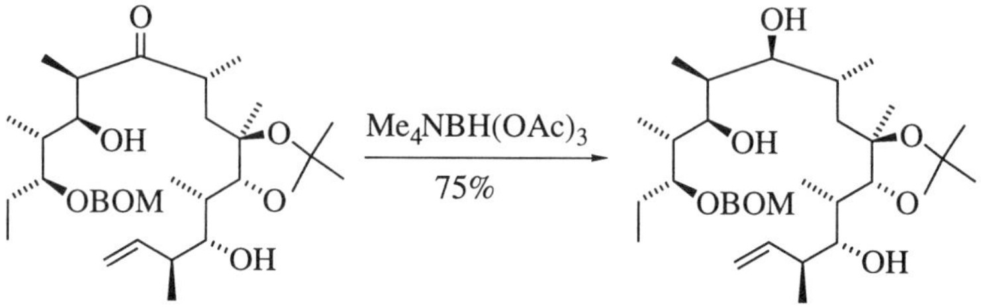

$$\text{(108)}$$

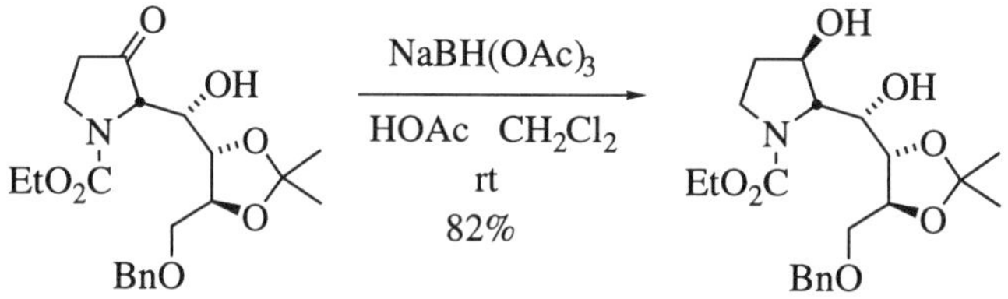

$$\text{(109)}$$

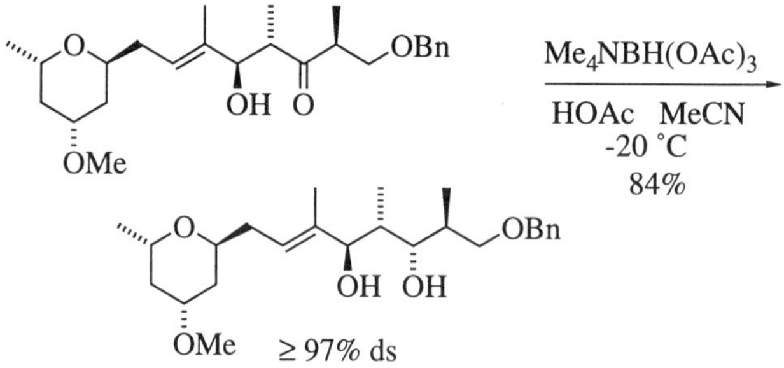

$$\text{(110)}$$

Numerous other examples of this stereoselective β-hydroxyketone reduction have been described in recent years (*175-193*), including the use of MBH(OAc)$_3$ in syntheses or synthetic approaches to verrucosidin (*194*), phorbol (*195*), streptenol B (*196*), calyculin A (*197*), lepicidin (*198*), muamvatin (*199*), mintlactone (*200*), rhizoxin (*201, 202*), *myo*-inositol derivatives (*203*), discodermolide (*204*), acutiphycin (*205*), and FK-506 (*206, 207*). Despite the enormous success of this β-hydroxyketone stereoselective reduction, Me$_4$NBH(OAc)$_3$ showed no improvement over conventional methods in at least one case (*208*).

Interestingly, several examples of apparent stereoselective hydroxyl-mediated reductions of α-hydroxyketones have been reported (equations 111-113) (*209-211*).

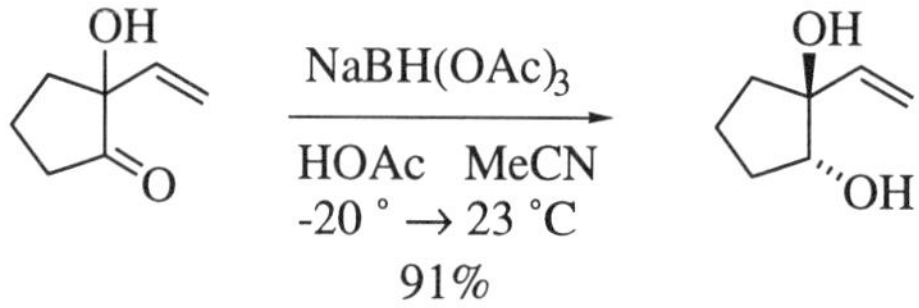

$$\text{(111)}$$

$$\text{(112)}$$

$$\text{(113)}$$

Other notable examples of the stereoselective reduction of α-hydroxyketones with MBH(OAc)$_3$ include the final step in the total syntheses of rocaglamide (*212*, *213*) and pancracine (*214*).

**Conclusions.**

Over the past 20 years, the combination of NaBH$_4$ and RCO$_2$H has developed into an amazingly versatile and efficient set of reducing and amine alkylating agents. These acyloxyborohydride species have rapidly emerged as the preeminent reagents of choice for many chemical transformations. The ability to control chemoselectivity, regioselectivity, and stereoselectivity by adjusting the carboxylic acid, hydride reagent, stoichiometry, solvent, temperature and time has no parallel in the arsenal of chemical reagents available to the organic chemist. Nevertheless, despite the extraordinary scope of acyloxyborohydrides in organic transformations, much work remains to be done in understanding the mechanisms of some of the reactions, such as *N*-alkylation, and in applying these reagents to asymmetric synthesis.

**References**

1. Dolby, L.J.; Gribble, G.W. *J. Heterocycl. Chem.* **1966**, *3*, 124.
2. Moore, M.L. *Org. React. V* **1949**, 301.
3. Hinman, R.L.; Whipple, E.B. *J. Am. Chem. Soc.* **1962**, *84*, 2534.
4. Marshall, J.A.; Johnson, W.S. *J. Am. Chem. Soc.* **1962**, *84*, 1485.
5. Marshall, J.A.; Johnson, W.S. *J. Org. Chem.* **1963**, *28*, 421.
6. Billman, J.H.; McDowell, J.W. *J. Org. Chem.* **1961**, *26*, 1437.
7. Gribble, G.W.; Lord, P.D.; Skotnicki, J.; Dietz, S.E.; Eaton, J.T.; Johnson, J.L. *J. Am. Chem. Soc.* **1974**, *96*, 7812.
8. Gribble, G.W. *Eastman Organic Chemical Bulletin* **1979**, *51*, No. 1, 1.
9. Gribble, G.W.; Nutaitis, C.F. *Org. Prep. Proc. Int.* **1985**, *17*, 317.
10. Nutaitis, C.F. *J. Chem. Ed.* **1989**, *66*, 673.
11. Gribble, G.W.; Wright, S.W. *Heterocycles* **1982**, *19*, 229.
12. Gribble, G.W.; Nutaitis, C.F.; Leese, R.M. *Heterocycles* **1984**, *22*, 379.
13. Gribble,G.W.; Hoffman, J.H. *Synthesis* **1977**, 859.
14. Kumar, Y.; Florvall, L. *Synth. Commun.* **1983**, *13*, 489.
15. Siddiqui, M.A.; Snieckus, V. *Tetrahedron Lett.* **1990**, *31*, 1523.
16. Toyota, M.; Fukumoto, K. *J. Chem. Soc., Perkin Trans. 1*, **1992**, 547.
17. Rawal, V.H.; Jones, R.J.; Cava, M.P. *J. Org. Chem.* **1987**, *52*, 19.
18. Brown, D.W.; Graupner, P.R.; Sainsbury, M.; Shertzer, H.G. *Tetrahedron* **1991**, *47*, 4383.
19. Yao, X.L.; Nishiyama, S.; Yamamura, S. *Chem. Lett.* **1991**, 1785.
20. Iwao, M.; Kuraishi, T. *Heterocycles* **1992**, *34*, 1031.
21. Dhanoa, D.S., *et al.*, *J. Med. Chem.* **1993**, *36*, 4232.

22. Bolton, R.E.; Moody, C.J.; Rees, C.W.; Tojo, G. *Tetrahedron Lett.* **1987**, *28*, 3163.
23. Boger, D.L.; Coleman, R.S.; Invergo, B.J. *J. Org. Chem.* **1987**, *52*, 1521.
24. Bolton, R.E.; Moody, C.J.; Rees, C.W.; Tojo, G. *Tetrahedron Lett.* **1987**, *28*, 3163.
25. Meghani, P.; Street, J.D.; Joule, J.A. *J. Chem. Soc., Chem. Commun.* **1987**, 1406.
26. Sundberg, R.J.; Hamilton, G.S.; Laurino, J.P. *J. Org. Chem.* **1988**, *53*, 976.
27. Martin, P. *Helv. Chim. Acta* **1989**, *72*, 1554.
28. Djerassi, C.; Monteiro, H.J.; Walser, A.; Durham, L.J. *J. Am. Chem. Soc.* **1966**, *88*, 1792.
29. Aimi, N.; Yamanaka, E.; Endo, J.; Sakai, S.; Haginiwa, J. *Tetrahedron* **1973**, *29*, 2015.
30. Thielke, D.; Wegener, J.; Winterfeldt, E. *Chem. Ber.* **1975**, *108*, 1791.
31. Wanner, M.J.; Koomen, G.J.; Pandit, U.K. *Tetrahedron* **1983**, *39*, 3673.
32. Gribble, G.W.; Johnson, J.L.; Saulnier, M.G. *Heterocycles* **1981**, *16*, 2109.
33. Takayama, H.; Seki, N.; Kitajima, M.; Aimi, N.; Seki, H.; Sakai, S. *Heterocycles* **1992**, *33*, 121.
34. Kucherova, N.F.; Sipilina, N.M.; Novikova, N.N.; Silenko, I.D.; Rozenberg, S.G.; Zagorevski, V.A. *Khim. Getero. Soed.* **1980**, 1383.
35. Lanzilotti, A.E.; Littel, R.; Fanshawe, W.J.; McKenzie,T.C.; Lovell, F.M. *J. Org. Chem.* **1979**, *44*, 4809.
36. Repic, O.; Long, D.J. *Tetrahedron Lett.* **1983**, *24*, 1115.
37. Maryanoff, B.E.; McComsey, D.F. *J. Org. Chem.* **1978**, *43*, 2733.
38. Maryanoff, B.E.; McComsey, D.F.; Nortey, S.O. *J. Org. Chem.* **1981**, *46*, 355.
39. Ketcha, D.M.; Lieurance, B.A. *Tetrahedron Lett.* **1989**, *30*, 6833.
40. Marchini, P.; Liso, G.; Reho, A.; Liberatore, F.; Moracci, F.M. *J. Org. Chem.* **1975**, *40*, 3453.
41. Oklobdzija, M.; Fajdiga, T.; Kovac, T.; Zonno, F.; Sega, A.; Sunjic, V. *Acta Pharm. Jugosl.* **1980**, *30*, 131; *Chem. Abstr.* **1981**, *94*, 121481.
42. Raftery, M.J.; Bowie, J.H. *Aust. J. Chem.* **1988**, *41*, 1477.
43. Thomas, E.W.; Nishizawa, E.E.; Zimmermann, D.C.; Williams, D.J. *J. Med. Chem.* **1985**, *28*, 442.
44. Fischer, J.W.; Nissan, R.A.; Lowe-Ma, C.K. *J. Heterocycl. Chem.* **1991**, *28*, 1677.
45. Gunzenhauser, S.; Balli, H. *Helv. Chim. Acta* **1989**, *72*, 1186.
46. Labaudiniére, R.; Dereu, N.; Cavy, F.; Guillet, M.-C.; Marquis, O.; Terlain, B. *J. Med. Chem.* **1992**, *35*, 4315.
47. Gribble, G.W.; Jasinski, J.M.; Pellicone, J.T.; Panetta, J.A. *Synthesis* **1978**, 766.
48. Thomas, E.W.; Cudahy, M.M.; Spilman, C.H.; Dinh, D.M.; Watkins,T.L.; Vidmar, T.J. *J. Med. Chem.* **1992**, *35*, 1233.
49. Katritzky, A.R.; Davis, T.L.; Rewcastle, G.W.; Rubel, G.O.; Pike, M.T. *Langmuir* **1988**, *4*, 732.
50. Cannon, J.G.; Walker, K.A.; Montanari, A.; Long, J.P.; Flynn, J.R. *J. Med. Chem.* **1990**, *33*, 2000.
51. James, L.J.; Parfitt, R.T. *J. Med. Chem.* **1986**, *29*, 1783.
52. Cannon, J.G.; Dushin, R.G.; Long, J.P.; Ilhan, M.; Jones, N.D.; Swartzendruber, J.K. *J. Med. Chem.* **1985**, *28*, 515.
53. Connor, D.T.; Unangst, P.C.; Schwender, C.F.; Sorenson, R.J.; Carethers, M.E.; Puchalski, C.; Brown, R.E.; Finkel, M.P. *J. Med. Chem.* **1989**, *32*, 683.
54. Cannon, J.G.; Jackson, H.; Long, J.P.; Leonard, P.; Bhatnagar, R.K. *J. Med. Chem.* **1989**, *32*, 1959.
55. Gribble, G.W.; Nutaitis, C.F. *Synthesis* **1987**, 709.
56. Gribble, G.W.; Ferguson, D.C. *J. Chem. Soc., Chem. Commun.* **1975**, 535.
57. Evans, D.A.; Chapman, K.T.; Carreira, E.M. *J. Am. Chem. Soc.* **1988**, *110*, 3560.
58. Rao, K.V.; Jackman, D. *J. Heterocycl. Chem.* **1973**, *10*, 213.

59. Gribble, G.W.; Heald, P.W. *Synthesis* **1975**, 650.
60. Johnson, J.V.; Rauckman, B.S.; Baccanari, D.P.; Roth, B. *J. Med. Chem.* **1989**, *32*, 1942.
61. Vigante, B.A.; Ozols, Ya.Ya.; Dubur, G.Ya. *Khim. Geter. Soed.* **1991**, 1680.
62. Carling, R.W.; Leeson, P.D.; Moseley, A.M.; Baker, R.; Foster, A.C.; Grimwood, S.; Kemp, J.A.; Marshall, G.R. *J. Med. Chem.* **1992**, *35*, 1942.
63. Uchida, M.; Chihiro, M.; Morita, S.; Yamashita, H.; Yamasaki, K.; Kanbe, T.; Yabuuchi, Y.; Nakagawa, K. *Chem. Pharm. Bull. Jpn.* **1990**, *38*, 534.
64. Ishii, H.; Ishikawa, T.; Ichikawa, Y.; Sakamoto, M.; Ishikawa, M.; Takahashi, T. *Chem. Pharm. Bull. Jpn.* **1984**, *32*, 2984.
65. Bergman, J.; Tilstam, U.; Törnroos, K.-W. *J. Chem. Soc., Perkin Trans. 1* **1987**, 519.
66. Bock, M.G.; DiPardo, R.M.; Rittle, K.E.; Evans, B.E.; Freidinger, R.M.; Veber, D.F.; Chang, R.S.L.; Chen, T.; Keegan, M.E.; Lotti, V.J. *J. Med. Chem.* **1986**, *29*, 1941.
67. Yadagiri, B.; Lown, J.W. *Synth. Commun.* **1990**, *20*, 175.
68. Moody, C.J.; Warrellow, G.J. *Tetrahedron Lett.* **1987**, *28*, 6089.
69. Evans, B.E., *et al.*, *J. Med. Chem.* **1987**, *30*, 1229.
70. Brown, D.W.; Mahon, M.F.; Ninan, A.; Sainsbury, M. *J. Chem. Soc., Perkin Trans. 1* **1995**, 3117.
71. Balaban, T.-S.; Balaban, A.T. *Tetrahedron Lett.* **1987**, *28*, 1341.
72. Balaban, T.-S.; Balaban, A.T. *Org. Prep. Proc. Int.* **1988**, *20*, 231.
73. Balaban, A.T.; Balaban, T.-S. *Rev. Roumaine Chim.* **1989**, *34*, 41.
74. Dinculescu, A.; Balaban, T.S.; Popescu, C.; Toader, D.; Balaban, A.T. *Bull. Soc. Chim. Belg.* **1991**, *100*, 665.
75. Ketcha, D.M.; Carpenter, K.P.; Zhou, Q. *J. Org. Chem.* **1991**, *56*, 1318.
76. Wasserman, H.H.; Rusiecki, V. *Tetrahedron Lett.* **1988**, *29*, 4977.
77. Bodar, N.; Koltai, E.; Pròkai, L. *Tetrahedron* **1992**, *48*, 4767.
78. Gribble, G.W.; Leiby, R.W.; Sheehan, M.N. *Synthesis* **1977**, 856.
79. Chiba, T.; Ishizawa, T.; Sakaki, J.; Kaneko, C. *Chem. Pharm. Bull. Jpn.* **1987**, *35*, 4672.
80. Kramer, J.B., *et al.*, *Bioorg. Med. Chem. Lett.* **1992**, *2*, 1655.
81. Waykole, L.M.; Shen, C.-C.; Paquette, L.A. *J. Org. Chem.* **1988**, *53*, 4969.
82. Wu, P.-L.; Chu, M.; Fowler, F.W. *J. Org. Chem.* **1988**, *53*, 963.
83. Cliffe, I.A.; Ifill, A.D.; Mansell, H.L.; Todd, R.S.; White, A.C. *Tetrahedron Lett.* **1991**, *32*, 6789.
84. Chaubet, F.; Duong, M.N.V.; Courtieu, J.; Gaudemer, A.; Gref, A.; Crumbliss, A.L. *Cand. J. Chem.* **1991**, *69*, 1107.
85. Guibourdenche, C.; Roumestant, M.L.; Viallefont, Ph. *Tetrahedron: Asym.* **1993**, *4*, 2041.
86. Williams, D.R.; Osterhout, M.H. *J. Am. Chem. Soc.* **1992**, *114*, 8750.
87. Pearson, W.H.; Schkeryantz, J.M. *J. Org. Chem.* **1992**, *57*, 6783.
88. Atarashi, S.; Tsurumi, H.; Fujiwara, T.; Hayakawa, I. *J. Heterocycl. Chem.* **1991**, *28*, 329.
89. Lewin, G.; Atassi, G.; Pierré, A.; Rolland, Y.; Schaeffer, C.; Poisson, J. *J. Nat. Prod.* **1995**, *58*, 1089.
90. Shono, T.; Matsumura, Y.; Uchida, K.; Kobayashi, H. *J. Org. Chem.* **1985**, *50*, 3243.
91. Klaver, W.J.; Hiemstra, H.; Speckamp, W.N. *J. Am. Chem. Soc.* **1989**, *111*, 2588.
92. Kreysel, M.; Vögtle, F. *Synthesis* **1992**, 733.
93. Hutchins, R.O.; Su, W.-Y.; Sivakumar, R.; Cistone, F.; Stercho, Y.P. *J. Org. Chem.* **1983**, *48*, 3412.
94. Petasis, N.A.; Lu, S.-P. *Tetrahedron Lett.* **1995**, *36*, 2393.
95. Cannon, J.G.; Amoo, V.E.D.; Long, J.P.; Bhatnagar, R.K.; Flynn, J.R. *J. Med. Chem.* **1986**, *29*, 2529.
96. Maiti, B.C.; Giri, V.S.; Pakrashi, S.C. *Heterocycles* **1990**, *31*, 847.

97. Rey, A.W.; Szarek, W.A.; MacLean, D.B. *Heterocycles* **1991**, *32*, 1143.
98. Kawecki, R.; Kozerski, L.; Urbánczyk-Lipkowska, Z.; Bocelli, G. *J. Chem. Soc., Perkin Trans. 1*, **1991**, 2255.
99. Bartoli, G.; Cimarelli, C.; Marcantoni, E.; Palmieri, G.; Petrini, M. *J. Org. Chem.* **1994**, *59*, 5328.
100. Audia, J.E.; Lawhorn, D.E.; Deeter, J.B. *Tetrahedron Lett.* **1993**, *34*, 7001.
101. Ainscow, R.B.; Brettle, R.; Shibib, S.M. *J. Chem. Soc., Perkin Trans. 1*, **1985**, 1781.
102. Comins, D.L.; Weglarz, M.A. *J. Org. Chem.* **1991**, *56*, 2506.
103. Jefford, C.W.; Wang, J.B. *Tetrahedron Lett.* **1993**, *34*, 2911.
104. Naito, T.; Shinada, T.; Miyata, O.; Ninomiya, I.; Ishida, T. *Heterocycles* **1988**, *27*, 1603.
105. Sankar, P.J.; Das, S.K.; Giri, V.S. *Heterocycles* **1991**, *32*, 1109.
106. Stoit, A.R.; Pandit, U.K. *Tetrahedron* **1988**, *44*, 6187.
107. Kuehne, M.E.; Zebovitz, T.C. *J. Org. Chem.* **1987**, *52*, 4331.
108. Umino, N.; Iwakuma, T.; Itoh, M. *Tetrahedron Lett.* **1976**, 763.
109. Cannon, J.G.; Chang, Y.; Amoo, V.E.; Walker, K.A. *Synthesis* **1986**, 494.
110. Sundaramoorthi, R.; Marazano, C.; Fourrey, J.-L.; Das, B.C. *Tetrahedron Lett.* **1984**, *25*, 3191.
111. Nutaitis, C.F. *Synth. Commun.* **1992**, *22*, 1081.
112. Miller, S.A.; Chamberlin, A.R. *J. Org. Chem.* **1989**, *54*, 2502.
113. Umino, N.; Iwakuma, T.; Itoh, N. *Tetrahedron Lett.* **1976**, 2875.
114. Ishikawa, F.; Saegusa, J.; Inamura, K.; Sakuma, K.; Ashida, S. *J. Med. Chem.* **1985**, *28*, 1387.
115. Mayer, J.P.; Cassady, J.M.; Nichols, D.E. *Heterocycles* **1990**, *31*, 1035.
116. Iyer, R.P.; Phillips, L.R.; Egan, W. *Synth. Commun.* **1991**, *21*, 2053.
117. Baldwin, J.E.; Otsuka, M.; Wallace, P.M. *Tetrahedron* **1986**, *42*, 3097.
118. Itsuno, S.; Hachisuka, C.; Ushijima, Y.; Ito, K. *Synth. Commun.* **1992**, *22*, 3229.
119. Marshall, J.A.; Johnson, W.S. *J. Org. Chem.* **1963**, *28*, 595.
120. Narayana, C.; Periasamy, M. *Tetrahedron Lett.* **1985**, *26*, 1757.
121. Gautam, V.K.; Singh, J.; Dhillon, R.S. *J. Org. Chem.* **1988**, *53*, 187.
122. Narayana, C.; Periasamy, M. *Tetrahedron Lett.* **1985**, *26*, 6361.
123. Gouzoules, F.H.; Whitney, R.A. *J. Org. Chem.* **1986**, *51*, 2024.
124. Gribble, G.W.; Leese, R.M.; Evans, B.E. *Synthesis* **1977**, 172.
125. Solberg, J.; Benneche, T.; Undheim, K. *Acta Chem. Scand.* **1989**, *43*, 69.
126. Rao, C.S.; Chakrasali, R.T.; Ila, H.; Junjappa, H. *Tetrahedron* **1990**, *46*, 2195.
127. Nutaitis, C.F.; Bernardo, J.E. *Synth. Commun.* **1990**, *20*, 487.
128. Kotha, S.; Kuki, A. *J. Chem. Soc., Chem. Commun.* **1992**, 404.
129. Olah, G.A.; Wu, A.; Farooq, O. *J. Org. Chem.* **1988**, *53*, 5143.
130. Bauder, C.; Ocampo, R.; Callot, H.J. *Tetrahedron* **1992**, *48*, 5135.
131. Kabalka, G.W.; Kennedy, T.P. *Org. Prep. Proc. Int.* **1989**, *21*, 348.
132. Nutaitis, C.F.; Patragnoni, R.; Goodkin, G.; Neighbour, B.; Obaza-Nutaitis, J. *Org. Prep. Proc. Int.* **1991**, *23*, 403.
133. Jensen, B.L.; Caldwell, M.W.; French, L.G.; Briggs, D.G. *Toxicol. Appl. Pharmacol.* **1987**, *87*, 1.
134. Lai, Y.-H.; Peck, T.-G. *Aust. J. Chem.* **1992**, *45*, 2067.
135. Currie, G.J.; Bowie, J.H.; Massy-Westropp, R.A.; Adams, G.W. *J. Chem. Soc., Perkin Trans. II* **1988**, 403.
136. Rajca, A.; Janicki, S. *J. Org. Chem.* **1994**, *59*, 7099.
137. Osuka, A.; Zhang, R.P.; Maruyama, K.; Yamazaki, I.; Nishimura, Y. *Bull. Chem. Soc. Jpn.* **1992**, *65*, 2807.
138. Nicholas, K.M.; Siegel, J. *J. Am. Chem. Soc.* **1985**, *107*, 4999.
139. McComsey, D.F.; Reitz, A.B.; Maryanoff, C.A.; Maryanoff, B.E. *Synth. Commun.* **1986**, *16*, 1535.
140. Sorgi, K.L.; Maryanoff, C.A.; McComsey, D.F.; Graden, D.W.; Maryanoff, B.E. *J. Am. Chem. Soc.* **1990**, *112*, 3567.

141. Gribble, G.W.; Kelly, W.J.; Emery, S.E. *Synthesis* **1978**, 763.
142. Lee, C.-M.; Parks, J.A.; Bunnell, P.R.; Plattner, J.J.; Field, M.J.; Giebisch, G.H. *J. Med. Chem.* **1985**, *28*, 589.
143. Daich, A.; Decroix, B. *J. Heterocycl. Chem.* **1992**, *29*, 1789.
144. Kurokawa, M.; Uno, H.; Itogawa, A.; Sato, F.; Naruto, S.; Matsumoto, J. *J. Heterocycl. Chem.* **1991**, *28*, 1891.
145. Ketcha, D.M.; Lieurance, B.A.; Homan, D.F.J.; Gribble, G.W. *J. Org. Chem.* **1989**, *54*, 4350.
146. Alazard, J.-P.; M.-Paillusson, C.; Boyé, O.; Guénard, D.; Chiaroni, A.; Riche, C.; Thal, C. *Bioorg. Med. Chem. Lett.* **1991**, *1*, 725.
147. Jacobi, P.A.; Cai, G. *Heterocycles* **1993**, *35*, 1103.
148. Jeandon, C.; Ocampo, R.; Callot, H.J. *Tetrahedron Lett.* **1993**, *34*, 1791.
149. Hutchins, R.O.; Natale, N.R. *J. Org. Chem.* **1978**, *43*, 2299.
150. Baeckström, P.; Li, L. *Synth. Commun.* **1990**, *20*, 1481.
151. Cho, B.T.; Yoon, N.M. *Synth. Commun.* **1985**, *15*, 917.
152. Suseela, Y.; Periasamy, M. *Tetrahedron* **1992**, *48*, 371.
153. Soucek, M.; Urban, J.; Saman, D. *Coll. Czech. Chem. Commun.* **1990**, *55*, 761.
154. Nutaitis, C.F.; Gribble, G.W. *Org. Prep. Proc. Int.* **1985**, *17*, 11.
155. Gribble, G.W.; Kelly, W.J.; Sibi, M.P. *Synthesis* **1982**, 143.
156. Gribble, G.W.; Saulnier, M.G.; Sibi, M.P.; Obaza-Nutaitis, J.A. *J. Org. Chem.* **1984**, *49*, 4518.
157. Koltai, E.; Horváth, G.; Botka, P.; Kórösi, J.; Tóth, G. *Acta Chim. Hung.* **1990**, *127*, 3.
158. Nutaitis, C.F.; Gribble, G.W. *Synthesis* **1985**, 756.
159. Nutaitis, C.F.; Gribble, G.W. *Tetrahedron Lett.* **1983**, *24*, 4287.
160. Odinokov, V.N.; Ishmuratov, G.Yu.; Ladenkova, I.M.; Tolstikov, G.A. *Chem. Nat. Cpds.* **1992**, 235.
161. Kanojia, R.M.; Chin, E.; Smith, C.; Chen, R.; Rowand, D.; Levine, S.D.; Wachter, M.P.; Adams, R.E.; Hahn, D. *J. Med. Chem.* **1985**, *28*, 796.
162. Barker, P.L.; Bahia, C. *Tetrahedron* **1990**, *46*, 2691.
163. Han, O.; Liu, H. *Tetrahedron Lett.* **1987**, *28*, 1073.
164. Yatagi, M.; Ohnuki, T. *J. Chem. Soc., Perkin Trans. 1* **1990**, 1826.
165. Hoffman, R.V.; Kim, H.-O. *J. Org. Chem.* **1991**, *56*, 6759.
166. Nutaitis, C.F.; Bernardo, J.E. *J. Org. Chem.* **1989**, *54*, 5629.
167. Saksena, A.K.; Mangiaracina, P. *Tetrahedron Lett.* **1983**, *24*, 273.
168. Evans, D.A.; Chapman, K.T. *Tetrahedron Lett.* **1986**, *27*, 5939.
169. Evans, D.A.; Gauchet-Prunet, J.A.; Carreira, E.M.; Charette, A.B. *J. Org. Chem.* **1991**, *56*, 741.
170. Evans, D.A.; Ng, H.P.; Clark, J.S.; Rieger, D.L. *Tetrahedron* **1992**, *48*, 2127.
171. Dondoni, A.; Perrone, D.; Merino, P. *J. Chem. Soc., Chem. Commun.* **1991**, 1313.
172. Martin, S.F.; Pacofsky, G.J.; Gist, R.P.; Lee, W.-C. *J. Am. Chem. Soc.* **1989**, *111*, 7634.
173. Thompson, S.H.J.; Subramanian, R.S.; Roberts, J.K.; Hadley, M.S.; Gallagher, T. *J. Chem. Soc., Chem. Commun.* **1994**, 933.
174. Paterson, I.; Cumming, J.G. *Tetrahedron Lett.* **1992**, *33*, 2847.
175. Farr, R.N.; Kwok, D.-I.; Daves, G.D., Jr. *J. Org. Chem.* **1992**, *57*, 2093.
176. Zhang, H.-C.; Daves, G.D., Jr. *J. Org. Chem.* **1992**, *57*, 4690.
177. Mori, Y.; Kohchi, Y.; Suzuki, M.; Furukawa, H. *Chem. Pharm. Bull. Jpn.* **1992**, *40*, 1934.
178. Sato, M.; Sugita, Y.; Abiko, Y.; Kaneko, C. *Tetrahedron: Asym.* **1992**, *3*, 1157.
179. Paterson, I.; Channon, J.A. *Tetrahedron Lett.* **1992**, *33*, 797.
180. Solladié, G.; Ghiatou, N. *Tetrahedron Lett.* **1992**, *33*, 1605.
181. Chen, C.; Crich, D. *Tetrahedron Lett.* **1992**, *33*, 1945.
182. Paterson, I.; Tillyer, R.D. *Tetrahedron Lett.* **1992**, *33*, 4233.
183. Casy, G. *Tetrahedron Lett.* **1992**, *33*, 8159.
184. Mori, Y.; Asai, M.; Furukawa, H. *Heterocycles* **1992**, *34*, 1281.

185. Shao, L.; Seki, T.; Kawano, H.; Saburi, M. *Tetrahedron Lett.* **1991**, *32*, 7699.
186. Paterson, I.; Lister, M.A.; Ryan, G.R. *Tetrahedron Lett.* **1991**, *32*, 1749.
187. Grabley, S.; Hammann, P.; Kluge, H.; Wink, J.; Kricke, P.; Zeeck, A. *J. Antibiot.* **1991**, *44*, 797.
188. Turner, N.J.; Whitesides, G.M. *J. Am. Chem. Soc.* **1989**, *111*, 624.
189. Gu, R.; Sih, C.J. *Tetrahedron Lett.* **1990**, *31*, 3283.
190. Robins, M.J.; Samano, V.; Johnson, M.D. *J. Org. Chem.* **1990**, *55*, 410.
191. Hill, R.K.; Nugara, P.N.; Holt, E.M.; Holland, K.P. *J. Org. Chem.* **1992**, *57*, 1045.
192. Shao, L.; Kawano, H.; Saburi, M.; Uchida, Y. *Tetrahedron* **1993**, *49*, 1997.
193. Enders, D.; Osborne, S. *J. Chem. Soc., Chem. Commun.* **1993**, 424.
194. Whang, K.; Cooke, R.J.; Okay, G.; Cha, J.K. *J. Am. Chem. Soc.* **1990**, *112*, 8985.
195. Wender, P.A.; Kogen, H.; Lee, H.Y.; Munger, J.D., Jr.; Wilhelm, R.S.; Williams, P.D. *J. Am. Chem. Soc.* **1989**, *111*, 8957.
196. Romeyke, Y.; Keller, M.; Kluge, H.; Grabley, S.; Hammann, P. *Tetrahedron* **1991**, *47*, 3335.
197. Evans, D.A.; Gage, J.R.; Leighton, J.L. *J. Am. Chem. Soc.* **1992**, *114*, 9434.
198. Evans, D.A.; Black, W.C. *J. Am. Chem. Soc.* **1992**, *114*, 2260.
199. Paterson, I.; Perkins, M.V. *J. Am. Chem. Soc.* **1993**, *115*, 1608.
200. Shishido, K.; Irie, O.; Shibuya, M. *Tetrahedron Lett.* **1992**, *33*, 4589.
201. Boger, D.L.; Curran, T.T. *J. Org. Chem.* **1992**, *57*, 2235.
202. Nakada, M.; Kobayashi, S.; Iwasaki, S.; Ohno, M. *Tetrahedron Lett.* **1993**, *34*, 1035.
203. Estevez, V.A.; Prestwich, G.D. *J. Am. Chem. Soc.* **1991**, *113*, 9885.
204. Smith, A.B., III; Qiu, Y.; Jones, D.R.; Kobayashi, K. *J. Am. Chem. Soc.* **1995**, *117*, 12011.
205. Smith, A.B., III; Chen, S. S.-Y.; Nelson, F.C.; Reichert, J.M.; Salvatore, B.A. *J. Am. Chem. Soc.* **1995**, *117*, 12013.
206. Jones, T.K.; Reamer, R.A.; Desmond, R.; Mills, S.G. *J. Am. Chem. Soc.* **1990**, *112*, 2998.
207. Fisher, M.J.; Chow, K.; Villalobos, A.; Danishefsky, S.J. *J. Org. Chem.* **1991**, *56*, 2900.
208. Burke, S.D.; Shankaran, K.; Helber, M.J. *Tetrahedron Lett.* **1991**, *32*, 4655.
209. Brown, M.J.; Harrison, T.; Herrinton, P.M.; Hopkins, M.H.; Hutchinson, K.D.; Mishra, P.; Overman, L.E. *J. Am. Chem. Soc.* **1991**, *113*, 5365.
210. Schulz, S. *Ann.* **1992**, 829.
211. Goldstein, S.W.; Overman, L.E.; Rabinowitz, M.H. *J. Org. Chem.* **1992**, *57*, 1179.
212. Trost, B.M.; Greenspan, P.D.; Yang, B.V.; Saulnier, M.G. *J. Am. Chem. Soc.* **1990**, *112*, 9022.
213. Davey, A.E.; Schaeffer, M.J.; Taylor, R.J.K. *J. Chem. Soc., Perkin Trans. 1*, **1992**, 2657.
214. Overman, L.E.; Shim, J. *J. Org. Chem.* **1991**, *56*, 5005.

# Chapter 12

# Use of Sodium Triacetoxyborohydride in Reductive Amination of Ketones and Aldehydes

### Ahmed F. Abdel-Magid and Cynthia A. Maryanoff

### Department of Chemical Development, R. W. Johnson Pharmaceutical Research Institute, Spring House, PA 19477

Herein we present an overview of the use of sodium triacetoxyborohydride in the reductive amination of ketones and aldehydes, with an emphasis on scope. In general, this is an extremely useful reagent and experimental conditions are convenient and simple. Alicyclic and cyclic ketones furnish excellent yields of secondary and tertiary amines. Where diastereomer formation is possible, this reagent is more sterically demanding than sodium cyanoborohydride, and higher selectivity is often observed, especially from bicyclic ketones. Acid sensitive groups in the molecule are unaffected under normal reaction conditions. Hydrazines are reductively alkylated with ketones to furnish monalkyated products. Ketoesters are reductively aminated with primary and secondary amines. The initial products from γ- and δ-ketoesters or acids cyclize to the lactams in a tandem reductive cyclization procedure. The combination of $NaBH(OAc)_3$ /$CF_3CO_2NH_4$ furnishes primary amines in high yields. Weakly basic amines demonstrate the clear advantage of this reagent, as they undergo reductive alkylation in high yields with ketones and aldehydes. Non-basic amines and sulfonamides furnish high yields of monoalkyl products with most aldehydes.

Amines occupy an important position in organic synthesis as target molecules of biological interest and as synthetic intermediates. Reductive amination of ketones and aldehydes is a cornerstone reaction for the synthesis of different kinds of amines. This reaction is also described in the literature as reductive alkylation of amines (*1*). In this chapter we limit our discussion to *direct* reductive amination reactions in which the carbonyl compound and the amine are mixed in the presence of the reducing agent. The reaction involves the initial formation of an intermediate carbinol amine which dehydrates to form an imine (with ammonia and primary amines) or an iminium ion (with secondary amines) (Scheme I). Reduction of any of these intermediates produces the amine product (*2*). The choice of the reducing agent is very critical to the success of the reaction, since it must reduce imines (or iminium ions) selectively over aldehydes or ketones under the reaction conditions.

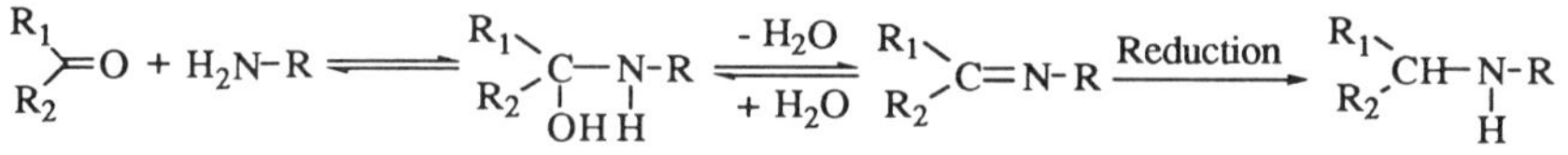

$$R_1\!\!\!\underset{R_2}{\diagup}\!\!\!=O + H_2N\text{-}R \rightleftharpoons \underset{R_2}{\overset{R_1}{\diagdown}}\!\!\underset{\underset{OH\ H}{|\ \ |}}{C}\!-N\text{-}R \xrightleftharpoons[+H_2O]{-H_2O} \underset{R_2}{\overset{R_1}{\diagdown}}C=N\text{-}R \xrightarrow{\text{Reduction}} \underset{R_2}{\overset{R_1}{\diagdown}}CH\!\!\underset{H}{\overset{}{-}}\!N\text{-}R$$

**Scheme I**

The reducing conditions utilized in this reaction may be catalytic hydrogenation or borohydride reagents. Catalytic hydrogenation has the advantage of being economical and easy to scale up (*1*). The borohydride reagents are more selective and can be used in the presence of C-C multiple bonds, divalent sulfur-containing compounds and in the presence of nitro and cyano groups (*2*). Sodium cyanoborohydride ($NaBH_3CN$) is the most commonly used borohydride reagent in reductive amination reactions (*3*). It is stable in hydroxylic solvents and has different selectivities at different pH values. At pH 1-3 it reduces aldehydes and ketones effectively. At pH 6-8, imines and iminium ions are reduced faster than aldehydes or ketones (*4*). On the other hand, it is highly toxic, (*5*) slow and sluggish with aromatic ketones and with weakly basic amines (*5a*) and its use may result in the contamination of the product with cyanide (*6*).

Over the past few years, we developed the use of the commercially available sodium triacetoxyborohydride [$NaBH(OAc)_3$] in the reductive amination of aldehydes and ketones (*7*). This borohydride is mild and exhibits remarkable selectivity as a reducing agent. It is very convenient and safe to handle. The use of this reagent, formed *in-situ* from sodium borohydride and glacial acetic acid, in reductive alkylation of aromatic amines was pioneered by Gribble et al (*8*). Our study of the scope and limitations of this reagent in the *direct* reductive amination of aldehydes and ketones with primary and secondary aliphatic and aromatic amines showed clearly that it is the reagent of choice in most cases (*9*). Since our earlier communication (*7*), we and other research groups have found it advantageous to use this procedure over other literature methods. The reactions are very easy to conduct and the isolated yields are good to excellent. This chapter will provide an overview of the utility of sodium triacetoxyborohydride as a reducing agent in reductive amination reactions.

## DISCUSSION

**(a) Reductive Amination of Ketones:**

Early reports on the use of sodium triacetoxyborohydride in reduction indicated its inertness towards ketones (*10*). We expected it to be ideal to use in reductive amination of ketones. A ketone would interact faster with an amine to form an imine or iminium ion without interference from the reducing agent. Under the reaction conditions, the more basic imine would be protonated and reduced faster than a ketone. This expectation was realized in the results obtained in our systematic study (*9*) which clearly showed the utility and wide scope of this reagent in reductive amination of different kinds of ketones. The ketone and amine are used in stoichiometric amounts or with a slight excess of the amine. The reducing agent is used in some excess, 1.5 - 2 equivalents. Most reactions were carried out in 1,2-dichloroethane (DCE) or in tetrahydrofuran (THF) in the presence of one equivalent of AcOH. These conditions are referred to herein as the standard conditions. In some cases they are modified to optimize the yield of a particular class of compounds.

Alicyclic ketones, ranging in size from cyclobutanone to cyclododecanone, give excellent yields in reductive amination reactions with primary and secondary amines. The small ring ketones are usually more reactive than the larger ones and all react efficiently under the standard conditions. Figure 1 features some amines obtained from representative reductive amination reactions of alicyclic ketones (*9*).

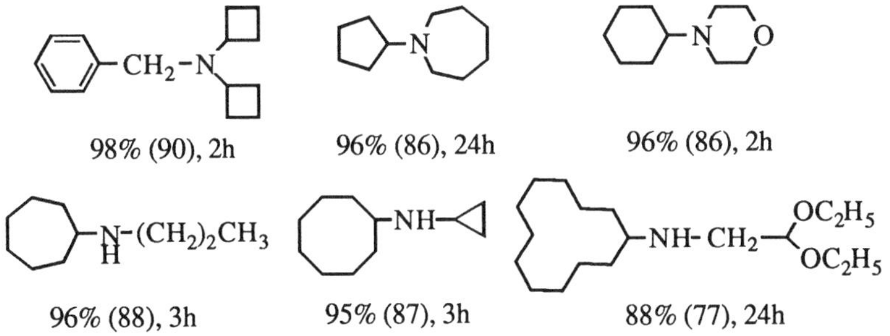

**Figure 1**: Reductive Amination of Alicyclic Ketones.
Values in parentheses are yields of recrystallized isolated salts.

The reductive amination of cyclic ketones such as β-tetralone and 2-indanone with aniline and NaBH(OAc)$_3$ (9) gives high isolated of the corresponding products and is superior to hydrogenation methods (11).

Other examples of forming carbon-nitrogen bonds by reductive amination of cyclic ketones with NaBH(OAc)$_3$ are reported in literature such as the preparation of the $^{14}$C-labeled amine **1** (12), the fused ring didemnin analogue intermediate **2** (13), the (+)-papuamine intermediate **3** (14) and the intermediate for photoactivatable analogues of substance P non-peptidic antagonist **4** (15).

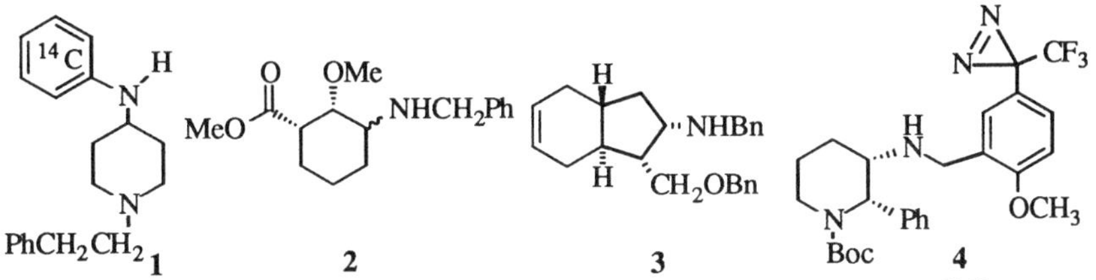

In the total synthesis of the marine alkaloids (-)-papuamine and (-)-Haliclonadiamine, the intermediate **6** was obtained in 58% (92% based on recovered starting material) as a 3.4:1 mixture of diastereomers from the reductive amination of the ketone **5** with 1,3-diaminopropane using NaBH(OAc)$_3$ (16).

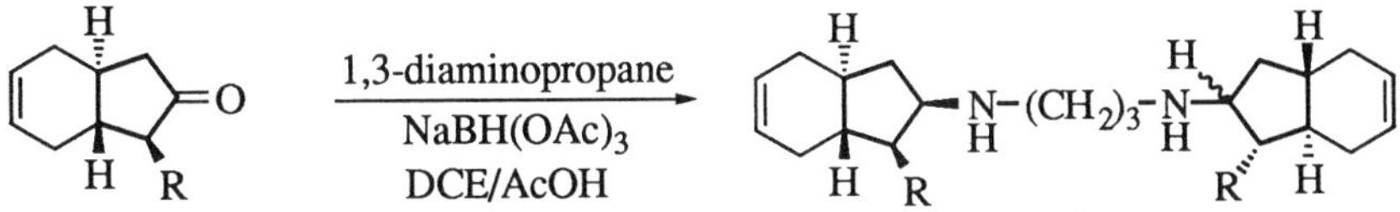

In the synthesis of compounds such as **2**, **3**, **4** and **6**, where diastereomer formation is possible, variable degrees of diastereoselectivity occur, from non-selective formation of **2** to exclusive formation of one diastereomer in **4**. A comparison between $NaBH_3CN$ and $NaBH(OAc)_3$ in reduction of vinylogous urethanes such as **7** to the diastereomeric tertiary amines **8** and **9** showed that *cis*-selectivity (formation of **8**) was dramatically enhanced with sodium triacetoxyborohydride (*17*). This reagent appears to be sterically more demanding in reducing the intermediate iminium ion than sodium cyanoborohydride (Scheme II). This finding agrees with earlier studies on reduction of 4-*tert*-butylcyclohexanone imines (*18*) and direct reductive amination of 4-*tert*-butylcyclohexanone (*9*).

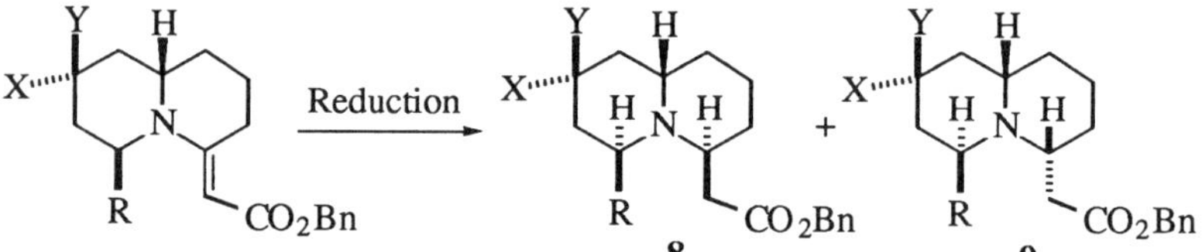
Y
H
X
N
R
CO2Bn
Reduction
H
H
+


**7** **8** **9**

| | $NaBH(OAc)_3$ | $NaBH_3CN$ |
|---|---|---|
| a: R = *m*-OMe-$C_6H_4$-; X = Y = H | 92% (>50 : 1) | 70% (5 : 1) |
| b: R = (*E*)-TMS-CH=CH-, X = H, Y = OCOPh | 72% (10 : 1) | 77% (1 : 2) |

**Scheme II**

Saturated acyclic ketones also undergo clean reductive amination with both primary and secondary amines, however, the reactions may be slower and the isolated yields may be lower than the alicyclic ketones, particularly with hindered secondary amines. Some of the slow reactions are accelerated by adding 1-2 equivalents of AcOH, the use of about 5 - 10% excess of the amine and up to 2 equivalents of sodium triacetoxyborohydride. Several products obtained by reductive amination of saturated ketones are shown in Figure 2 (*9*).

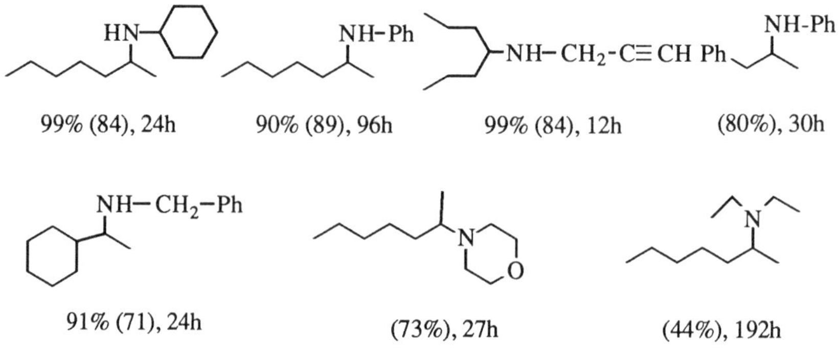
HN
NH-Ph
NH-CH2-C≡CH
Ph
NH-Ph
99% (84), 24h
90% (89), 96h
99% (84), 12h
(80%), 30h
NH-CH2-Ph
N
O
N
91% (71), 24h
(73%), 27h
(44%), 192h

**Figure 2**: Reductive Amination of Acyclic Aliphatic Ketones.
Values in parentheses are yields of recrystallized isolated salts.

Bicyclic ketones such as norcamphor and tropinone are successfully reductively aminated with primary and secondary amines in good to excellent yields. The reactions involving these systems usually show high levels of diastereoselectivity towards the *endo*- products. Products from norcamphor and both primary and secondary amines are exclusively *endo*- while those obtained from tropinone and primary amines show at least a 20 : 1 ratio of the *endo*- to *exo*- products. The exception is the reaction of tropinone with piperidine which is very slow and gives a 1 : 1 ratio of the *endo*- and *exo*- products (Figure 3) (9). A related example is the reductive amination of the diketone **10** to give the symmetric diamine **11** in 96% yield with slight contamination of two diastereomers (19).

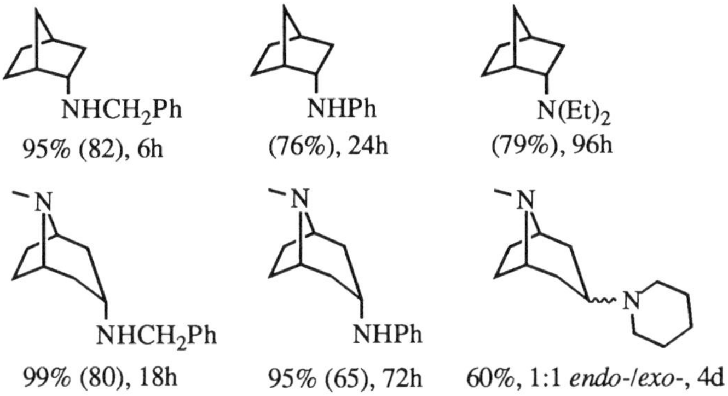

NHCH$_2$Ph
95% (82), 6h

NHPh
(76%), 24h

N(Et)$_2$
(79%), 96h

NHCH$_2$Ph
99% (80), 18h

NHPh
95% (65), 72h

60%, 1:1 *endo-/exo-*, 4d

**Figure 3**: Reductive Amination of Bicyclic Ketones.
Values in parentheses are yields of recrystallized isolated salts.

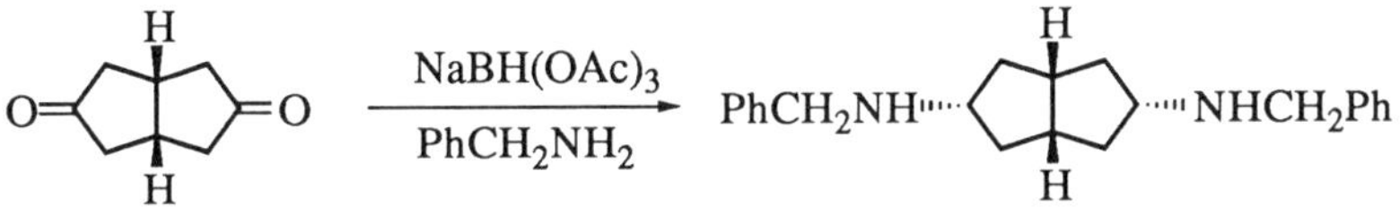

**10** → **11**
NaBH(OAc)$_3$, PhCH$_2$NH$_2$

The reaction conditions are modified in another study to use NaBH(OAc)$_3$ in reductive amination of polymer bound bicyclo[2,2,2]octanones (**12**). While the standard conditions give only a trace of product (**13**), the use of excess amine, addition of Na$_2$SO$_4$, and application of ultrasound effects efficient reductive amination with various amines (20).

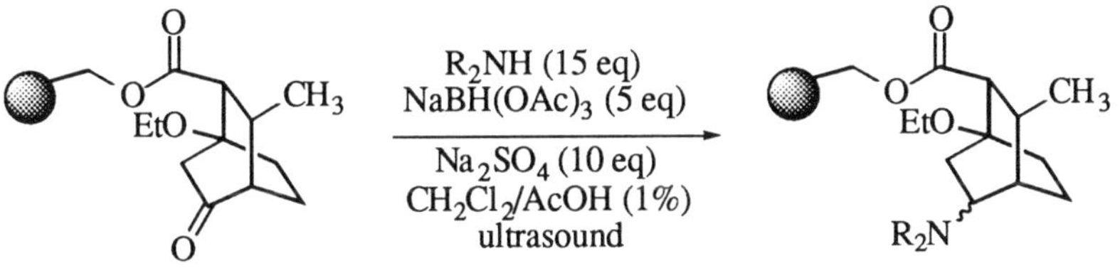

**12** → **13**
R$_2$NH (15 eq)
NaBH(OAc)$_3$ (5 eq)
Na$_2$SO$_4$ (10 eq)
CH$_2$Cl$_2$/AcOH (1%)
ultrasound

The reaction conditions have been applied to substrates containing acid sensitive functionalities such as acetals and ketals on either reactant. With the use of AcOH or no acid, the products are stable to aqueous work up conditions. The reductive amination of cyclohexanedione monoethylene ketal with primary and secondary amines affords good to excellent isolated yields of the corresponding amines. Similar results are obtained from reactions of aminoacetaldehyde diethyl acetal with ketones (Figure 4) (9).

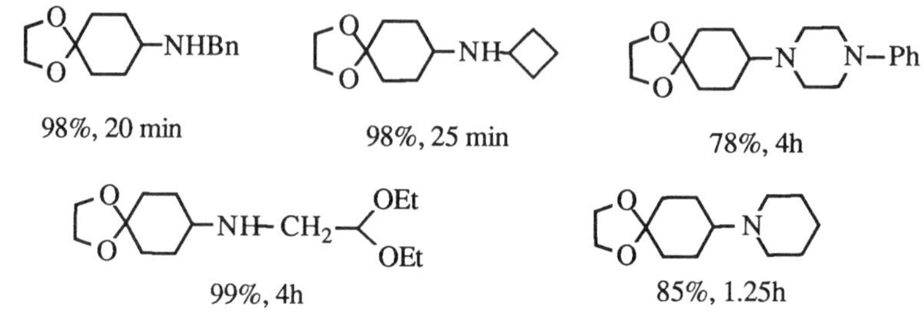

[structures with yields:]
- 98%, 20 min (NHBn)
- 98%, 25 min (NH-cyclopropyl)
- 78%, 4h (N-piperazine-N-Ph)
- 99%, 4h (NH-CH₂-CH(OEt)₂)
- 85%, 1.25h (N-piperidine)

**Figure 4**: Reductive Amination in The Presence of Ketals and Acetals.

This reaction has found an application in the synthesis of the potential dopamine receptor agonist **14** in a recent publication (21). In this case, $NaBH(OAc)_3$ gives a better result than $NaBH_3CN$ or the $Ti(OiPr)_4/NaBH_3CN$ combination.

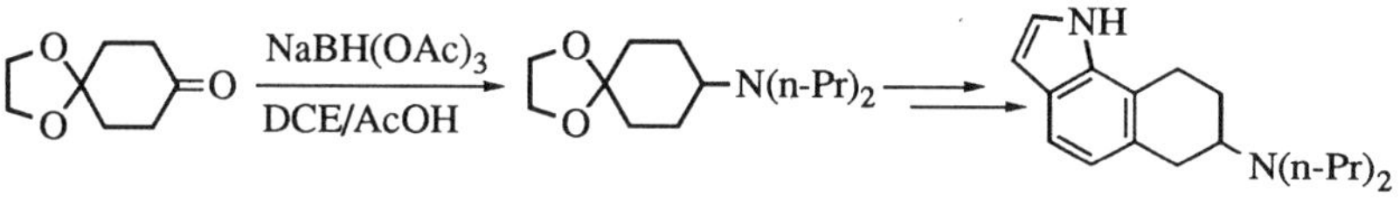

[reaction scheme: ketal-ketone → NaBH(OAc)₃, DCE/AcOH → ketal-N(n-Pr)₂ (crude quantitative yield) → compound **14** with N(n-Pr)₂]

Other related examples showing the use of $NaBH(OAc)_3$ in reductive amination in the presence of ketals are reported for the synthesis of compounds **15 - 17** (22-24).

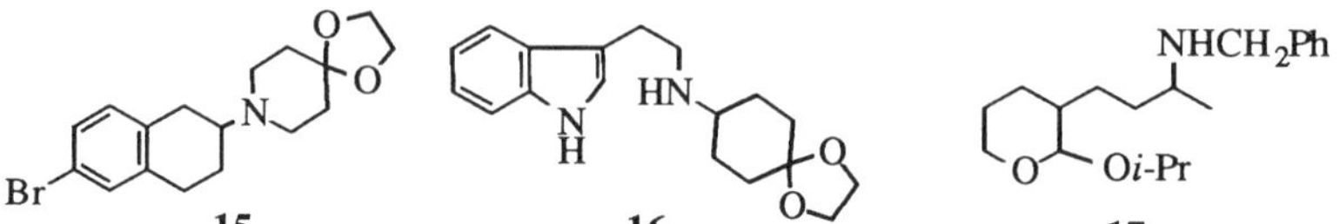

**15** — 85%, intermediate in synthesis of MK-0499 (ref. 22)

**16** — 87%, intermediate in synthesis of Melinonine-E (ref. 23)

**17** — mixture of diastereomers (ref. 24)

Substituted hydrazines such as phenylhydrazine and 1-aminopiperidine are reductively alkylated with ketones giving rise to monoalkylation products. The reaction of phenylhydrazine with cyclohexanedione monoethylene ketal is a very clean reaction, however, those with cyclohexanone and 2-heptanone show variable ratios of byproducts and give lower yields (9). The reactions of 1-aminopiperidine with either 2-heptanone or cycloheptanone are very clean giving exclusively the expected N-monoalkyl products (Figure 5) in nearly quantitative yields (unpublished results).

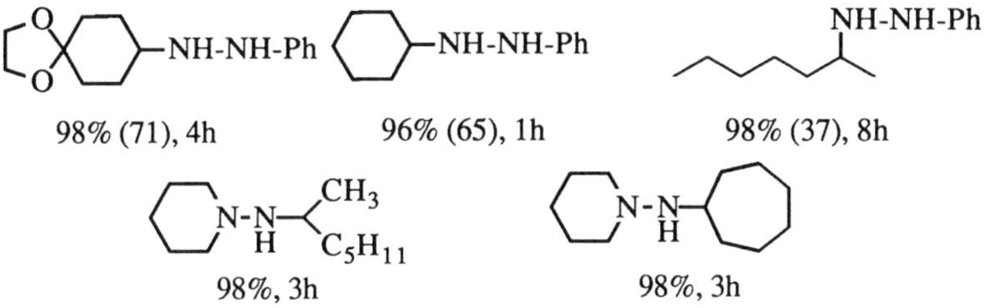

[Structures with yields:]

- cyclohexanone ethylene ketal –NH-NH-Ph, 98% (71), 4h
- cyclohexyl –NH-NH-Ph, 96% (65), 1h
- 2-hexyl NH-NH-Ph, 98% (37), 8h
- piperidine N-N(H)-CH(CH₃)(C₅H₁₁), 98%, 3h
- piperidine N-N(H)-cycloheptyl, 98%, 3h

**Figure 5**: Reductive Amination with Hydrazines.
Values in parentheses are yields of recrystallized isolated salts.

Ketoesters are a special class of ketones. The relative position of the keto group to the ester may effect the outcome of the reaction chemically or stereochemically or may result in a secondary reaction. The reductive amination of α-ketoesters with primary and secondary amines gives the corresponding *N*-substituted α-aminoesters. The reductive amination of various α-ketoesters with benzylamine affords the α-benzylamino esters (**18 - 20**) in good to excellent yields (*9*). Reactions involving other amines, such as aniline or morpholine, are not as efficient and give variable amounts of ketone reductions. Alternatively, *N*-substituted α-aminoesters may be prepared by the reductive alkylation of α-aminoesters with aldehydes and ketones. This is demonstrated in the reductive amination of 2-butanone with methyl glycinate (*9*) which results in the clean formation of methyl *N*-(2-butyl)glycinate (**21**).

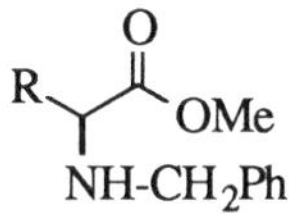

R–CH(NH-CH₂Ph)–C(O)OMe

**18** R = Me, 90%
**19** R = i-Pr, 82%
**20** R = Ph, 58%

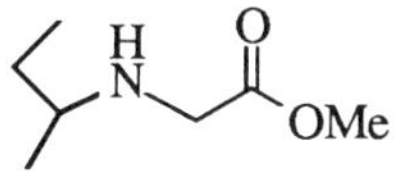

sec-Bu–NH–CH₂–C(O)OMe

**21**
88%

The reductive amination of cyclic β-ketoesters such as methyl cyclohexanone-2-carboxylate gives almost exclusively the *cis*-product (**22**) (*25*). There is evidence that the intermediate is an enamine rather than an imine. This reaction is currently under further investigation to evaluate its generality in different systems and its mechanistic pathway.

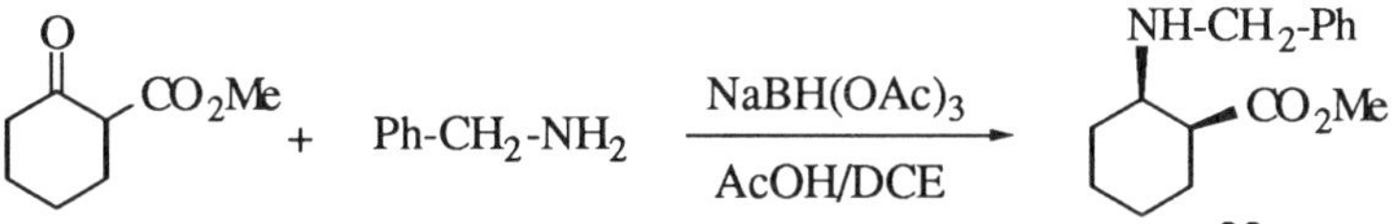

2-oxocyclohexane-CO₂Me + Ph-CH₂-NH₂ →[NaBH(OAc)₃ / AcOH/DCE]→ 2-(NH-CH₂-Ph)cyclohexane-CO₂Me

**22**

The reductive amination of γ- and δ-ketoesters or acids with primary amines is a special case (*26*). The initial product, an *N*-substituted γ- or δ-aminoester or acid cyclizes to the corresponding lactam (such as **23** and **24**) under the reaction conditions (Scheme III). This tandem two-step procedure, which we termed "reductive lactamization" is a convenient method for the synthesis of *N*-substituted butyro- and valerolactams under mild conditions.

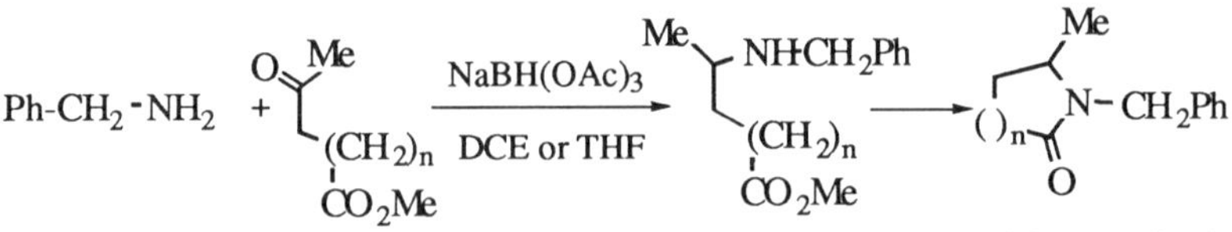

**Scheme III**

23 n = 1, 84%
24 n = 2, 80%

The same products are obtained from reductive alkylation of γ- or δ-aminoacids or esters with carbonyl compounds. As in the above case, the initial reductive amination product, cyclizes to the corresponding lactam under the reaction conditions (26). Some representative examples are shown in Scheme IV. The reaction could not be applied to ε-aminoesters or larger homologues to obtain lactams; these reactions result only in reductive amination but no cyclization.

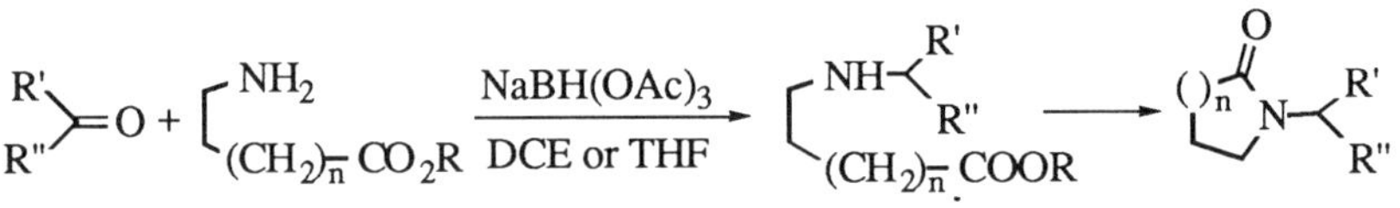

n = 1 or 2

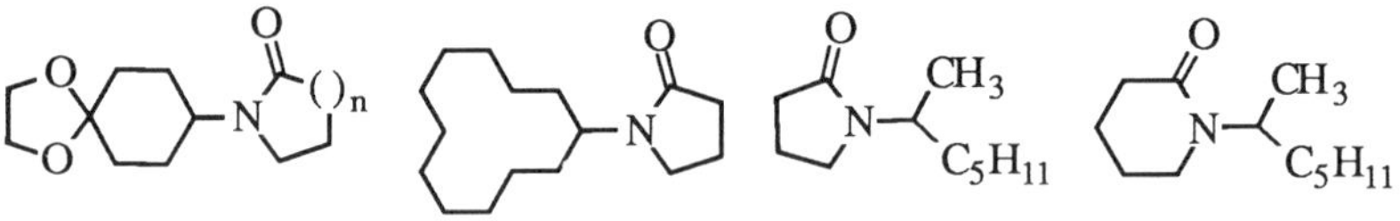

n = 1, 92%
n = 2, 85%

91%

96%

55%

**Scheme IV**

Our attempts to develop a practical procedure for the synthesis of primary amines by reductive amination of ketones with NaBH(OAc)$_3$ were hindered by the poor solubility of ammonium acetate, the most common and convenient source of ammonia, in THF or DCE. The reaction gives dialkylamines exclusively. Even the use of a large excess, up to 10 equivalents, of ammonium acetate in DCE, THF or CH$_3$CN, still gives the dialkylamines. This reaction can thus be used for the effective preparation of symmetric dialkylamines, such as dicycloheptylamine (9). Our search for better conditions that may be used in preparation of primary amines via reductive amination in aprotic solvents led to ammonium trifluoroacetate. This salt is soluble in THF and can be used effectively in reductive amination reactions. The reaction with cycloheptanone and cyclododecanone (Scheme V) gives the corresponding primary amines as the major products with 7% or less of the dialkylamines (unpublished results). This reaction is currently under further investigation and will be the subject of a future report.

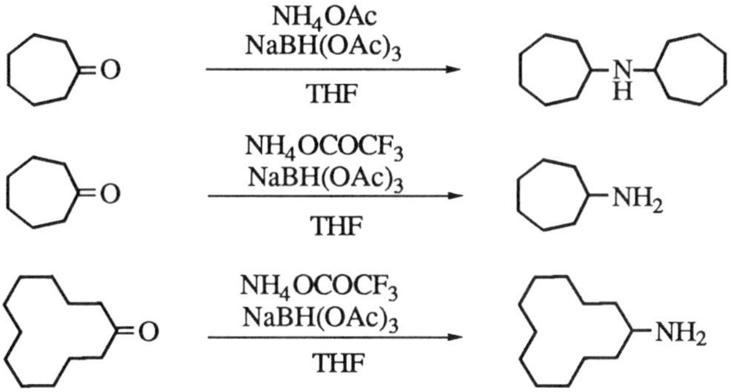

## Scheme V

Perhaps, the results that best demonstrate the superior advantage of using NaBH(OAc)$_3$ over other reagents are those obtained from reactions with weakly basic amines. These amines are weak bases as well as poor nucleophiles. The pK$_a$ values for the amines used in this study range from 3.98 for *p*-chloroaniline to -4.26 for 2,4-dinitroaniline (27). The reductive amination of ketones with anilines substituted with electron withdrawing groups is usually sluggish and slow and the ketone may be reduced preferentially with most reagents. The use of sodium triacetoxyborohydride in reductive amination with several of the monosubstituted anilines in stoichiometric quantities or in the presence of excess ketones gives the corresponding *N*-alkyl products in very good isolated yields. However, the efficiency of these reactions decreases considerably with less basic amines such as *o*-nitroaniline, 2,6-dibromoaniline and 2,4,6-trichloroaniline which react slow or result in no reaction. Some of the products obtained from reductive amination of ketones with some weakly basic amines are listed in Figure 6 with their isolated yields and reaction times.

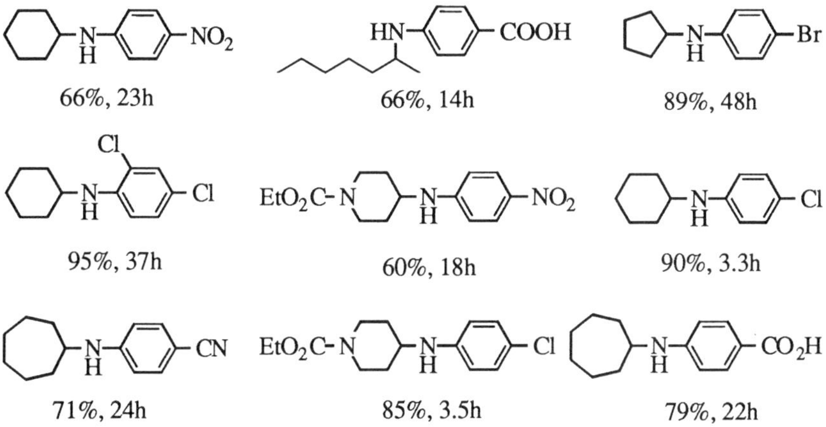

**Figure 6**: Reductive Amination of Ketones with Weakly Basic Amines.

Aromatic, α,β-unsaturated, and sterically hindered saturated ketones react very slowly or show no reaction under the standard reaction conditions (9). In competition studies, featuring cyclohexanone *vs.* acetophenone and acetylcyclohexane *vs.* 1-acetylcyclohexene, the saturated ketones reacted faster and were selectively reductively aminated in the presence of the slow reacting aromatic and α,β-unsaturated ketones (Scheme VI) (9).

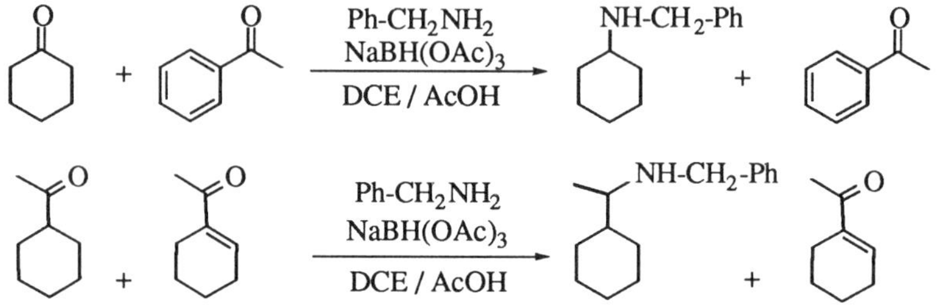

**Scheme VI**

### (b) Reductive Amination of Aldehydes:

Unlike ketones, aldehydes can be reduced with sodium triacetoxyborohydride (28). However, under the standard reaction conditions the reductive aminations with aldehydes occur very effectively and result in fast reactions with no aldehyde reductions in most cases. Both aliphatic and aromatic aldehydes are very reactive and give reductive amination products with nearly all kinds of primary and secondary amines. In most reactions, the aldehyde and amine are mixed in stoichiometric amounts in DCE or THF with about 1.4 - 1.5 equivalents of $NaBH(OAc)_3$. The reaction times range from 20 minutes to 24 hours. These conditions are mild and allow for a convenient reaction and an easy work up and isolation of products. Some representative examples from our work (9) are illustrated in Figure 7.

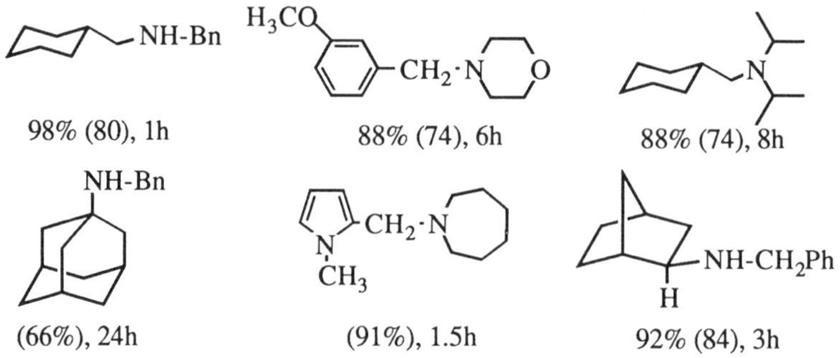

**Figure 7**: Reductive Amination of Aldehydes.
Values in parentheses are yields of recrystallized isolated salts.

When compared to most other reducing agents, $NaBH(OAc)_3$ does not cause significant aldehyde reduction when used in reductive amination reactions. Most aldehyde reactions do not require the use of acid activation as is the case with ketones. This seems to minimize the chance of aldehyde reduction. For example, the formation of *N,N*-diisopropylcyclohexylmethylamine (Figure 7) from cyclohexane carboxaldehyde and diisopropylamine is accompanied by about 25% aldehyde reduction in the presence of AcOH, but only 5% in its absence. A literature example has shown, in the construction of the internucleoside -$CH_2$-$CH_2$-NH- linkage in the thymidine dimer (**25**), that the use of $NaBH(OAc)_3$ gives a high yield of product and very little reduction of aldehyde. The use of $NaBH_3CN$ causes significant aldehyde reduction and gives a lower yield (*29*).

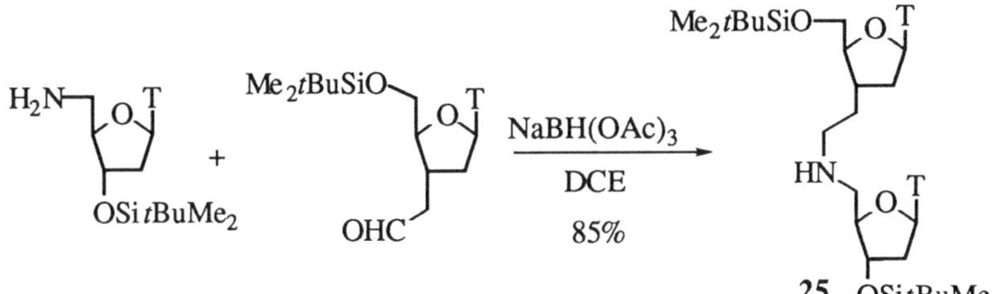

The reductive amination of aromatic aldehydes with ethyl 2-carboxypiperidine under the standard conditions (*9*) shows no aldehyde reduction and is superior to other literature procedures (*6b*).

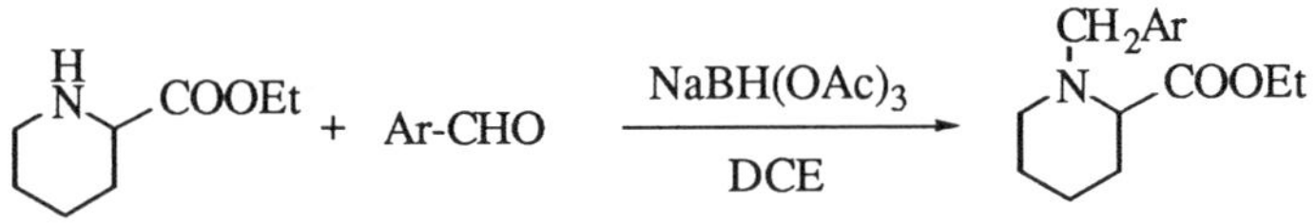

Ar = 4-pyridyl, 95%
Ar = 3-$NO_2$-$C_6H_4$-, 96%

The mild nature of this reducing agent is well represented in the reductive amination of the 1,1',2'-tris-nor-squalene aldehyde. The aldehyde was cleanly converted to the corresponding amines in high yields with no detectable aldehyde reduction or other side reactions (*9*). This is a significant improvement over results obtained from $NaBH_3CN$ (*30*).

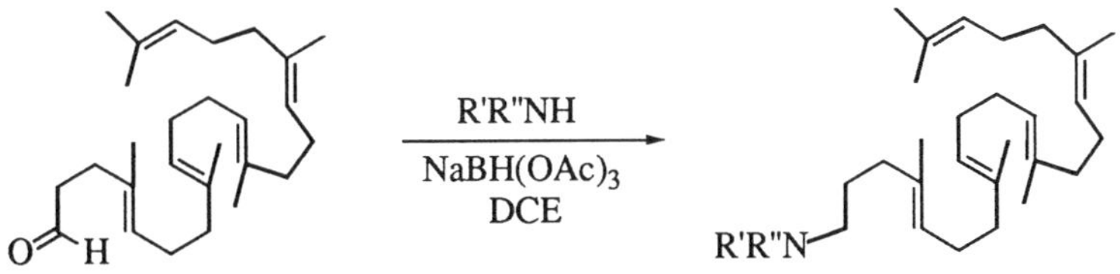

R' = H, R" = cyclohexyl, 94%
R' = R" = i-Pr, 91%
R' = R" = Et, 98%

The mildness of the reaction conditions is also demonstrated in the synthesis of the allenic amines **26a-c** and **27** from hexa-4,5-dienal and the appropriate amine (*31*).

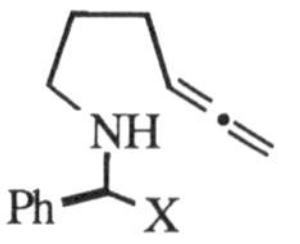

**26a** X = CH$_2$Ph
**26b** X = CH$_2$Sph
**26c** X = CH$_2$SePh

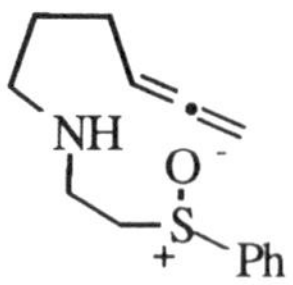

**27**

The reductive alkylation of aminoacids or esters with aldehydes is achieved effectively by using NaBH(OAc)$_3$. The *N*-benzylation of *N*-ε-Cbz-L-lysine with benzaldehyde and NaBH(OAc)$_3$ in DCE at 50 °C gives the desired product (**28**) in 82% isolated yield (unpublished result).

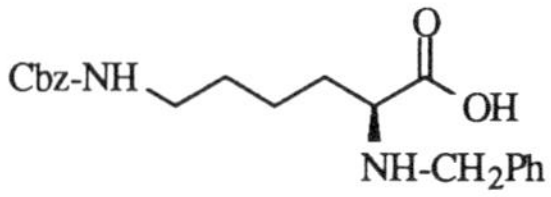

**28**

The reductive amination of aldehydes with aminoesters and NaBH(OAc)$_3$ under the standard conditions was found to be more convenient than other methods in the preparation of *N*-alkyl aminoesters (*33*). Various *N*-alkylated amino acid esters were prepared according to our standard procedure in good isolated yields as shown in Scheme VII (*32, 33*).

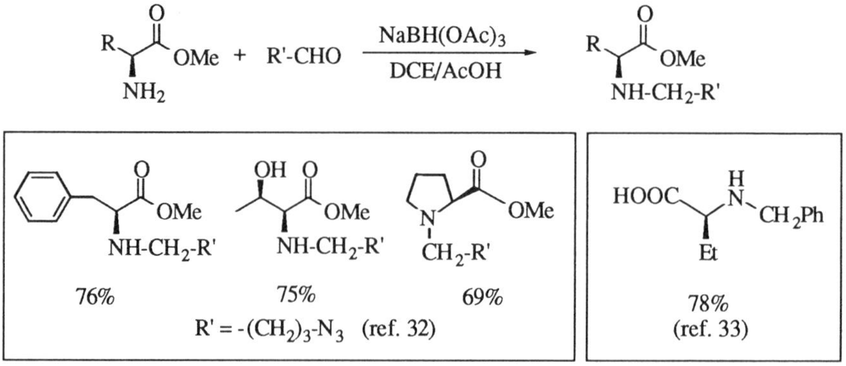

76%   75%   69%   78%
R' = -(CH$_2$)$_3$-N$_3$   (ref. 32)   (ref. 33)

**Scheme VII**

Similar to ketones, aldehydes react with primary γ- and δ-aminoacids or esters, to give the initial *N*-alkyl products which cyclize to the corresponding lactams under the reaction conditions. A particularly interesting aldehyde is *o*-carboxybenzaldehyde which gives the benzolactams with either simple amines or aminoesters. The cyclization occurs preferentially on the *o*-carboxyl group even in the presence of a δ-ester group on the chain (Scheme VIII) (*26*).

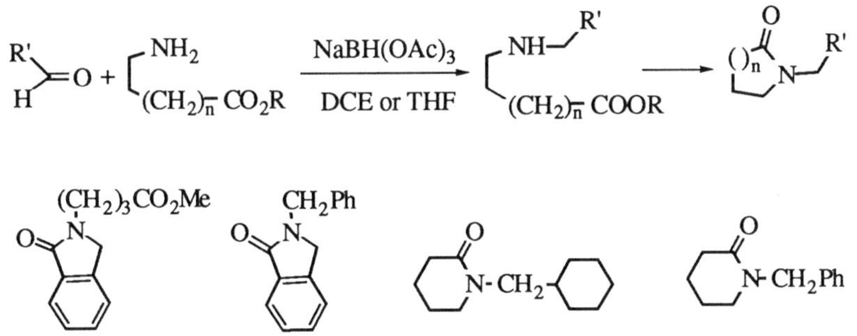

**Scheme VIII**

The *N*-methylation of amines can be carried out using formaldehyde under the standard conditions. This reaction, however, is not selective with primary amines; it gives only the *N,N*-dimethyl derivative. Paraformaldehyde may be used as a source of formaldehyde as in the synthesis of compound **29** (*34*) which was isolated in 89% yield.

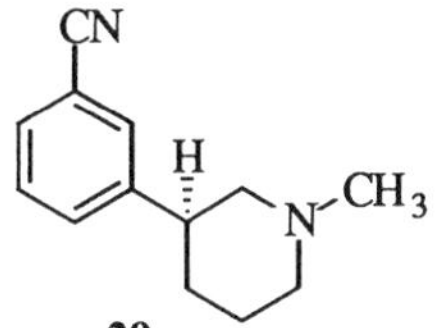

Formalin may also be used in reductive amination reactions on a small scale (10 - 20 mmol) with excess sodium triacetoxyborohydride. The restriction on the scale results from the exothermic decomposition of the hydride reagent by water. About 5 equivalents of the hydride reagent are used in the reaction which may be impractical in larger scale reactions. Aqueous glutaraldehyde may similarly be used in these reactions. Reactions of formalin and glutaraldehyde with 1-phenylpiperazine give nearly quantitative yields of the corresponding amines **30** and **31** (*9*).

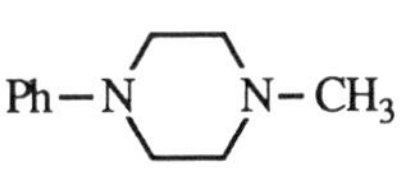

**30** (98%) 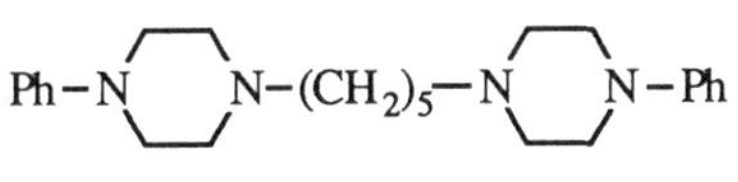     **31** (98%)

The reductive amination of aldehydes with weakly basic amines is faster and has wider scope than ketones. Most reactions with anilines monosubstituted by electron-withdrawing groups are carried out under the standard conditions with undetectable aldehyde reduction. The products are obtained effectively and in high yields, such as the synthesis of the compounds listed in Figure 8 (*9*).

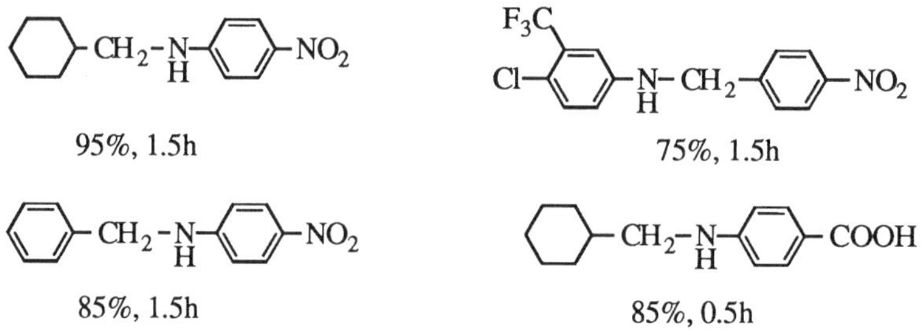

**Figure 8:** Reductive Amination of Aldehydes with Weakly Basic Amines

As the basicity and nucleophilicity of the amines decrease, the reductive amination becomes slow and aldehyde reduction becomes a competing reaction. The reductive amination of aldehydes with amines such as *o*-nitroaniline, 2,4-dichloroaniline, 2-aminothiazole and iminostilbene are accompanied by about 10 - 30% aldehyde reduction. The reactions are modified to use the amines as limiting agents and up to 1.5 equivalents of aldehyde to compensate for this side reaction. These reductive aminations are very efficient and give isolated yields ranging from 60 to 96% (Figure 9) (*9*). These reactions expand the scope of reductive amination reactions to limits never before achieved with any of the commonly used reducing agents.

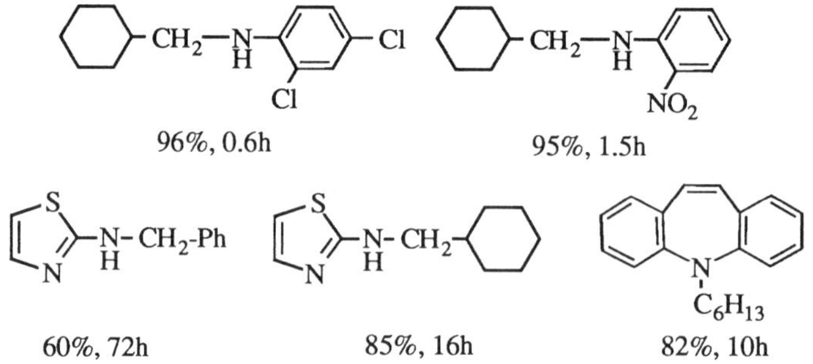

**Figure 9:** Reductive Amination of Aldehydes with More Weakly Basic Amines.

The non-basic amines such as 2,4-dinitroaniline and 2,4,6-trichloroaniline are the least reactive. There is no detectable reaction with aromatic aldehydes such as benzaldehyde. However, the reductive amination of cyclohexane carboxaldehyde with either amine progresses slowly and is accompanied by considerable aldehyde reduction. The reaction is carried out in the presence of 3 - 5 equivalents of AcOH and requires the occasional addition of excess aldehyde and reducing agent up to 5 equivalents each over 2 - 4 days to effect complete reaction of the amines. The products (**32** and **33**) are isolated by chromatography in 61% and 58% yield respectively (*9*). We believe that these reactions proceed via initial formation of enamines which will be reduced under the reaction conditions. This also explains the lack of reactivity toward aromatic aldehydes which can not form enamines.

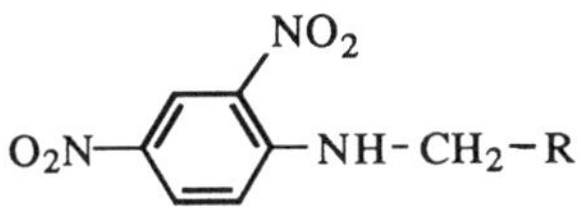

**32 (61%)**   R = cyclohexyl   **33 (58%)**

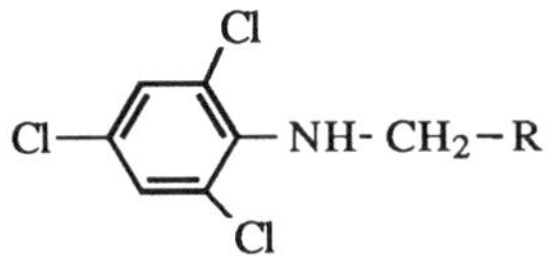

This procedure has been extended to substrates that never before used in reductive amination reactions, namely, sulfonamides. The reaction of *p*-toluenesulfonamide with aldehydes affords the corresponding *N*-alkyl sulfonamides, e.g., **34** and **35** in good isolated yields (9).

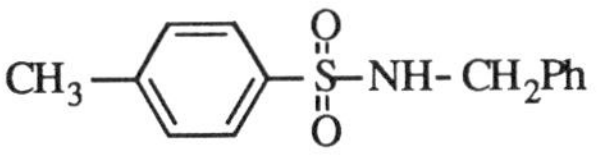

**34 (80%), 48h** 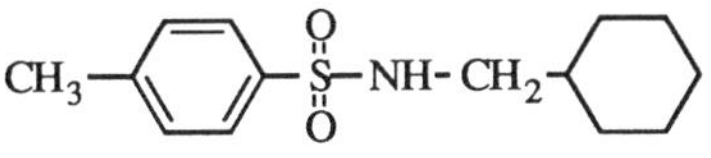            **35 (85%), 28h**

**Conclusion:**

This overview shows clearly that sodium triacetoxyborohydride is a synthetically useful reagent that can be used effectively in reductive amination of ketones and aldehydes with most amines. While the wide applications mentioned here make it an attractive choice for reductive amination, the most attractive features of this reagent are the convenience of use, the ease of work up and the simplicity of product isolation.

## References

1. Emerson, W. S. *Organic Reactions* **1948**, *4*, 174 and references therein.
2. Schellenberg, K. A. *J. Org. Chem.*, **1963**, *28*, 3259.
3. (a) Hutchins, R. O.; Natale, N. R. *Org. Prep. Proced. Int.* **1979**, *11*(5), 201; (b) Lane, C. F. *Synthesis*, **1975**, 135.
4. (a) Borch, R. F., Bernstein, M. D.; Durst, H. D. *J. Am. Chem. Soc.* **1971**, *93*, 2897. (b) Borch, R. F., Durst, H. D. *J. Am. Chem. Soc.* **1969**, *91*, 3996.
5. The *Sigma-Aldrich Library of Chemical Safety Data*, Edition I, p 1609, Lenga, R. E., Ed., Sigma-Aldrich Corp. **1985**.
6. (a) We detected cyanide contamination in large scale reduction of imines with sodium cyanoborohydride. (b) A similar result was reported: Moormann, A. E. *Synth. Commun.* **1993**, *23*, 789.
7. (a) Abdel-Magid, A. F.; Maryanoff, C. A.; Carson, K. G. *Tetrahedron Lett.* **1990**, *31*, 5595. (b) Abdel-Magid, A. F.; Maryanoff, C. A. *Synlett* **1990**, 537.
8. (a) Gribble, G. W.; Lord, P. D.; Skotnicki, J.; Dietz, S. E.; Eaton, J. T.; Johnson, J. L. *J. Am. Chem. Soc.* **1974**, *96*, 7812; (b) Gribble, G. W.; Jasinski, J. M.; Pellicone, J. T.; Panetta, J. A. *Synthesis* **1978**, 766.
9. Abdel-Magid, A. F.; Carson, K. G.; Harris, B. D.; Maryanoff, C. A.; Shah, R. D. *J. Org. Chem.*, **1996**, *61*, 3849.
10. (a) Gribble, G. W. and Ferguson, D. C. *J. C. S. Chem. Commun.* **1975**, 535; (b) Nutaitis, C. F.; Gribble, G. W. *Tetrahedron Lett.* **1983**, *24*, 4287.
11. Campbell, J. B.; Lavagnino, E. P. In *Catalysis in Organic Syntheses*; Jones, W. H. Ed.; Academic Press, New York, 1980; p 43.

12. Bagley, J. R.; Wilhelm, J. A. *J. Labelled Comp. Radiopharm.* **1992**, *31*, 945.
13. Mayer, S. C., Pfizenmayer, A. J.; Cordova, R.; Li, W-R; Joullié, M. M. *Tetrahedron: Asymm.* **1994**, *5*, 519.
14. Barrett, A. G. M.; Boys, M. L.; Boehm, T. L. *J. Chem. Soc., Chem. Commun.* **1994**, 1881.
15. Kersey, I. D.; Fishwick, C. W. G.; Findlay, J. B. C.; Ward, P. *Bioorg. Med. Chem. Lett.* **1995**, *5*, 1271.
16. McDermott, T. S.; Mortlock, A. A.; Heathcock, C. H. *J. Org. Chem.*, **1996**, *61*, 700.
17. Hart, D. J.; Leroy, V. *Tetrahedron* **1995**, *51*, 5757.
18. (a) Wrobel, J. E.; Ganem, B. *Tetrahedron Lett.* **1981**, *22*, 3447; (b) Hutchins, R. O.; Markowitz, M. *J. Org. Chem.* **1981**, *46*, 3571; (c) Hutchins, R. O.; Su, W-Y.; Sivakumar, R.; Cistone, F.; Stercho, Y. P. *J. Org. Chem.* **1983**, *48*, 3412; (d) Hutchins, R. O.; Adams, J.; Rutledge, M. C. *J. Org. Chem.* **1995**, *60*, 7396.
19. Camps, P.; Muñoz-Torrero, D.; Pérez, F. *J. Chem. Res. (S)* **1995**, 232.
20. Ley, S. V.; Mynett, D. M.; Koot, W-J. *Synlett* **1995**, 1017.
21. Demopoulos, V. V.; Gavalas, A.; Rekatas, G.; Tani, E. *J. Heterocyclic Chem.* **1995**, *32*, 1145.
22. Tschaen, D. M.; Abramson, L.; Cai, D.; Desmond, R.; Dolling, U-H.; Frey, L.; Karady, S.; Shi, Y-J.; Verhoeven, T. R. *J. Org. Chem.* **1995**, *60*, 4324.
23. Quirante, J.; Escolano, C.; Bosch, J.; Bonjoch, J. *J. Chem. Soc., Chem. Commun.* **1995**, 2141.
24. Duhamel, P.; Deyine, A.; Dujardin, G.; Plé, G.; Poirier, J-M. *J. Chem. Soc., Perkin Trans. 1* **1995**, 2103.
25. (a) Abdel-Magid, A. F.; Carson, K. G.; Harris, B. D.; Maryanoff, C. A. The 33rd ACS National Organic Symposium, Bozeman, Mo, June **1993**, Paper A-4. (b) Vanderplas, B.; Murtiashaw, C. W.; Sinay, T.; Urban, F. *J. Org. Prep. Proced. Int.* **1992**, *24*, 685.
26. Abdel-Magid, A. F.; Harris, B. D.; Maryanoff, C. A. *Synlett* **1994**, 81.
27. (a) Albert, A.; Serjeant, E. P. In *The Determination of Ionization Constants;* Chapman and Hall: London **1971**, p 91; (b) Yates, K; Wai, H. *J. Am. Chem. Soc.* **1964**, *86*, 5408.
28. (a) Gribble, G. W.; Ferguson, D. C. *J. Chem. Soc., Chem. Commun.* **1975**, 535. (b) Nutaitis, C. F.; Gribble, G. W. *Tetrahedron Lett.* **1983**, *24*, 4287. (c) Gribble, G. W. In *Encyclopedia of Reagents for Organic Synthesis*; Paquette, L. A. Ed., John Wiley and Sons, New York, vol. 7, p 4649, **1995**.
29. Caulfield, T. J.; Prasad, C. V. C.; Prouty, C. P.; Saha, A. K.; Sardaro, M. P.; Schairer, W. C.; Yawman, A.; Upson, D. A.; Kruse, L. I. *Bioorg. Med. Chem. Lett.* **1993**, *3*, 2771.
30. (a) Duriatti, A.; Bouvier-Nave, P.; Benveniste, P.; Schuber, F.; Delprino, L.; Balliano, G.; Cattel, L. *Biochem. Pharmacol.* **1985**, *34*, 2765. (b) Ceruti, M.; Balliano, G.; Viola, F.; Cattel; L.; Gerst, N.; Schuber, F. *Eur. J. Med. Chem.* **1987**, *22*, 199.
31. Davies, I. W.; Gallagher, T.; Lamont, R. B.; Scopes, D. I. C. *J. Chem. Soc., Chem. Commun.* **1992**, 335.
32. Ramanjulu, J. M.; Joullié, M. M. *Synthetic Commun.* **1995**, *26*,
33. Kubota, H.; Kubo, A.; Takahashi, M.; Shimizu, R.; Da-te, T.; Okamura, K.; Nunami, K. *J. Org. Chem.* **1995**, *60*, 6776.
34. Sonesson, C.; Lin, C-H.; Hansson, L.; Waters, N.; Svensson, K.; Carlsson, A.; Smith, M. W.; Wikstrom, H. *J. Med. Chem.* **1994**, *37*, 2735.

# INDEXES

# Author Index

Abdel-Magid, Ahmed F., 127,201
Adams, Jeffrey, 127
Beck, Albert K., 52
Blake, James F., 112
Brown, Herbert C., 1,84
Costanzo, Michael J., 138
Dahinden, Robert, 52
Fisher, Gary B., 153
Genet, Jean Pierre, 31
Godjoian, Gayane, 153
Goralski, Christian T., 153
Gribble, Gordon W., 167
Harris, Bruce D., 138
Hutchins, MaryGail K., 127
Hutchins, Robert O., 127
King, Anthony O., 98
Kühnle, Florian N. M., 52
Maryanoff, Bruce E., 138
Maryanoff, Cynthia A., 138,201
Mathre, David J., 98
Oskay, Enis, 127
Quallich, George J., 112
Ramachandran, P. Veeraraghavan, 1,84
Rao, Samala J., 127
Seyden-Penne, Jacqueline, 70
Shinkai, Ichiro, 98
Singaram, Bakthan, 153
Tschaen, David M., 98
Woodall, Teresa M., 112
Zhang, Han-Cheng, 138
Zhu, Qi-Cong, 127

# Affiliation Index

Dartmouth College, 167
Dow Chemical Company, 153
Drexel University, 127
ETH-Zentrum, 52
Ecole Nationale Supérieure de Chimie de Paris, 31
Merck Research Laboratories, 98
Pfizer, Inc., 112
Purdue University, 1,84
R. W. Johnson Pharmaceutical Research Institute, 138,201
Université Paris Sud, 70
University of California—Santa Cruz, 153

# Subject Index

**A**

Acetals
  reductive cleavage by sodium acyloxyborohydrides, 190–191
  role in reductive amination by sodium triacetoxyborohydride, 206
Acetoxyazetidinone, synthesis, 48
Acidic reducing agents, use in reductions, 13–19
Acridine, reduction by sodium acyloxyborohydrides, 176
Acyclic aliphatic ketones, reductive amination by sodium triacetoxyborohydride, 204–206
Acyclic diastereocontrol in reductions of $1,n$-hydroxy ketones, remote, *See* Remote acyclic diastereocontrol in reductions of $1,n$-hydroxy ketones
Acyloxyborohydrides, reduction, 8

2-Acylpyridines, asymmetric reductions with B-chlorodiisopinocampheylborane, 94
Air-stable reducing agents, *See* Lithium aminoborohydrides
Alane, reductions, 13
Alane–amine complex, reductions, 13–14
Alcohols, reduction by sodium acyloxyborohydrides, 186–188
Aldehydes
  reduction by lithium aminoborohydrides, 156–157
  reductive amination
    with sodium triacetoxyborohydride, 210–215
    with weakly basic amines by sodium triacetoxyborohydride, 213–215
  selective reduction by sodium acyloxyborohydrides, 191–193
Alicyclic ketones, reductive amination by sodium triacetoxyborohydride, 202–204
Aliphatic azides, reduction by lithium aminoborohydrides, 161–162
Aliphatic ketones, enantioselective lithium aluminum hydride reduction, 54–55
Alkenes
  hydroboration by sodium acyloxyborohydrides, 185
  reduction by sodium acyloxyborohydrides, 185–186
Alkoxyaluminohydrides, reduction, 6–7
Alkoxyboranes, reductions, 18
Alkoxyborohydrides, reduction, 7–8
Alkyl halides, reduction by lithium aminoborohydrides, 160
Alkyl-substituted borohydrides, synthesis, 162–163
Alkylaluminoborohydrides, reduction, 9–12
Alkylborane, reduction, 16–18,19*f*
Alkylchloroboranes, reduction, 20
Alkylidene lactones, C=C bond hydrogenation, 39
α-Alkynyl ketones, reduction, 54–55
Allylic alcohols, C=C bond hydrogenation, 35–36
Alpine-Borane, *See* B-Isopinocampheyl-9-borobicyclo[3.3.1]nonane

Aluminohydrides, electrophilic assistance in reduction of six-membered cyclic ketones, 70–81
Amides, reduction by sodium acyloxyborohydrides, 182–184
Amine(s)
  N-alkylation by sodium acyloxyborohydrides, 171–175
  importance in organic synthesis, 201
  synthesis, 127–137
Amine enantiomers, synthesis, 136–137
Amine nitrogen, role in reduction, 143–146
Amino acids, synthesis, 127–137
Amino ketones, asymmetric reductions with B-chlorodiisopinocampheylborane, 93–94
Aminoborohydrides, reduction, 8
Anti-β-hydroxy-α-hydrazino esters, synthesis, 46
Anti-inflammatory arylacetic acids, enantioselective synthesis, 37
Aralkyl ketones, asymmetric reductions with B-chlorodiisopinocampheylborane, 85–86
Arenes, Friedel–Crafts alkylation by sodium acyloxyborohydrides, 191
Aromatic azides, reduction by lithium aminoborohydrides, 161–162
Aryl alkyl ketones, reduction, 54–55
Asymmetric hydrogenation, use of chiral ruthenium(II) catalysts, 31–48
Asymmetric lithium aluminum hydride reduction, pursuit of high enantioselectivities, 54–58
Asymmetric reduction
  borohydride reagents, 23
  chiral lithium triethylborohydrides, 23–24
  lithium aluminum hydride modified reagents, 22
  oxazaborolidines, 24
  prochiral ketones
    advantages of oxazaborolidines, 112
    limitations with stoichiometric reagents, 112
  sodium aluminum hydride modified reagents, 22

# INDEX

Asymmetric reduction—*Continued*
  with *B*-chlorodiisopinocampheylborane
    advantages, 96
    applications
      BMS–181100 analogue synthesis, 95
      (*R,S*)-eprozinol synthesis, 95
      *N*-methyl-γ-[4-(trifluoromethyl)-
        phenoxy]benzenepropanamine
        hydrochloride synthesis, 95
    aralkyl ketone reduction, 85–86
    asymmetric amplification, 89
    *B*-chlorodiisopinocampheylborane
      synthesis and reaction, 85
    development, 84
    diketone reduction, 90–91
    double asymmetric synthesis, 89–90
    α-hindered reduction, 87
    *B*-isopinocampheyl-9-borabicyclo-
      [3.3.2]nonane, 85
    kinetic resolution, 89
    mechanism, 87
    modified workup procedure, 86
    neighboring group effects
      2-acylpyridines, 94
      amino ketones, 93–94
      hydroxy ketones, 91–92
      keto acids, 93
    perfluoroalkyl ketone reduction, 87–88
    perfluoroalkyloxirane reduction, 88
Asymmetric synthesis, studies, 31
Azides, reduction by lithium
  aminoborohydrides, 160–162

## B

Bicyclic carbapenem analogues, synthesis,
  47–48
Bicyclic chelation control, *See* Remote
  acyclic diastereocontrol in reductions
  of 1,*n*-hydroxy ketones
Bicyclo ring derivatives, diastereoselective reductions, 130–132
BINOL, enantioselective lithium
  aluminum hydride reduction, 52–64
Biphenanthrol, enantioselective lithium
  aluminum hydride reduction, 52–64

Biphenol, enantioselective lithium
  aluminum hydride reduction, 52–64
1,4-Bis(dimethylamino)-(2*S*,3*S*)-butane-
  2,3-diol, synthesis, 53
(*R*)-2,2′-Bis(diphenylphosphino)-1,1′-
  binaphthyl ruthenium(II) catalysts,
  synthesis, 33
(*R*)-2,2′-Bis(diphenylphosphino)-1,1′-
  binaphthyl ruthenium(arene) catalysts,
  synthesis, 34
BMS–181100, synthesis, 95
Borane, reductions, 14–16
Borane–amines, reductions, 16
Borane–dimethyl sulfide complex, 15–16
Borane–tetrahydrofuran, reductions, 14–15
Borohydride(s)
  alkyl substituted, synthesis, 162–163
  electrophilic assistance in reduction of
    six-membered cyclic ketones, 70–81
  sulfurated, reduction, 8
Borohydride reagents, reductions, 22

## C

$C_2$-symmetric diols, synthesis, 122–124
C=C bond hydrogenation, use of chiral
  ruthenium(II) catalysts, 35–39
C=N bond hydrogenation, use of chiral
  ruthenium(II) catalysts, 45
Carbonyl group hydrogenation, use of
  chiral ruthenium(II) catalysts, 39–42
Carbonyl reduction, use of diborane, 2
Carboxylic acid(s) and esters, reduction
  by sodium acyloxyborohydrides, 190
Carboxylic acid(s) and sodium borohydride,
  *See* Sodium acyloxyborohydrides
Carboxylic acid esters, reduction by
  lithium aminoborohydrides, 158
Catalysts, chiral ruthenium(II), use for
  asymmetric hydrogenation, 31–48
Cation, role in reductions, 5–6
Chelation-controlled mechanism, remote
  acyclic diastereocontrol in reductions
  of 1,*n*-hydroxy ketones, 138–151
Chemoselectivity, reductions, 20

Chiral diphosphine ruthenium(2-methylallyl)$_2$ complexes, general synthesis, 33–34
Chiral ketone reduction with B-chlorodiisopinocampheylborane
  comparison to oxaborolidine-catalyzed enantioselective reductions, 108–109
  isolation, 107–108
  MK–0287, 103–104
  MK–0476, 105–107,108*t*
Chiral ligands, use in asymmetric hydrogenation, 32
Chiral lithium triethylborohydrides, reductions, 23–24
Chiral reducing agent, role in reduction, 143
Chiral ruthenium(II) catalysts for asymmetric hydrogenation
  C=C bond hydrogenation
    α,β-unsaturated acids, 36–38
    alkylidene lactones, 39
    allylic alcohols, 35–36
    dehydroamino acids, 38–39
  C=N bond hydrogenation, 45
  carbonyl group hydrogenation
    conditions, 39,40*t*
    dynamic kinetic resolution, 42–45
    generality, 41
    screening of chiral diphosphines and catalysts, 39–41
    sense of enantioselectivity, 42
  catalyst preparation, 33–35
  chiral ligands used for preparation, 32
  nonchelating C=O group hydrogenation, 45
  synthetic applications, 45–48
Chiral ruthenium(II) containing atropisomeric ligands, synthesis, 34
B-Chlorodiisopinocampheylborane
  asymmetric reductions, 84–96
  enantioselective reduction of prochiral ketones, 98–109
  synthesis and reactions, 85,86*f*
[2.1.1]Cryptand, role in reduction of ketones, 70–71
Cyanoborohydride, reduction, 11–12

Cyclic β-keto esters, reductive amination by sodium triacetoxyborohydride, 207–208
Cyclic ketones, electrophilic assistance in reduction, 70–81
Cyclohexyl azide, reduction by lithium aminoborohydrides, 161–162

D

Dehydroamino acids, C=C bond hydrogenation, 38–39
Dialkylalanes, reductions, 14
Dialkylborohydrides, reduction, 11
Diastereocontrol in reductions of 1,*n*-hydroxy ketones, *See* Remote acyclic diastereocontrol in reductions of 1,*n*-hydroxy ketones
Diastereoselective reductions of ketone phosphinylimines to phosphinylamines, synthesis of protected amines and amino acids, 127–137
Diborane, use for carbonyl reduction, 2
Dihalide ruthenium catalysts, in situ synthesis, 34–35
2,2-Diphenylcyclopentanone, enantioselective reduction, 102
Diphenyloxazaborolidine for enantioselective reduction of ketones
  ab initio calculations, 119–120
  enantioselective alkynylation, 124
  in situ oxazaborolidines, 120–122
  synthesis of $C_2$-symmetric diols, 122–124
  synthetic evaluation of oxazaborolidine structure and enantioselectivity relationships, 115–119
*N*-Diphenylphosphinylimine reduction to phosphinylamines
  diastereoselective reductions of cyclic *N*-diphenylphosphinylimines
  bicyclo ring derivatives, 130–132
  imine preparation, 128
  substituted cyclohexyl derivatives, 128–130
  substituted cyclopentyl derivatives, 130–132

N-Diphenylphosphinylimine reduction to phosphinylamines—*Continued*
  experimental description, 128
  α-imino ester derivatives with chiral hydride reagents to afford protected phenyl glycines, 133–136
  use of chiral phosphorus auxiliary group for generation of chiral phosphinyl imines and subsequent reduction to protected amine enantiomers, 136–137
Disubstituted acrylic acids, asymmetric hydrogenation, 36–37

E

Electrophilic assistance in reduction of six-membered cyclic ketones
  competition between saturated and α,β-unsaturated ketones, 75–78
  regioselectivity in reduction of α-enones, 71–75
  stereoselectivity of reduction, 79–81
Electrophilic catalysis in ketone reduction, evidence, 70
Enamides, reduction by sodium acyloxyborohydrides, 182
Enamines, reduction by sodium acyloxyborohydrides, 181–182
Enantiomerically enriched alcohols, synthesis, 41–42, 44–45
Enantiomerically pure secondary alcohols
  catalysts, 52–53
  synthetic approaches, 52
Enantioselection, use of stoichiometric amounts of chirally modified lithium hydride, 53
Enantioselective lithium aluminum hydride reduction of ketones, tartrate-derived ligands, 52–64
Enantioselective reductions
  ketone(s), diphenyloxazaborolidine, 112–124
  ketone phosphinylimines to phosphinylamines, synthesis of protected amines and amino acids, 127–137

Enantioselective reductions—*Continued*
  prochiral ketones
    chiral ketone reduction with *B*-chlorodiisopinocampheylborane, 103–109
    comparison of methods, 108–109
    oxazaborolidine catalyzed, *See* Oxazaborolidine-catalyzed enantioselective reductions
α-Enones, regioselectivity in reduction, 71–75
Epoxides, reduction by lithium aminoborohydrides, 160
(*R*,*S*)-Eprozinol, synthesis, 95
Ethers, reductive cleavage by sodium acyloxyborohydrides, 190–191
Ethyl benzoate, in-air reduction, 158

F

Free hydroxyl, role in reductions, 146–148
Functionalized β-ketones with atropisomeric ligands, hydrogenation, 42

H

Haloboranes, reductions, 18–19
Heterocycles
  saturated, reduction by sodium acyloxyborohydrides, 179
  synthesis, 47
α-Hindered ketones, asymmetric reductions with *B*-chlorodiisopinocampheylborane, 87
Hydrazines, role in reductive amination by sodium triacetoxyborohydride, 206–207
1,*n*-Hydroxy ketones, remote acyclic diastereocontrol in reductions, 138–151
4(*R*)-Hydroxycyclopentenone, synthesis, 35–36
Hydroboration, discovery, 24–25
Hydrogenation, asymmetric, use of chiral ruthenium(II) catalysts, 31–48
Hydroxy ketones, asymmetric reductions with *B*-chlorodiisopinocampheylborane, 91–92

Hydroxyl-directed carbonyl reduction, sodium acyloxyborohydrides, 193–195

### I

(S)-Ibuprofen, synthesis, 38
Imines
  reduction by sodium acyloxyborohydrides, 177–178,180–181
  synthesis, 128
Indoles, reduction by sodium acyloxyborohydrides, 168–171
Indoline alkaloids, synthesis, 38–39
B-Isopinocampheyl-9-borobicyclo[3.3.1]-nonane
  dehydroboration as cause of achiral reduction, 85
  synthesis and reactions, 85
Isoquinoline, reduction by sodium acyloxyborohydrides, 175–176

### K

Ketals
  reductive cleavage by sodium acyloxyborohydrides, 190–191
  role in reductive amination by sodium triacetoxyborohydride, 206
Keto acids, asymmetric reductions with B-chlorodiisopinocampheylborane, 93
β-Keto esters, asymmetric hydrogenation, 39–45
Ketone(s)
  cyclic, electrophilic assistance in reduction, 70–81
  diphenyloxazaborolidine for enantio-selective reduction, 112–124
  1,n-hydroxy, See 1,n-Hydroxy ketones
  prochiral, enantioselective reduction, 98–109
  reduction
    by lithium aminoborohydrides, 156–157
    by sodium acyloxyborohydrides, 188–189

Ketone(s)—*Continued*
  reductive amination
    with sodium triacetoxyborohydride, 202–210
    with weakly basic amines by sodium triacetoxyborohydride, 208–210
  tartrate-derived ligands for enantio-selective lithium aluminum hydride reduction, 52–64
Ketone phosphinylimines to phosphinyl-amines, diastereo- and enantio-selective reductions, 127–137

### L

Lithium aluminum hydride
  reduction
    ketones, tartrate-derived ligands, 52–64
    reaction, 4
    search for safe and convenient alternatives, 153
Lithium aluminum hydride–1,4-bis(dimethyl-amino)-(2S,3S)-butane-2,3-diol complex, selectivities in reduction of aryl alkyl ketones, 53–54
Lithium aluminum hydride modified reagents, reductions, 22
Lithium aminoborohydrides
  applications, 163–165
  characterization
    physical properties, 154,156
    reducing properties, 156
    spectral properties, 154
  discovery, 154
  reactivity, 164
  reduction
    aldehydes, 156–157
    alkyl halides, 160
    azides, 160–162
    carboxylic acid esters, 158
    epoxides, 160
    tertiary amides, 158–160
    α,β-unsaturated aldehydes and ketones, 157–158

# INDEX

Lithium aminoborohydrides—*Continued*
  safety considerations during synthesis, 155–156
  synthesis
    solid hydrides, 155
    tetrahydrofuran solutions, 155
    transfer reactions, 162–163
Lithium (*N,N*-diethylamino)borohydride, solid, synthesis, 155
Lithium piperidinoborohydride, synthesis, 155
Lithium triethylborohydride
  chiral, reductions, 23–24
  reduction, 9–10

## M

1β-Methylcarbapenem intermediate, synthesis, 36
*N*-Methyl-γ-[4-(trifluoromethyl)-phenoxy]benzenepropanamine hydrochloride, synthesis, 95
MK–0287, synthesis, 103–104
MK–0476, synthesis, 105–107,108*t*
MK–0499, enantioselective reduction for synthesis, 102–103
Monoalkoxyaluminum trihydride, reduction, 7
Monoalkylborohydrides, reduction, 11

## N

Nebivolol, 139
Neighboring group, role in asymmetric reductions with *B*-chlorodiisopinocampheylborane, 91–94
Nitrile, reduction by sodium acyloxyborohydrides, 184–185
Nonchelating C=O group hydrogenation, use of chiral ruthenium(II) catalysts, 45

## O

Optically active phenyl glycidates, synthesis, 46
Optically pure compounds, asymmetric synthesis, 31

Optically pure sultam, synthesis, 45
Oxazaborolidine(s)
  advantages for asymmetric reduction of prochiral ketones, 112
  applications, 113
  enantioselective reduction of ketones, 113–114
  reductions, 24
  structure, 112–113
Oxazaborolidine-catalyzed enantioselective reductions
  comparison to chiral ketone reduction with *B*-chlorodiisopinocampheylborane, 108–109
  examples of chiral oxazaborolidines, 99
  mechanism, 101–103
  *B*-methyloxazaborolidine synthesis, 99–101
  reaction, 98
  synthetic evaluation of structure and enantioselectivity relationships, 115–119
Oximes, reduction by sodium acyloxyborohydrides, 179–180

## P

Perfluoroalkyl ketones, asymmetric reductions with *B*-chlorodiisopinocampheylborane, 87–88
Perfluoroalkyloxiranes, asymmetric reductions with *B*-chlorodiisopinocampheylborane, 88
Phenyl glycidates, optically active, synthesis, 46
Phenylglycines, synthesis, 133–136
*N*-(Phenylsulfonyl)pyrroles, reduction by sodium acyloxyborohydrides, 178–179
Primary amines, synthesis, 127
Prochiral ketones, enantioselective reduction, 98–109
Protected amines and amino acids, synthesis, 127–137
Prozac, 95
Pyrrolidinooctanamide, reduction, 160
Pyrylium salt, reduction by sodium acyloxyborohydrides, 178

## Q

Quinoline, reduction by sodium acyloxyborohydrides, 175–177

## R

Reductions
  acidic reducing agents
    alane, 13
    alane–amine complex, 13–14
    alkoxyboranes, 18
    alkylborane
      from borohydrides, 18,19f
      from olefins via hydroboration, 16–17
    alkylchloroboranes, 20
    borane
      borane–amines, 16
      borane–dimethyl sulfide complex, 15–16
      borane–tetrahydrofuran, 14–15
      preparation, 14
    dialkylalanes, 14
    haloboranes, 18–19
    rate of reaction, 13
  asymmetric reduction
    borohydride reagents, 23
    chiral lithium triethylborohydrides, 23–24
    lithium aluminum hydride modified reagents, 22
    oxazaborolidines, 24
    sodium aluminum hydride modified reagents, 22
  chemoselectivity, 20
  development, 2
  diborane for carbonyl reductions, 2
  $1,n$-hydroxy ketones, remote acyclic diastereocontrol, 138–151
  ketone(s), tartrate-derived ligands, 52–64
  ketone phosphinylimines to phosphinylamines, 127–137
  landmarks, 25
  lithium aluminum hydride, 4
  pre-borohydride era, 2

Reductions—*Continued*
  prochiral ketones, enantioselective, *See* Enantioselective reduction of prochiral ketones
  selective reductions
    cation effect, 5–6
    extremes, 4–5
    solvent effect, 5
    substituent effect
      acyloxyborohydrides, 8
      alkoxyaluminohydrides, 6–7
      alkoxyborohydrides, 7–8
      alkylaluminoborohydrides
        cyanoborohydride, 11–12
      dialkylborohydrides, 11
      lithium triethylborohydride, 9–10
      monoalkylborohydrides, 11
      Selectrides, 10–11
      trialkylborohydrides, 9
    aminoborohydrides, 8
    monoalkoxyaluminum trihydride, 7
    sulfurated borohydride, 8
  six-membered cyclic ketones, electrophilic assistance, 70–81
  spectrum of reducing agents, 24,25t
  stereoselectivity, 20
  World War II research and preparation of sodium borohydride, 3–4
Reductive alkylation of amines, *See* Reductive amination of ketones and aldehydes
Reductive amination
  aldehydes, 210–215
  conditions, 202
  ketones, 202–210
  reaction, 201–202
Regioselectivity, reduction of α-enones, 71–75
Remote acyclic diastereocontrol in reductions of $1,n$-hydroxy ketones
  models, 148–151
  reasons for interest, 138–139
  reductions
    amine nitrogen effect, 143–146
    chiral reducing agent effect, 143
    free hydroxyl effect, 146–148

Remote acyclic diastereocontrol in reductions of 1,*n*-hydroxy ketones—*Continued*
reductions—*Continued*
reducing agents, 140–141
substrate effect, 141–143
Ruthenium(II) catalysts for asymmetric hydrogenation, chiral, *See* Chiral ruthenium(II) catalysts for asymmetric hydrogenation

**S**

Saturated heterocycles, reduction by sodium acyloxyborohydrides, 179
Saturated ketones, competition with α,β-unsaturated ketones, 75–78
Secondary alcohols, enantiomerically pure, 52–53
Selective reductions, description, 4–8
Selectrides, reduction, 10–11
Six-membered cyclic ketones, electrophilic assistance in reduction, 70–81
Sodium acyloxyborohydrides
  N-alkylation, amines, 171–175
  development, 167
  Friedel–Crafts alkylation of arenes, 191
  hydroboration, alkenes, 185
  hydroxyl-directed carbonyl reduction, 193–195
  reduction
    acridine, 176
    alcohols, 186–188
    alkenes, 185–186
    amides, 183–184
    enamides, 182
    enamines, 181–182
    imines, 177–178, 180–181
    indoles, 168–171
    isoquinoline, 175–176
    ketones to hydrocarbons, 188–189
    nitrile, 184–185
    oximes, 179–180
    *N*-(phenylsulfonyl)pyrroles, 178–179
    pyrylium salt, 178
    quinoline, 175–177
    saturated heterocycles, 179
    vinylogous amides, 182–183

Sodium acyloxyborohydrides—*Continued*
  reductive cleavage of acetals, ketals, and ethers, 190–191
  selective reduction of aldehydes, 191–193
Sodium aluminum hydride modified reagents, reductions, 22
Sodium aminoborohydride, use as reducing agent, 153
Sodium bis(2-methoxy)aluminum hydride, advantages and disadvantages, 153
Sodium borohydride, World War II research and preparation, 3–4
Sodium borohydride and carboxylic acids, *See* Sodium acyloxyborohydrides
Sodium triacetoxyborohydride in reductive amination
  aldehydes, 210–215
  ketones, 202–210
Solid lithium (*N*,*N*-diethylamino)borohydride, synthesis, 155
Solvent, role in reductions, 5
Stereochemical control of reactions involving conformationally flexible, acyclic molecules, importance, 138
Stereoselectivity, reductions, 20
Substituent, role in reductions, 6–12
Substituted cyclohexyl derivatives, diastereoselective reductions, 128–130
Substituted cyclopentyl derivatives, diastereoselective reductions, 130–132
Substrate, role in reductions, 141–143
Sulfurated borohydride, reduction, 8
Sultam, optically pure, synthesis, 45
Super Hydride, *See* Lithium triethylborohydride

**T**

Tartrate-derived ligands for enantioselective lithium aluminum hydride reduction of ketones
  comparison of mechanistic proposals, 61–64
  enhancement of enantiopurity by clathrate formation in crude product mixture, 60–61
  experimental description, 58

Tartrate-derived ligands for enantioselective lithium aluminum hydride reduction of ketones—*Continued*
  lithium aluminum hydride–α,α,α',α'-Tetraaryl-1,3-dioxolane-4,5-dimethanol
  preparation, 58–59
  reduction of ketones, 59–60
Tertiary amides, reduction by lithium aminoborohydrides, 158–160
α,α,α',α'-Tetraaryl-1,3-dioxolane-4,5-dimethanol
  structure, 54–55
  synthesis, 54
  use for enantioselective lithium aluminum hydride reduction of ketones, 52–64
1,2,3,4-Tetrahydrocarbazole, reduction via Leuckart-type reaction, 167
Tetrahydrofuran solutions of lithium aminoborohydrides, synthesis, 155
Tetrahydrolistatin, synthesis, 47
1-Tetralone, enantioselective reduction, 102
Transfer reactions, lithium aminoborohydrides, 162–163
Trialkylborohydrides, reduction, 9
(3*R*,7*R*)-3,7,11-Trimethyldodecanol, synthesis, 35

## U

α,β-Unsaturated acids, C=C bond hydrogenation, 36–38
α,β-Unsaturated aldehydes and ketones, reduction by lithium aminoborohydrides, 157–158
α,β-Unsaturated ketones, competition with saturated ketones, 75–78

## V

Vinylogous amides, reduction by sodium acyloxyborohydrides, 182–183
Vitride, 153

## W

Weakly basic amines, role in reductive amination by sodium triacetoxyborohydride, 208–210, 213–215
World War II research and preparation, sodium borohydride, 3–4

# Highlights from ACS Books

*Good Laboratory Practice Standards: Applications for Field and Laboratory Studies*
Edited by Willa Y. Garner, Maureen S. Barge, and James P. Ussary
ACS Professional Reference Book; 572 pp; clothbound ISBN 0–8412–2192–8

*Silent Spring Revisited*
Edited by Gino J. Marco, Robert M. Hollingworth, and William Durham
214 pp; clothbound ISBN 0–8412–0980–4; paperback ISBN 0–8412–0981–2

*The Microkinetics of Heterogeneous Catalysis*
By James A. Dumesic, Dale F. Rudd, Luis M. Aparicio, James E. Rekoske,
and Andrés A. Treviño
ACS Professional Reference Book; 316 pp; clothbound ISBN 0–8412–2214–2

*Helping Your Child Learn Science*
By Nancy Paulu with Margery Martin; Illustrated by Margaret Scott
58 pp; paperback ISBN 0–8412–2626–1

*Handbook of Chemical Property Estimation Methods*
By Warren J. Lyman, William F. Reehl, and David H. Rosenblatt
960 pp; clothbound ISBN 0–8412–1761–0

*Understanding Chemical Patents: A Guide for the Inventor*
By John T. Maynard and Howard M. Peters
184 pp; clothbound ISBN 0–8412–1997–4; paperback ISBN 0–8412–1998–2

*Spectroscopy of Polymers*
By Jack L. Koenig
ACS Professional Reference Book; 328 pp;
clothbound ISBN 0–8412–1904–4; paperback ISBN 0–8412–1924–9

*Harnessing Biotechnology for the 21st Century*
Edited by Michael R. Ladisch and Arindam Bose
Conference Proceedings Series; 612 pp;
clothbound ISBN 0–8412–2477–3

*From Caveman to Chemist: Circumstances and Achievements*
By Hugh W. Salzberg
300 pp; clothbound ISBN 0–8412–1786–6; paperback ISBN 0–8412–1787–4

*The Green Flame: Surviving Government Secrecy*
By Andrew Dequasie
300 pp; clothbound ISBN 0–8412–1857–9

---

For further information and a free catalog of ACS books, contact:
American Chemical Society
Customer Service & Sales
1155 16th Street, NW, Washington, DC 20036
Telephone 800–227–9919

# Bestsellers from ACS Books

*The ACS Style Guide: A Manual for Authors and Editors*
Edited by Janet S. Dodd
264 pp; clothbound ISBN 0–8412–0917–0; paperback ISBN 0–8412–0943–X

*Understanding Chemical Patents: A Guide for the Inventor*
By John T. Maynard and Howard M. Peters
184 pp; clothbound ISBN 0–8412–1997–4; paperback ISBN 0–8412–1998–2

*Chemical Activities* (student and teacher editions)
By Christie L. Borgford and Lee R. Summerlin
330 pp; spiralbound ISBN 0–8412–1417–4; teacher ed. ISBN 0–8412–1416–6

*Chemical Demonstrations: A Sourcebook for Teachers,
Volumes 1 and 2,* Second Edition
Volume 1 by Lee R. Summerlin and James L. Ealy, Jr.;
Vol. 1, 198 pp; spiralbound ISBN 0–8412–1481–6;
Volume 2 by Lee R. Summerlin, Christie L. Borgford, and Julie B. Ealy
Vol. 2, 234 pp; spiralbound ISBN 0–8412–1535–9

*Chemistry and Crime: From Sherlock Holmes to Today's Courtroom*
Edited by Samuel M. Gerber
135 pp; clothbound ISBN 0–8412–0784–4; paperback ISBN 0–8412–0785–2

*Writing the Laboratory Notebook*
By Howard M. Kanare
145 pp; clothbound ISBN 0–8412–0906–5; paperback ISBN 0–8412–0933–2

*Developing a Chemical Hygiene Plan*
By Jay A. Young, Warren K. Kingsley, and George H. Wahl, Jr.
paperback ISBN 0–8412–1876–5

*Introduction to Microwave Sample Preparation: Theory and Practice*
Edited by H. M. Kingston and Lois B. Jassie
263 pp; clothbound ISBN 0–8412–1450–6

*Principles of Environmental Sampling*
Edited by Lawrence H. Keith
ACS Professional Reference Book; 458 pp;
clothbound ISBN 0–8412–1173–6; paperback ISBN 0–8412–1437–9

*Biotechnology and Materials Science: Chemistry for the Future*
Edited by Mary L. Good (Jacqueline K. Barton, Associate Editor)
135 pp; clothbound ISBN 0–8412–1472–7; paperback ISBN 0–8412–1473–5

---

For further information and a free catalog of ACS books, contact:
American Chemical Society
Customer Service & Sales
1155 16th Street, NW, Washington, DC 20036